Coal Preparation Conference & Exhibition 2019

Secunda, South Africa
19-22 August 2019

ISBN: 978-1-7138-2393-3

Printed from e-media with permission by:

Curran Associates, Inc.
57 Morehouse Lane
Red Hook, NY 12571

Some format issues inherent in the e-media version may also appear in this print version.

Printed with permission by Curran Associates, Inc. (2022)

For permission requests, please contact Southern African Coal Processing Society
at the address below.

Southern African Coal Processing Society
P O Box 3028
Glenvista, Gauteng, 2058

Phone: +27 79 872 6403

https://www.sacoalprep.co.za/

Additional copies of this publication are available from:

Curran Associates, Inc.
57 Morehouse Lane
Red Hook, NY 12571 USA
Phone: 845-758-0400
Fax: 845-758-2633
Email: curran@proceedings.com
Web: www.proceedings.com

TABLE OF CONTENTS

A PERFORMANCE COMPARISON BETWEEN POLYURETHANE AND CERAMIC DENSIFYING CYCLONES

KC Naidoo

Exxaro (Metallurgical Engineer); ECSA (Candidate Engineer)

SUMMARY/ABSTRACT

There are two main types of densifying cyclones currently available in the market, namely polyeurethane (PU) and ceramic. The PU densifying cyclones are currently used throughout the Grootegeluk complex and coal beneficiation in general. It has been identified that PU densifiers wear relatively fast depending on the operating pressure and need to be replaced frequently as a result. The motivation for this project is to achieve an increase in the densifying cyclone lifetime through less wear and an increase in efficiency resulting in the ability to build higher relative densities (RD) quicker. This will allow for a reduction in the downtimes due to replacing densifying cyclones as well as building RD and a reduction in the operating costs associated with replacing worn densifying cyclones.

Some of the major findings of the study indicated that the ceramic cyclone theoretically undergoes less wear than the PU cyclone, but since wear is application specific and was evaluated, the wear on the polyurethane cyclone was found to be minimal.

The RD differential is higher on the PU cyclones than the ceramic cyclones, showing that the PU cyclones perform better in a densifying application. This is the first test conducted on this application and type of cyclones. It should therefore be noted that having the PU cyclone significantly outperform the ceramic cyclone was not expected, due to the only change being the material of construction. This gives guidance to take the material of construction into account when selecting cyclones as it can have a significant impact on the cyclone performance in specific applications.

1. BACKGROUND

1.1 Dense medium separation

Material in the run-of-mine (ROM) feed contains the desired (product) as well as an undesired (discard) material. It is therefore necessary to separate the product from the discard in the feed. The product is distinguished from the discard through the composition (ash, sulphur, moisture, fixed carbon) and is based on the clients' specification.

Dense medium separation (DMS) is one of the most commonly used practices for the beneficiation of coal. DMS utilises gravity as a method of separating particulate matter based on density. The dense medium utilised in industrial coal separations is a thick suspension of heavy solid in water, which behaves as a heavy liquid. In the case of coal separation for this study, magnetite is used as the dense solid in the medium. The dense medium is mixed accordingly to achieve the required relative density (RD) in order for the desired separation to take place.

The separation takes place as follows: material of densities higher than that of the medium will sink whereas material with densities lower than that of the medium will float. The product reports to the floats and has a lower ash content than the discard which reports to the sinks. The relative density at which the dense medium is set is called the cut-point and indicates the density at which the feed material must be separated (i.e. report to either the product or discard streams).

In order to achieve the desired cut-point, it is necessary to control the RD. There are a few types of equipment that assist in RD control such as orifices, magnetic separators and densifying cyclones. This project relates to investigating the trend in performance of a polyurethane against a ceramic lined densifying cyclone.

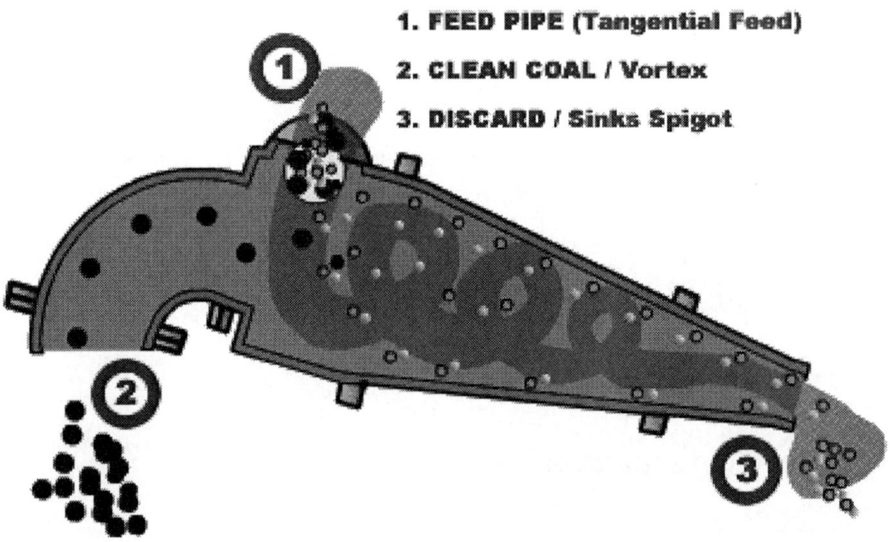

Figure 1: DMS Cyclone Operability

The need for this project arises from the difficulty of achieving the desired cut-point RD's in the GG2 plant of the Grootegeluk Complex. The desired RD's are seemingly too high for the system to reach. The RD control system used in the GG2 plant comprises of magnetic separators and densifying cyclones. The magnetic separators separate the magnetite from the coal and water, and the densifying cyclones help further increase the RD by taking out excess water from the dense medium going back into the correct cone using centrifugal forces.

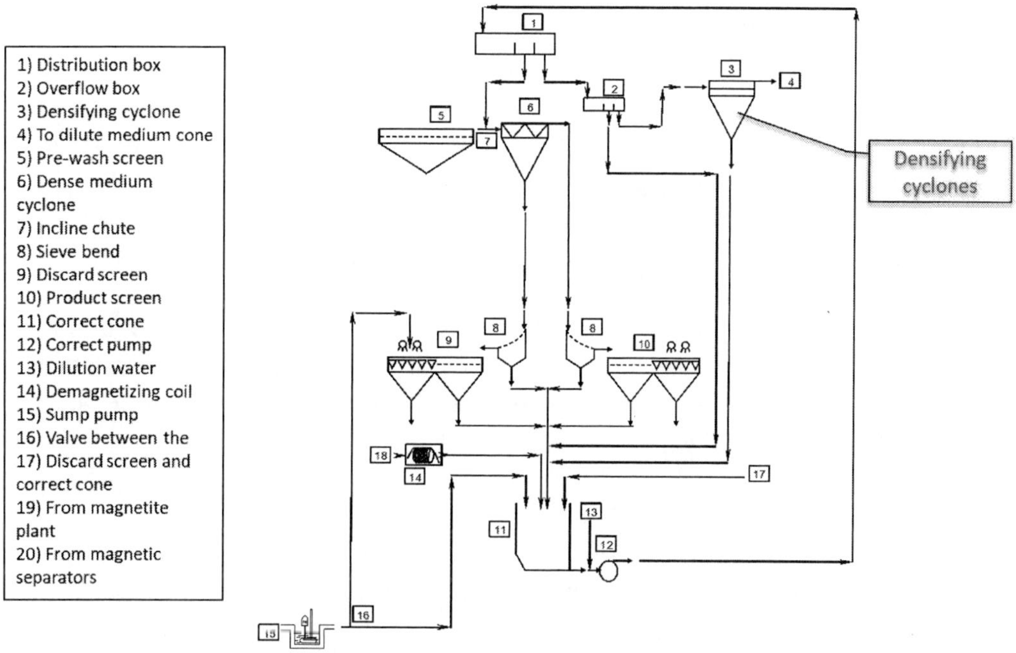

Figure 2: GG2 Cyclone Modules Dense Medium Flowline

There are two types of densifying cyclones available in the GG2 cyclone plant, namely polyurethane (PU) and ceramic. The PU densifying cyclones are currently used throughout the GG2 cyclone plant. It has been found that PU densifiers wear easily and need to be replaced frequently as a result. This forms another basis for carrying out research into comparing the performance of the different densifying cyclones.

The reasons for experimental work include being able to determine the following key performance indicators that will enable us to establish which the better performing densifying cyclone is are:

- Percentage wear of densifying cyclone
- Rate of wear of densifying cyclone

- RD of overflow/ underflow of densifying cyclone
- Percentage magnetics in overflow/ underflow.

1.2 Ceramic as a material for densification

Dense medium separation is one of the most complex unit processes in mineral processing today especially if cyclones are used. It is also the most efficient gravity concentrating process available to process engineers. The complexity of the process and large number of design and operating variables means that the process has largely remained more an art than a science (Bornman, 2018). Dense medium separation is well established in the diamond, coal and iron ore industries but is rising in popularity as a pre-concentration step in other industries. The maintenance issues that affect the efficiency of the operation range throughout the unit process equipment used (Bornman, 2018). With densifying cyclones being part of the range of equipment used in the dense medium process, it comes with its own set of maintenance issues. The duty of densifying cyclones is to offer a mechanism to adjust the relative densities (RD's) to the desired set point, in order to achieve the specified product quality, yield and throughput from the dense medium cyclones. Maintenance issues such as worn cyclones affect the efficiency dealing with the RD set point adjustment, associated product qualities and yield.

A dynamic model of the dense medium cyclone process can be derived from mass flow balances. An automated controller can be set to adjust the density of medium used to enhance separation in the DMC process according to measurements on percentages of different components in raw coal. The first objective of the control is to maintain the carbon content in the clean coal to a set level based on the other constituents. The second purpose is to attain the highest yield while satisfying the quality restrictions (Zhang, et al., 2013). The controller solves an optimization problem formulated to determine the density of medium whenever new measurements are available. It ensures that a precise change in the RD set points is achieved. This is an element of advanced process control and densifying cyclones form the basis or hardware for the success of such initiatives.

A densifying cyclone can be described as a "cyclo-thick" apparatus, which combines features of the hydrocyclone and the thickener in a single piece of equipment. Field tests conducted with the "cyclo-thick" demonstrated that the unit is remarkably simple and clean in design, and can effectively separate, dewater, and densify fine particulates. It is a unit that should be considered as a viable alternative when evaluating potential solutions to a given thickening problem (Lin, 2006). It is not limited to the fine particulates that may be processed.

Current background knowledge and the gathered literature indicate that ceramics have a very high melting point, are resistant to oxidation and corrosion and are very hard. They are however extremely brittle and prone to thermal shock failure (Khalil & Rosner, 1996). Polyurethane on the other hand is flexible providing a high load bearing capacity which allows it to return to its original shape if deformed under load. Polyurethane is also resistant to temperature/ thermal extremes (Metcalf, 1985).

Since previous years, ceramics have been consistently gaining a position in wear resistant applications as an avant-garde, usable wear resistant lining. The traditional use of ceramics in industry has however been avoided in any high impact situation. This attitude has retarded the application of this superior lining material. Increasingly of late, however, ongoing research and development (R&D), and test work on plant wear and impact applications has led to the conclusion that, with careful engineering, unique impact resistant ceramics can be used in high impact zones using modern ceramic wear resistant linings and ballistic armour technology (Viljoen, 2010).

The material used in the concerned application is dense medium consisting of magnetite, water and fine coal particles at a fairly constant temperature. The material is highly abrasive and flows at a relatively high rate through the densifying cyclones (Khalil & Rosner, 1996). Applying the background knowledge and literature data it is evident that ceramic cyclones would be more suited to processing the medium associated with the densifying cyclone application. This finding is due to the absence of thermal shock (constant temperature) as well as ceramics being resistant to oxidation and corrosion which aid in combatting the abrasive effects of the dense medium. In addition, the usually advantageous property of polyurethane which is its flexibility is not necessary in this application as there is an absence of high loads.

Figure 3 indicates that polyurethane, grouped under polymers and elastomers possess an exponentially higher wear rate than ceramics, especially aluminium ceramics which are used in the industrial linings on the test densifying cyclones. The hardness of ceramics is however much harder than that of polyurethane, again confirming that it should not be used in high impact situation where momentum of large particles is of concern.

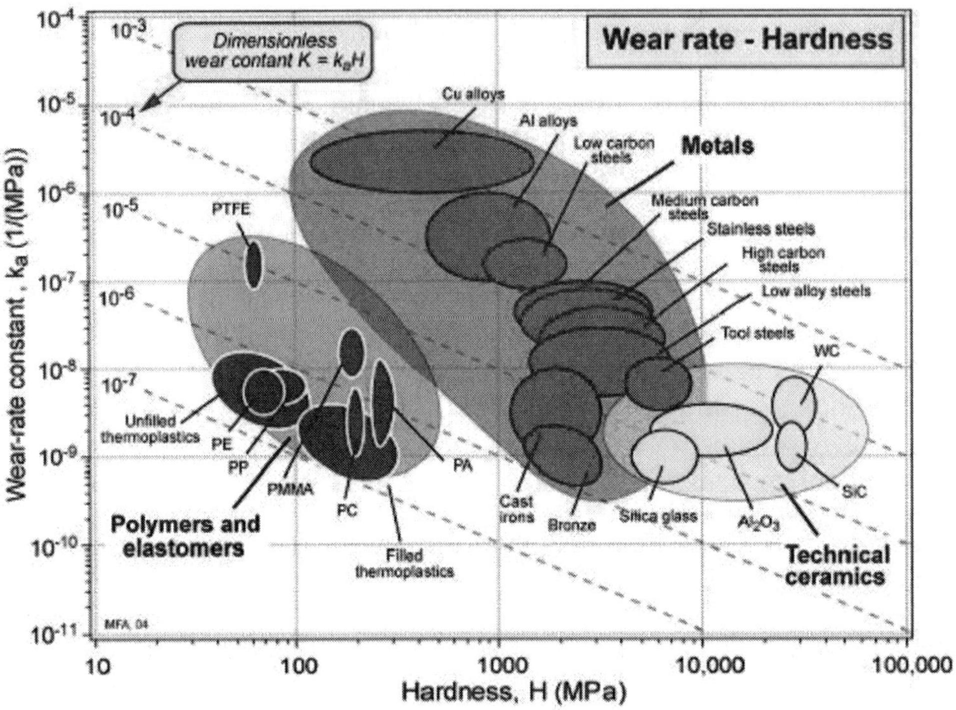

Figure 3: Wear Rate versus Hardness (Cambridge University, 2010)

Therefore, in terms of wear, ceramics come out on top as per literature when compared to polyurethane. The cost factor on the other hand paints a different picture as described by Figure 4.

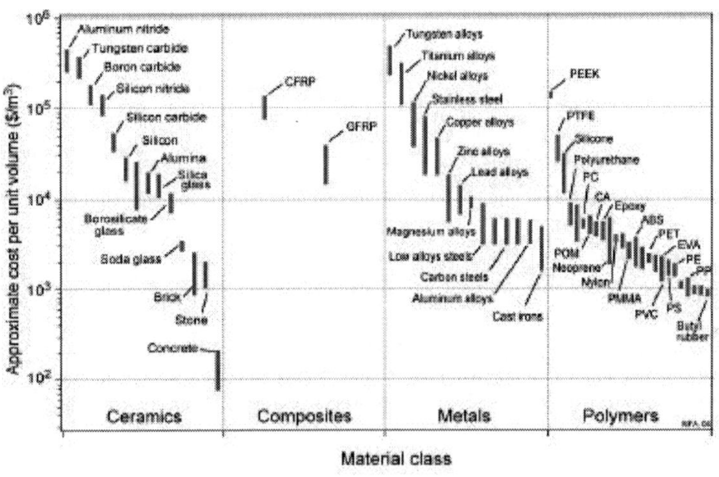

Figure 4: Cost Relationship per Unit Volume (Cambridge University, 2010)

The cost of ceramics is significantly higher than that of polyurethane. This may however be overcome by the extended lifetime of the ceramic products due to their high wear rate, reducing the frequent need for replacement of the densifying cyclones. The test-work should therefore cover the economic aspect of evaluating the benefit of the ceramic cyclones.

1.3 Motivation

The motivation for this project is to achieve an increase in the densifying cyclone lifetime through less wear and an increase in efficiency resulting in the ability to build higher RD's quicker. This will allow for a reduction in the downtimes due to replacing densifying cyclones as well as building RD and a reduction in the operating costs associated with replacing worn densifying cyclones.

2. PROJECT OBJECTIVES

- Identify the wear of PU and ceramic densifying cyclones
- Determine the percentage improvement in RD that can be achieved from the PU and ceramic densifying cyclones
- Establish the replacement cost of the PU and ceramic cyclones and draw up an annual budget using the wear rate
- Identify the better performing densifying cyclone of the two through comparing the key performance indicators mentioned
- Make recommendations on the standard of densifying cyclones in the GG2 plant
- Improve the RD control and hence overall plant control in the GG2 plant
- Maximise the yield where possible through being able to reach higher set-point RD's
- Reduce the downtimes in the GG2 plant through reducing the frequency of replacing worn densifying cyclones/ building RD
- Reduce the operating cost (R/ton) in the GG2 plant through reducing the frequency of replacing worn densifying cyclones and magnetite pumped

3. HYPOTHESIS

Assessment and interpretation of the gathered literature has led to the following hypothesis being generated: The ceramic cyclones will undergo less wear and therefore last longer than the polyurethane densifying cyclones. The ceramic cyclones will also offer better performance in terms of magnetite recovered to the correct medium system due to the structural integrity of the cyclones being maintained to a greater extent than that of the polyurethane cyclones (spigot/ vessel). The ceramic cyclones will offer a cost benefit in terms of a longer replacement plan as well as from the increased performance in the ability to pick up the relative densities as required consistently (yield and magnetite consumption benefit).

4. EXPERIMENTAL PLANNING

The experimental plan that will be used to test the hypothesis is outlined in detail. The ceramic cyclone was a *AA165-8-1/A-B/30, spigot from the 7-HS spigot range*. The PU cyclone was a *VV165-8-1/A-B/30, spigot from the 50-HS spigot range*. The PU densifying cyclone used for the test did not possess a barrel section as do the other production PU densifying cyclones. This is due to the footprint of the barrel section making the PU cyclone too large to fit into the cyclone box, hence the barrel section is removed. The experimental plan consists of 3 different experiments as mentioned below:

4.1 Experiment 1 – Image Analysis

Image analysis of the polyurethane test densifying cyclones for the evaluation of wear at the University of Pretoria.

The spigots on the cyclones were scanned using micro CT tomography at NECSA (South African Nuclear Energy Corporation). The reason why only the spigots were scanned were due to:

- Size limitations of sample for micro CT scan
- The wear is the highest in most cases at the lower part of the cyclone and is therefore a good representative sample of the wear throughout the hydrocyclone.

The amount of wear is measured using the volume of the void within the spigot. The larger the void, the more the wear will be relative to a new spigot (a new spigot will have a smaller void if wear occurred).

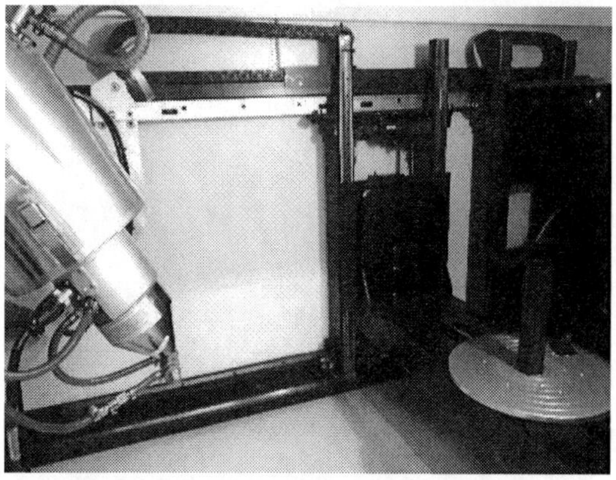

Figure 5: Micro CT scanner at NECSA. The X-ray source is at the left. The spigot is at the right and is rotated and the detector is at the back. Micro CT tomography works on the principle of density differences to create a 3D image of the object.

4.2 Experiment 2 – Spigot Measurements

Measurements of the spigot diameters of each of the test cyclones in order to measure the wear on one of the most important components of a cyclone. A worn spigot can result in spraying thus prematurely leading to cyclone failure and the need for replacement through reduced efficiency/ performance. The alpha water (water in feed stream that reports to the underflow) increases incrementally. Alpha water is a good proxy for efficiency as the fine particle that should have reported to the overflow gets dragged by the water and misplaced to the underflow (Herbst, 2019). The spigot diameters will therefore be measured at specified time intervals.

4.3 Experiment 3 – Sample Analyses

Samples of the feed, overflow and underflow of the test densifying cyclones were collected and analysed for density.

For good operation of densifying cyclone, the density difference between the material in the underflow and the material in the overflow must be greater than 5%.

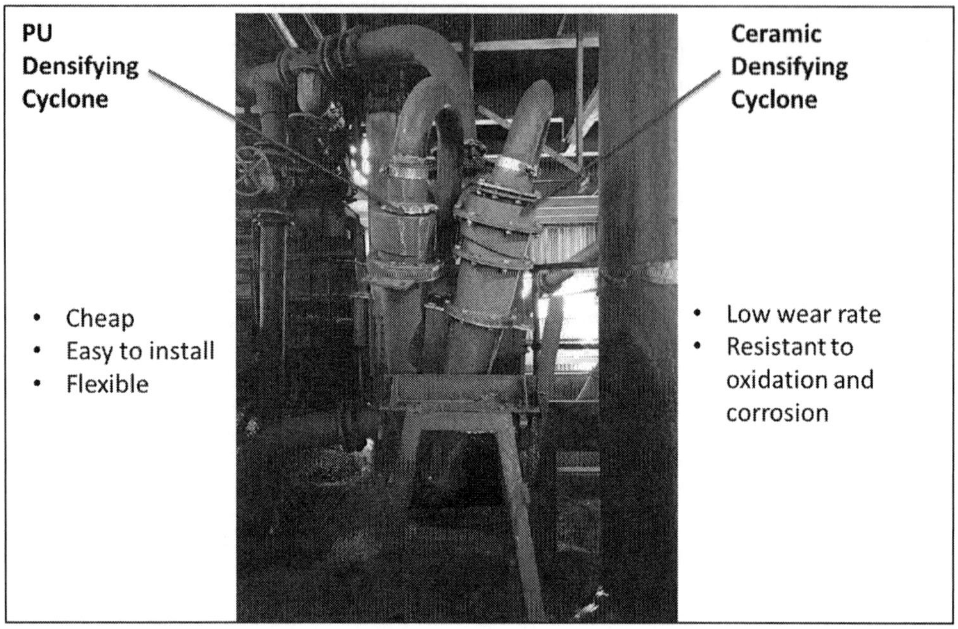

Figure 6: Test Setup

The potential cost benefit was evaluated based on the cyclone that has a lower reduction rate in performance and hence a longer replacement plan.

5. SAFETY

5.1 Personal Protective Equipment (PPE)

The correct PPE was utilised as per the SWP and included:

- Gloves
- Safety Goggles
- Hard Hat
- Ear Plugs
- Reflective Overalls
- Safety Boots.

Figure 7: PPE used for the task

5.2 Mini Hira

Mini Hira was completed to identify and put in place effective controls to mitigate any identified risks.

5.3 Equipment

The correct equipment was selected and used for the task.

6. RESULTS AND DISCUSSION

6.1 Image Wear Analysis

Image analysis of the polyurethane densifying cyclone (VV165) spigots were conducted using micro CT tomography at NECSA. Spigots were scanned at specified intervals. The intervals were as follows:

- 0 weeks (new)
- 2 weeks
- 8 weeks
- 3 months.

Figure 8 describes the test polyurethane cyclone used (left) as well as the CT tomography scan of a new spigot indicated by the red colour.

VV165 PU Densifying Cyclone

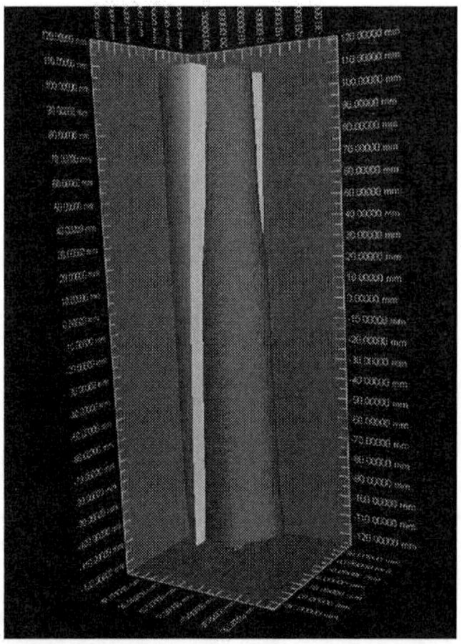

New Spigot Image Scan (Red)

Figure 8: PU Densifying Cyclone and New Spigot Image Scan

The new spigot has inherently undergone no wear. It was therefore overlapped with the used spigots for each of the time intervals in order to make a comparison in terms of wear and the wear rate.

Figure 9 shows scans of spigots that have been in operation for 2, 8 and 3 months, respectively, each overlapped onto a new spigot scan. The last image in figure 9 shows all of the spigots overlapped onto each other. The legend regarding the time intervals are as follows:

Table 1: Legend for Spigot Operating Time Intervals

Time Interval	Indicator Colour
0 weeks (new)	Red
2 weeks	Blue
8 weeks	Yellow
3 months	Green

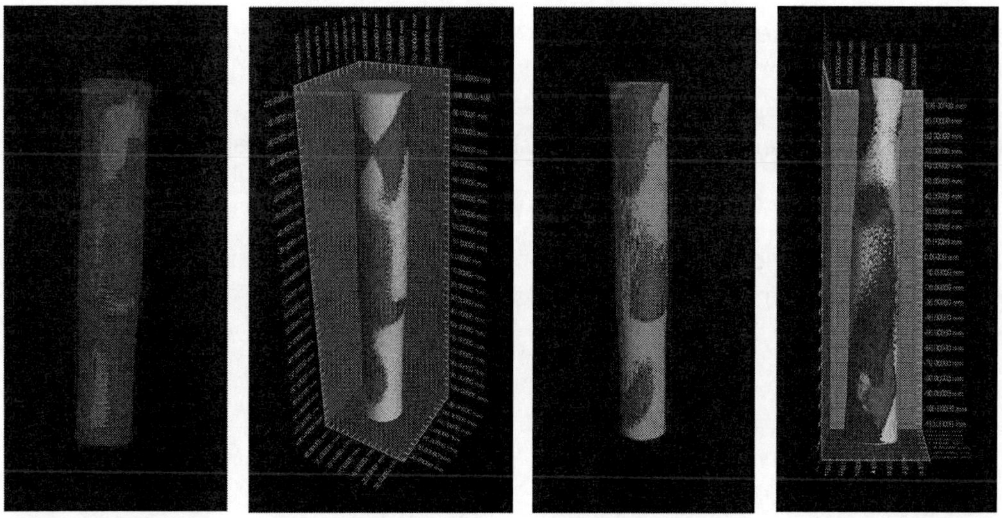

2 Week Spigot Image scan (Blue) | 8 Week Spigot Image Scan (Yellow) | 3 month Spigot Image Scan (Green) | New, 2 Week, 8 Week, 3 Month Spigot Overlay

Figure 9: 2 weeks, 8 weeks, 3 Months Scans of Spigots

The visual representation of wear among the cyclone spigots that were in operation compared to that of a new spigot in Figure 9 show some regions of wear. It was therefore necessary to measure and quantify the wear among the spigots so that a comparison could be made. This exercise was executed and the results are as shown in Table 2.

Table 2: Wear Analysis on VV165 Cyclones

Wear Analysis on Exarro VV165 Hydrocyclone				
		Volume of void (mm^3)	Radius at outlet (mm)	% Difference relative to new spigot
void of new	red	197 926	15.58	0
void of 2 weeks	blue	199 771	16.20	3.98
void of 8 Weeks	yellow	198 422	16.01	2.76
void of 3 months	green	199 076	16.24	4.24

From Table 1 it is evident that there is no significant wear among the spigots that have been in operation when compared to that of a new spigot. This is show by the highest percentage difference relative to the new spigot being only 4.24% after being in operation for 3 months. This finding is justifiable since wear is application based.

Figure 10 describes the change in the erosion test wear rate with a change in particle size for four different materials including polyurethane.

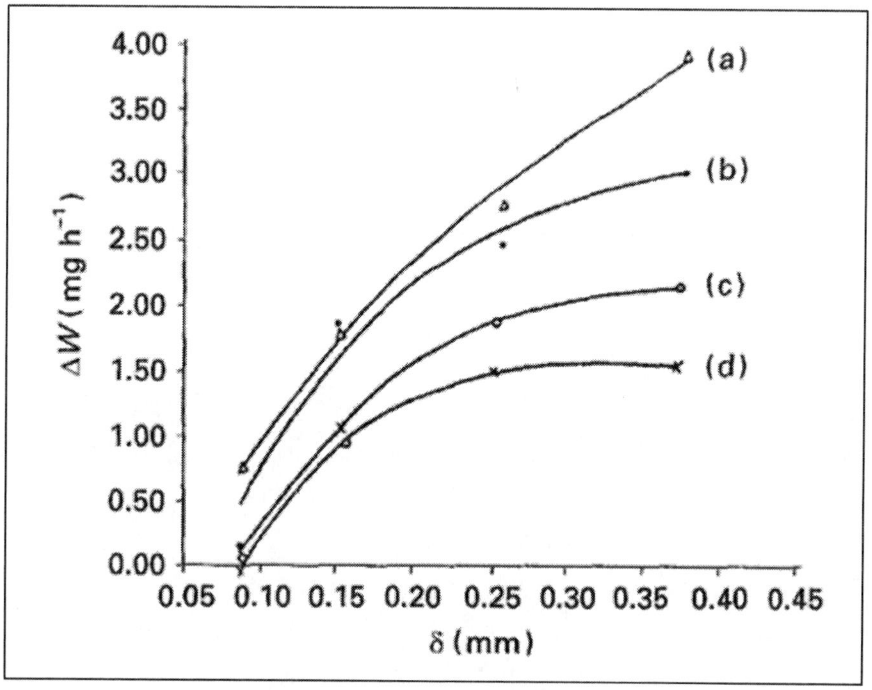

Figure 10: Erosion test wear rate with a change in particle size for (a) PTFE, (b) Nylon, (c) SBR, (d) Polyurethane at concentration of 10 % Sm and particle velocity of 11 m/s. (Zhang et al. 1995)

Figure 10 offers confirmation regarding the findings made as the application the cyclones were used in was for magnetite slurry densification, where the average particle size is 45 microns or 0.045mm. From Figure 10, it is evident that wear on polyurethane only has a significant impact when the particle size exceeds 0.07mm, thus explaining the minimal wear recognized on the cyclone spigots used for a 0.045mm application.

6.2 Sampled RD and Pressure Comparisons

Relative density (RD) samples were collected from both the test densifying cyclones (ceramic and polyurethane). The test PU cyclone used at the point of sampling was in operation for a period of 8 weeks. The samples were collected 15 minutes after any changes were made to the system. A total of three sample sets were collected at each sampling point for the respective cyclones. The results shown comprise of the average of the three sample sets. The feed RD remained constant and comparable across all three sample sets. The RD samples were collected from the feed, overflow and underflow points of each of the cyclones. These samples were collected for two instances, when two cyclones out of the four were opened and when only one cyclone of the four was opened. This enabled the difference in effective pressure and the impact posed on the cyclone performance to be evaluated and compared. Two open cyclones produced a low effective pressure scenario since the pressure from the feed distributor is split between two open cyclones instead of just going straight through to one cyclone which was the second and high effective pressure scenario.

A summary of the results is given by Figures 11 and 12. Figure 11 shows the results for when two cyclones are open or the low effective pressure scenario.

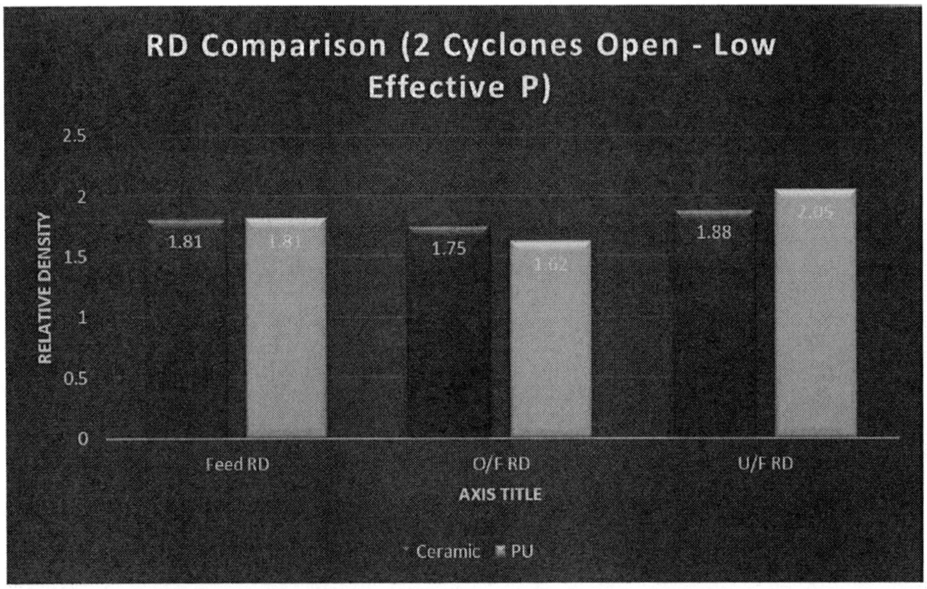

2 Cyclones				
	Feed RD	O/F RD	U/F RD	% Difference Between O/F and U/F
Ceramic	1.81	1.75	1.88	7.09
PU	1.81	1.62	2.05	21.10

Figure 11: RD Comparison when 2/4 Cyclones are Open (Low Effective Pressure)

From figure 11 it is evident that the polyurethane cyclone performs significantly better than the ceramic cyclone with the % difference in RD between the overflow and underflow being three times that of the ceramic cyclone. This allows for more densification at a faster rate, allowing to build higher RD's as well as reducing the time required to reach higher set point RD's thus reducing building RD downtimes. According to the densities, the ceramic cyclone has a solids split (Θ) of 50.1% compared to the PU cyclone of 57.3%.

Figure 12 shows the results for when one cyclone is open or the high effective pressure scenario.

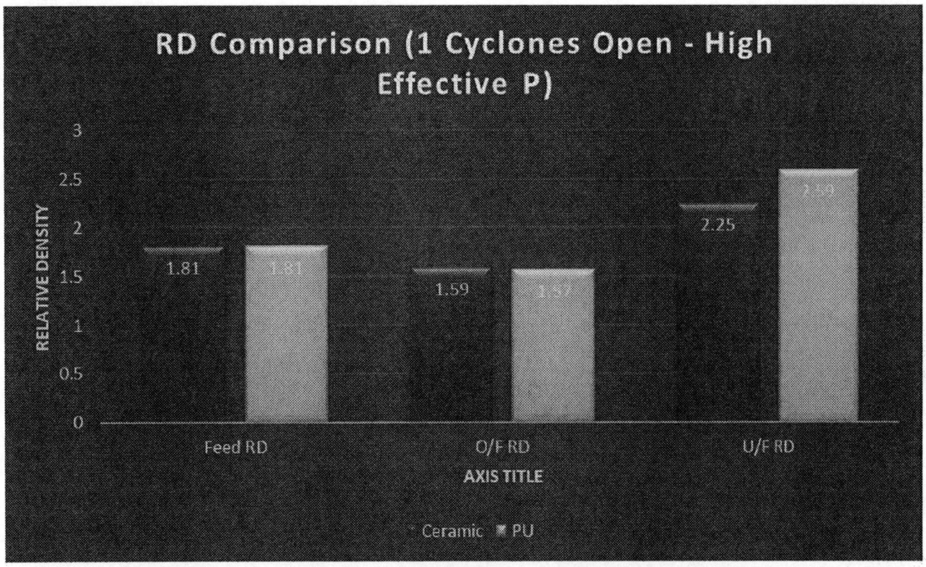

1 Cyclone				
	Feed RD	O/F RD	U/F RD	% Difference Between O/F and U/F
Ceramic	1.81	1.59	2.25	29.29
PU	1.81	1.57	2.59	39.59

Figure 12: RD Comparison when 1/4 Cyclones are Open (High Effective Pressure)

It can be seen that the ceramic cyclone performs much better when one cyclone is open yielding a higher effective pressure with a % difference in RD between the overflow and underflow of 29.29%. This is however still overshadowed by the % difference in RD between the overflow and underflow of the polyurethane cyclone of 39.59%. According to the densities, the ceramic cyclone has a solids split (ϴ) of 51.4% compared to the PU cyclone of 46.2%. Although a higher underflow density is achieved, more solids (magnetite) is lost to the overflow.

The polyurethane densifying cyclone has proved to perform better than the ceramic cyclone in terms of solids split (ϴ) and underflow density for the low effective pressure scenario and better in terms of underflow density but not solids split (ϴ) for the high effective pressure scenario. The reason for this finding points to the uniformity of the material polyurethane which is cast as one piece instead of the ceramic material which is made up of a few monolithic components.

The feed pipe into the pulp distributor (PD) could lead to a bias in solids distribution as shown in the picture below. Due to the direction of the slurry flow (black arrow), bending into the PD, more solids and the more coarse solids can report to the ceramic cyclone.

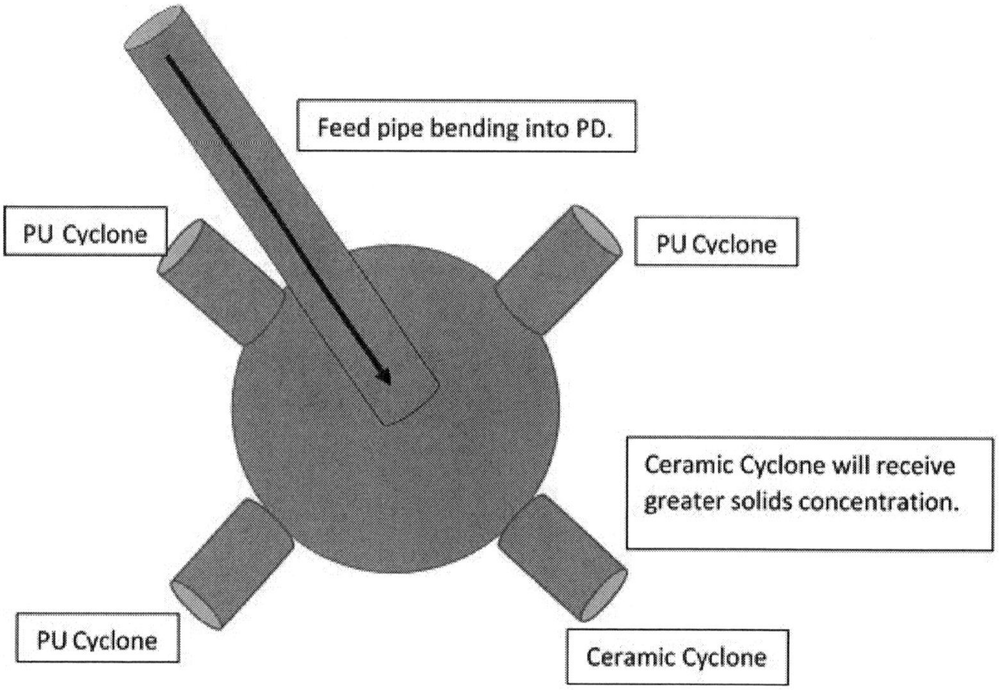

Figure 13: Test Pulp Distributor and Cyclone Configuration

The actual pressure experienced by the 165mm densifying cyclones is outlined by Table 3. The cyclones are gravity fed.

Table 3: Actual Pressure at 165mm Densifying Cyclones in the GG2/6 Plant

	Head (m)	P (kPa)
Actual	5.0	93.1

It is evident that the actual pressure is significant when looking at the size of the cyclone and accompanying spigots. The high pressure may cause the spigots to come off the cyclone as well as contribute to wear on the overflow pipes of the cyclones, which is currently being experienced in the GG2/6 plant.

6.3 Alternative RD Control Methods

A few alternative methods for RD control are shown in Figure 14. The comprise of orifices as well as adjustable densifying chutes.

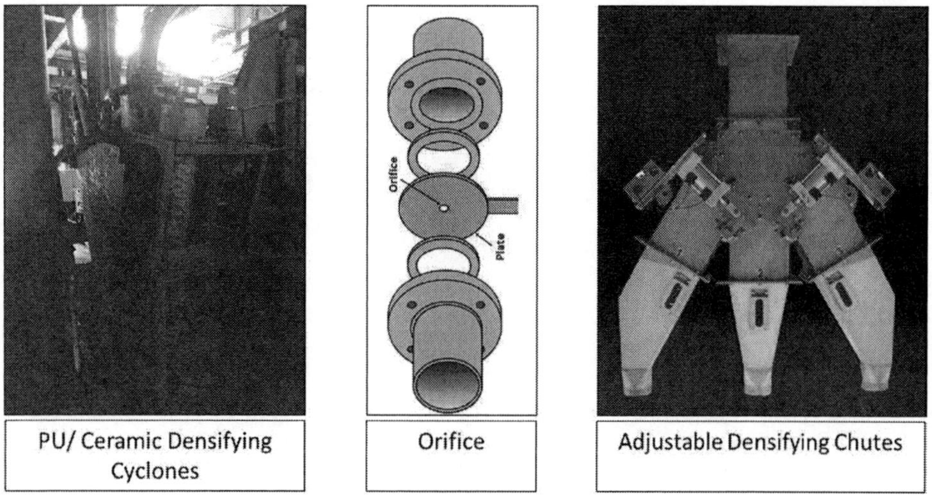

Figure 14: Alternative RD Control Equipment

The operating philosophy of the orifice and the adjustable densifying chutes are briefly described via Figure 15.

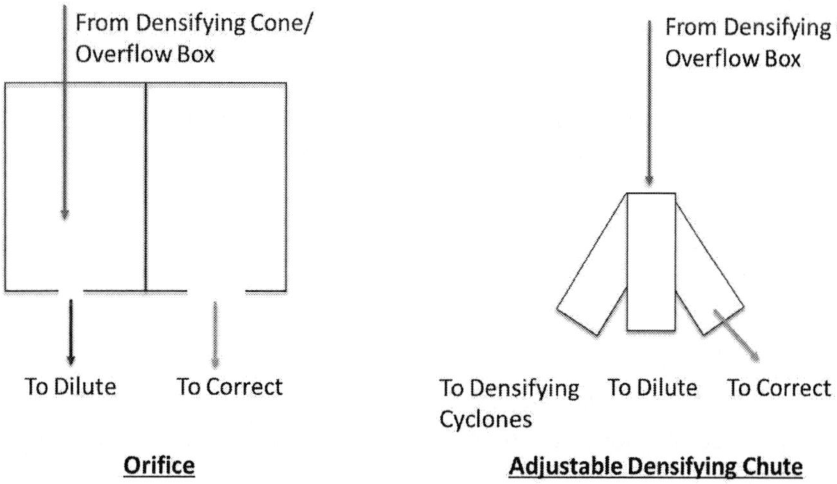

Figure 15: Orifice and Adjustable Densifying Chute Operating Philosophy

The orifice works by allowing correct/ dense medium to flow at a slow rate back into the dilute system where it reports to the magseps and back into the correct medium system increasing the effective correct RD. The adjustable densifying chutes on the other hand allows you to switch to where you send correct/ dense medium according to the plant's needs. It offers more options for control.

The RD control methods utilised by the large-scale plants at Grootegeluk are summarised in Table 4.

Table 4: Various Utilised RD Control Methods

GG1	GG26	GG8
Densifying Cyclones	Densifying Cyclones	Densifying Cyclones
Orifice	Orifice	-
Adjustable Densifying Chute	-	Adjustable Densifying Chute
Additional Water	Additional Water	Additional Water

7. CONCLUSIONS

The results and findings from the project/ investigation into a performance comparison between PU and ceramic densifying cyclones have led to the following conclusions being made:

- The cost of the ceramic cyclone (R 67 289) is more than 4 times that of the PU cyclone (R 14 687).
- The ceramic cyclone theoretically undergoes less wear than the PU cyclone but since wear is application specific and was evaluated, the wear on the polyurethane cyclone is minimal.
- The RD differential is higher on the PU cyclones than the ceramic cyclones.
- The cyclone spigots on the PU cyclones come off due to the high pressure.

- The mass of the ceramic cyclones are significantly greater than that of the PU cyclones (4 operators are required to remove ceramic cyclone to unblock instead of 1 operator for the PU cyclone).
- The ceramic cyclones are brittle and can break if dropped, damaging the cyclone leading to an increase in costs.
- The installation of ceramic cyclones require the installation of a bigger cyclone housing which costs approximately R 300 000 each and may require civil work due to the current footprint.

It must be noted that the ceramic cyclone delivered and installed at Exxaro - Grootegeluk GG2, is still a prototype in development and not a production unit. The aim of the ceramic unit was to achieve a longer wear life, due to complaints from site concerning the wear life of the polyurethane (PU) cyclones. Performance of the cyclone did not form part of the scope of the development work on the ceramic cyclone. It was confirmed after the test that cyclones were changed out due to underflow density decreases, assumed to be because of wear. This was then proven wrong, as the density decreases were due to spigots 'popping' out of the cone extensions (Herbst, 2019)

8. RECOMMENDATIONS

- Use larger diameter cyclones, and evaluate cyclone performance at this pressure.
- Investigate solution to secure spigot to cyclone more effectively, use the spigot coupling instead of the currently used older clamp design.
- Test the ceramic cyclone over a longer period (2 years) and monitor performance, in order to verify repeatability of results.
- Continue to use PU densifying cyclones instead of changing to ceramic cyclones due to the conclusions made as a result of this project

9. ACKNOWLEDGEMENTS

- Simeon Combrink – University of Pretoria, Multotec
- Dr. Natasia Naude – University of Pretoria
- Erick Herbst – Multotec

- Justin McDonald – Multotec
- Lindie Steyn – University of Pretoria, Advanced Coal Preparation Co-Ordinator
- Willie Pieterse – Exxaro, Grootegeluk
- Frikkie De Beer – Necsa
- Joyce Pieters – Exxaro, Grootegeluk

10. REFERENCES

Bornman , F. 2013. *Practical Operational Aspects of Dense Medium Cyclone Separation.* Kempton Park : Multotec Process Equipment, 2013.

Cambridge University. 2010. *Material and Process Selection Charts.* 2010.

Khalil, Y F and Rosner, D E. 1996. *Erosion Rate Prediction and Correlation Technique for Cermamic Surfaces Exposed to High Speed Flows of Abrasive Suspensions.* s.l. : Wear, 1996.

Lin, I J. 2006. *Hydrocycloning Thickening: Dewatering and Densification of Fine Particulates .* s.l. : Separation Science and Technology, 2006.

Metcalf , R L. 1985. *Cyclone Separator having Abrasion Resistant Cone Covered by a Plastic SLeeve with Flexible Seal Regions.* 1985.

Viljoen, S D. 2010. *Reduced Maintenance Costs Resulting from the use of Wear Resistant Materials.* s.l. : SAIMM, 2010.

Zhang, L and Xia, X. 2013. *Medium Density Control for Coal Washing Dense Medium Cyclone Circuits.* s.l. : IEEE, 2013.

Presentation 2019 Southern African Coal Processing Society Conference.

A practical proven fines circuit.

M. Craddock Mineral Processing Technologies (MPTEQ) (Pty) Ltd

ABSTRACT

Fines Coal with a nominal – 3 mm +0.1 mm due to its liberation has the potential to be upgraded to a quality higher than the larger size fractions in a wash plant. This paper describes a circuit that has proven itself as a practical simple to operate circuit that can achieve these objectives. It will describe the circuit and practical considerations when designing the circuit.

Coal fines are generated by the increase in mechanization on coal mines and plants, it has the potential to be used as a blend material to sweeten larger size fractions in limited quantities due to contractual obligations regarding the percentage fines in the product or it all can be added if the product is agglomerated or briquetted although this is potentially a more expensive option due to binder costs.

The circuit is simple, requires minimal operator intervention and utilizes proven technology with no reagent addition. It is not a new unproven design, Landau had one in the early 2000's.

Practical design considerations are described as well as operational considerations based on a plant recently designed and built by Obsideo Consulting.

INTRODUCTION

Coal fines are generated by the increase in mechanization on coal mines and plants, and are often considered a waste product sold as low value duff coal, however with this circuit it can become a high value product, especially if agglomeration is used.

There are some key considerations when designing this fines plant:

- - 80 μm material is typically poison and every opportunity to remove it must be taken advantage of.
- Guard screen 10 mm aperture. Contrary to common design parameters an elutriator can float coal larger than 3 mm in small quantities.
- Vibrating screens are preferable to static wedge wire sieve bends for removal of fines product / spiral feed from the large fines product due to the greater cut accuracy achieved.
- Water quality is important, together with the flow rate and pressure.

The ideal proven circuit has been used globally and within Southern Africa for a number of years but if one considers the number of coal plants in operation the adoption is still relatively low, percentage wise.

CIRCUIT

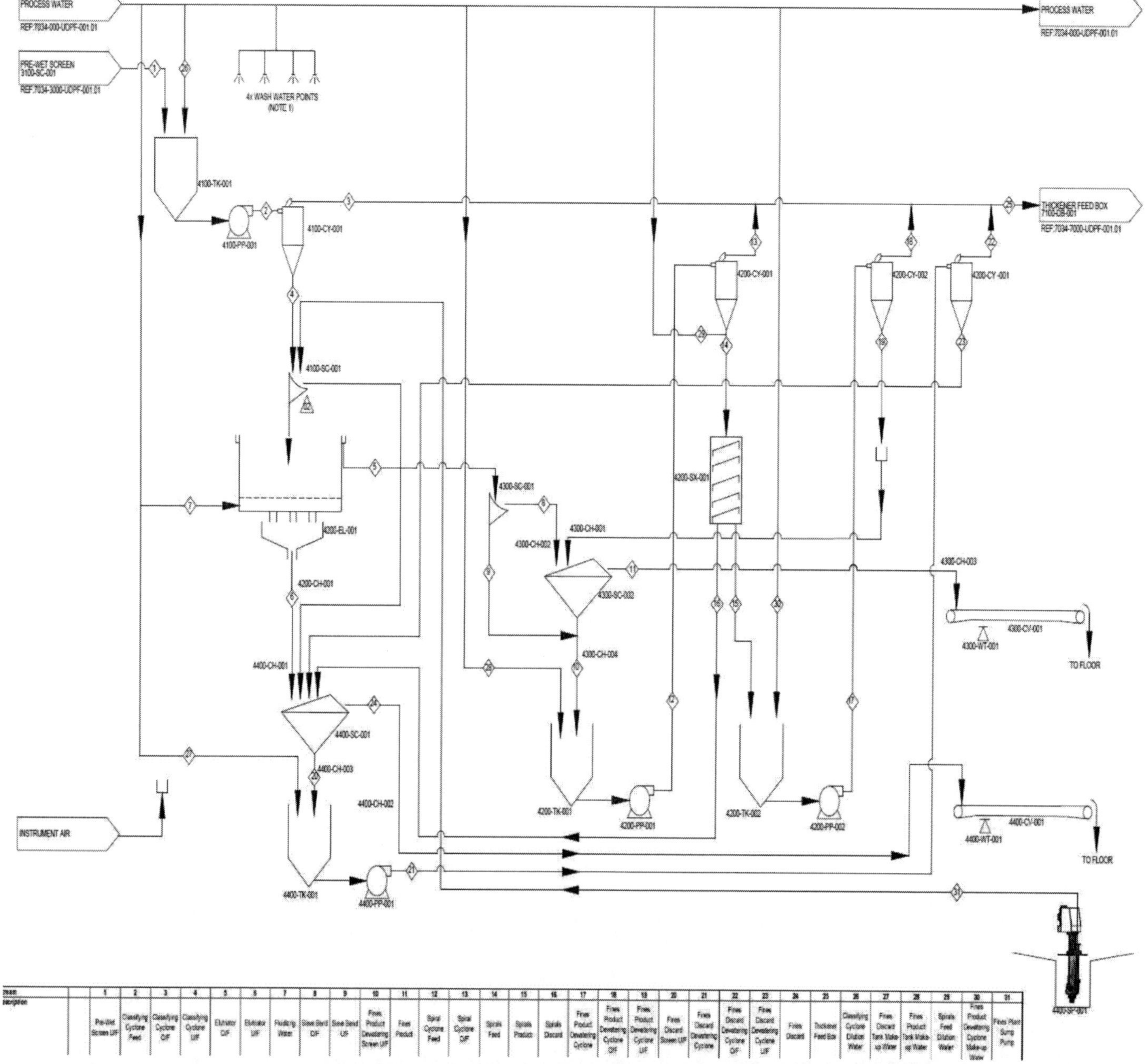

Fines from the plant sizing screen with a 2 mm aperture are pumped to a classifying cyclone, this cyclone's task is to de-slime the – 80 µm material and to optimally produce an underflow with 60 % solids. It is often stated that the feed percent solids should be 50 percent but in reality the higher the percent solids the better. The optimal is a wet – 3 mm feed from a conveyor. The reason for this is that the water in the cyclone underflow increases the volume of water in the upper zone of the elutriator, increasing the velocity in this zone in a potentially negative way.

If you have a feed rate of say 50 tons per hour the increase in water volume to elutriator between 50 and 60 percent solids is 16.67 m^3/hour. A 50 ton per hour elutriator is sized with an area of approximately 2.5 square meters. A good design water flow rate is 60 m^3 per hour, if you had 50% solids as opposed to 60% solids the velocity from below the feed well to the overflow will increase by almost 22%. This is obviously not optimal.

The feed water to the elutriator must have an accurate method of volume and pressure control, a header tank with an overflow box works well if you have the headroom within your plant, a VSD on the pump motor is also optimal. A local flow and pressure gauge is recommended. It is critical to have the feed pipe extend above the overflow launder by approximately 500 mm, this is to prevent the feed water creating a negative pressure in the feed water chamber that can suck solids into the water chamber blocking the water distribution system apertures.

A pressure probe located above the water distribution plate by approximately 500 to 1000 mm depending on the unit size. This pressure probe measures the hydrodynamic head above the probe and varies depending on the quantity of solids fluidized above the pressure transducer. This is used to control the opening and closing of the dart valves discharging the discard in coal applications.

This bed must always be fluidized, this is how you check the operation of an elutriator. Because the beds is fluidized you can use a wooden pole to test this and it should touch the water distribution system without any force apart from the poles buoyancy.

The underflow from the elutriator must be open to the atmosphere to prevent back-pressure and allow the free flow of solids. The underflow from the elutriator has a high percentage of solids varying between 35 to 70% solids by mass. The slopes used when designing the underflow piping should take this into account and have a minimum of a 30-degree slope steeper is better. The underflow reports to a dewatering screen. If the elutriator underflow is deposited onto the feed end of the screen first, it will form a bed on the screen allowing the spiral discard to be deposited on top of this bed reducing the quantity of fines lost through the screen.

A standard dewatering screen is very effective at dewatering both the product and discard, producing a conveyable product.

The overflow from the elutriator is collected in a peripheral launder and reports to a screen to remove the fine material that has not been upgraded within the elutriator. The aperture of this screen is interesting and is site specific. It typically ranges from 0.8 mm to 0.4 mm. Depending on the quality of the size fraction, a quick fractional analysis can indicate this aperture. 600 micron is a good starting point. The quality of the + 600 micron should thus be the target product quality.

The nominal – 600-micron fraction passes through the screen and reports to a sump where it is pumped to a classifying cyclone. The overflow reports to the MPTEQ high rate thickener feed box while the underflow reports to spirals. Spiral product joins the + 600 micron product from the elutriator while the spiral discard is layered on top of the elutriator discard and dewatered on a standard dewatering screen.

The dewatered product and discard are conveyed to stockpiles via conveyors.

PLANT PEFORMANCE AND DISCUSSION

The advantages of this flow sheet versus a pure spiral circuit can be listed as follows:

- The elutriator can absorb large fluctuations in plant feed tonnage.
- The spirals are fed a narrow size range at a reasonably constant feed tonnage.
- Due to this there is very little beaching on the spirals.
- Very little or no supervision of the circuit is required.
- Very little maintenance is required.
- Product qualities are exceptional.
- Low operating cost biggest power consumer is the elutriator water supply pump.

PERFORMANCE

Feed CV	Feed Ash	Product CV	Product Ash	Discard Ash	Discard CV
16.22	43.00	26.45	17.5	59.6	10.45
21.63	29.0	25.77	18.2	56.3	11.05
21.19	30.0	27.92	12.6	35.9	13.13
22.01	27.8	26.23	16.9	58.2	10.3
21.83	28.4	25.30	19.4	56.7	13.27
21.09	30.2	25.49	18.9	45.7	13.20
19.74	33.9	24.64	16.9	59.5	9.89
21.05	30.4	25.95	13.6	59.4	9.83

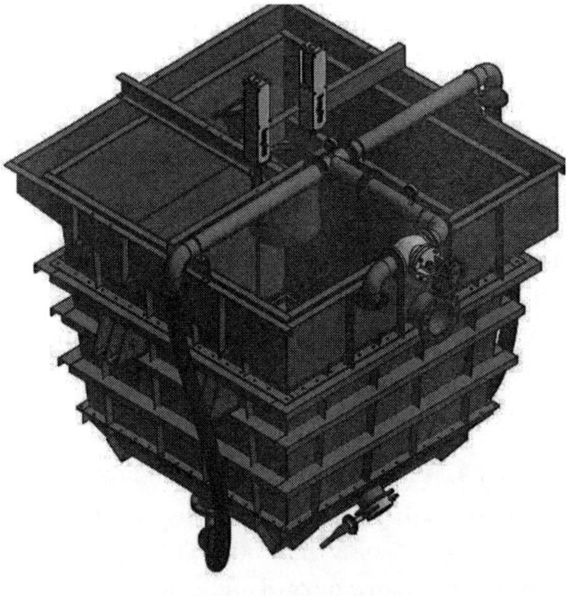

ISOMETRIC VIEW

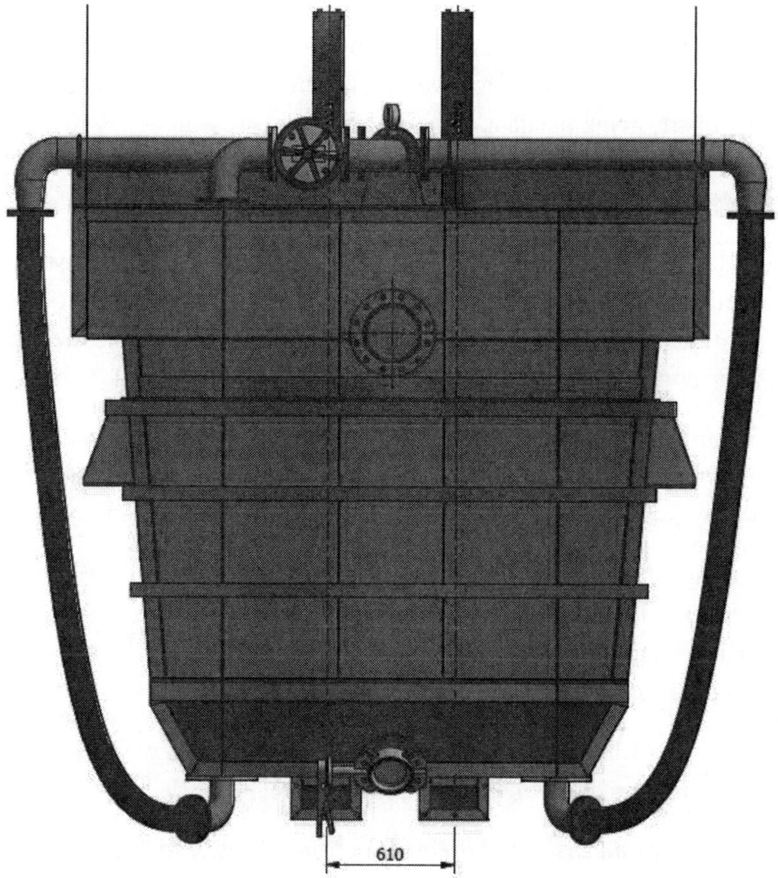

DISCUSSION

There are alternative circuits, the fine coal DMS circuit being built by Exxaro is very interesting but this circuit is so simple to operate and maintain with good yields and quality that it is hard to beat. It has proven robust and reliable, arguably the least visited part of the plant by the operators.

The elutriator derivatives are by now widely accepted in mineral processing industry due to its high separation efficiency, simplicity and low operating costs. It also has other advantages such as controllable density cut points, high solids handling capacity in a single unit, small footprint area, minimal feed slurry distribution problems. MPTEQ likes open tanks without added separation systems within the tank due to ease of access and maintenance. We also believe in unit operations each piece of equipment is used where it performs best.

The reason that two stage elutriators is not the optimal solution and spirals are selected for the – 400 to -600 + 100-micron material is due to the fact that the quantity of water required to fluidize the bed and to keep it fluidized gives an up-current velocity that is too great for efficient separation, a spiral however performs well with this material.

TECHNICAL DESIGN Elutriat

Upward water current at velocity v brings particles to the overflow launder on the basis of their relative terminal velocity. Under this condition it can be readily shown by the application of Newtons law that the terminal velocity of any given particle, v can be simply expressed as $v = k[d(Þ_{particle} - Þ_{liquid})\}^{0.5}$, where Þ is the density, d is particle diameter, and k is the proportional constant. The term $(Þ_{particle} - Þ_{liquid})$ is referred to as the effective density.

Typically the separation is carried out in water with $Þ_{liquid}$ equal to 1.0. Therefore the terminal velocity, v as well as the separation is governed by both particle diameter, d and the particle density, $Þ_{particle.}$ However, if the concentration of particles in the slurry is sufficiently high, the particles will be operating in the hindered settling zone and interfere with each others settling path.

The above formula thus needs to be modified. While this can become a very complicated theoretical calculation, it can be seen that, as a first approximation, that the equasion above needs to be modified as $v_{Hindered\ Settling} = k\ [d(Þ_{particle} - Þ_{liquid})\}^{0.5}$.

Under hindered settling conditions, it can be seen that the lower the density of the particle, the more pronounced the corresponding reduction in corresponding reduction in effective density term. This implies that the particle has a greater reduction in the interference on the settling velocity with increasing particle separation density.

In other words, hindered-settling velocity is reduced by the influence of the particle size, while increasing the effect of the particle density on separation[1].

In a Coal plant the separator allows the particle suspension density to build up within the elutriator against an upward fluidizing column of water until an autogenous dense medium bed is formed, in which the average particle separation is such that a perfect particle separation is formed.

CONCLUSION

The circuit Obsideo Consulting installed works, it produced a high CV and low ash product with a high ash discard. The plant runs without supervision. The product is blended carefully with the larger product sizes to sweeten the CV.

ADSORBENT ASSISTED DRYING OF SPIRAL AND FLOTATION COAL FINES

R. Erlank, M. le Roux*, Q.P. Campbell

School of Chemical and Minerals Engineering, North-West University, South Africa. Private Bag X6001, Potchefstroom 2520, South Africa

**Corresponding author: marco.leroux@nwu.ac.za*

ABSTRACT

Wet coal is difficult to handle, has a consequent lower heating value and relates to higher transportation costs and lower boiler efficiencies. Alumina-based adsorbent assisted drying of fine coal has been proven successful to dry run-of-mine coal to below $8\%_{wt}$ moisture content within 10 minutes.

In this study, the drying ability of spiral and flotation coal products was studied using alumina-based adsorbents. This was done on a laboratory-scale rotating bed with the coal being dried in a cascading motion inside cylindrical vessels.

The study showed that an increase in contact surface area between the coal and the adsorbents reduces the contact time for sufficient drying significantly. For both the spiral and flotation coal products, dewatering with new as well as used adsorbents was possible. Spiral and flotation product could be dried to below $8\%_{wt}$ moisture within 2 minutes using new adsorbents.

Keywords: coal processing, fine coal drying, fine coal dewatering

INTRODUCTION

In South Africa, coal has been mined for about 150 years and plays a key role in the economy (Venter & Naude, 2015). South Africa produces on average 224 Mt of marketable coal annually and exports about 25% of its marketable coal (Eskom, 2016). The majority of the domestic-use coal is used for electricity generation while coal-fired electricity generation accounts for 83.3% of the total South African energy mix (Eskom, 2017).

The increasing mechanised means of coal mining have caused an abundance of fine and ultra-fine coal. Coal with particle sizes of -500 μm can account for up to 15% of run-of-mine coal and coal smaller than 200 µm can account for up to 6% of run-of-mine coal (SACPS, 2015; Venter & Naude, 2015).

Fine coal has a larger surface area per mass unit than its coarser brethren, causing it to retain more moisture, which makes the dewatering process more challenging. High moisture levels in fine coal hold several problems including combustion, transportation, handling and marketability (Tao, et al., 2000). South African coal-fired power stations discard -200 µm coal fines, because of its usual higher moisture content and consequently lower heating value (Venter & Naude, 2015).

Some workers report that an economically appealing coal moisture content is $8\%_{wt}$ and below. In South Africa, thermal-grade coal market specifications call for a product moisture content of $9\%_{wt}$ or below (Bland & McDaniel, 2014; De Korte, 2019).

Current mechanical dewatering methods employed at coal preparation plants can lower fine coal's moisture content down to approximately $18\text{-}30\%_{wt}$ (Prat, 2012). Still, more drying is required to meet market specifications.

Two emerging drying technologies, Parsepco Drying Technolgy (PDT) and Nano Drying Technology (NDT) have shown the feasibility to yield the desired moisture reduction of fine coal from $18\%_{wt}$ to $9\%_{wt}$ (Mohanty, et al., 2012). Alumina-based adsorbent assisted drying of run-of-mine coal has been proven successful to dry coal in the range -2 mm +0.25 mm from moisture contents of $15\text{-}20\%_{wt}$ to below $8\%_{wt}$ in less than 10 minutes (Peters, 2016).

EXPERIMENTAL

Material

Coal

The origin of the coal being used in this study is No. 4 Seam of the Witbank Coalfield, South Africa. Spiral and flotation coal products are being investigated. The coal was sampled directly after filtration from the spiral and flotation plants of a colliery mining on No. 4 and No. 5 Seam of the Witbank Coalfield.

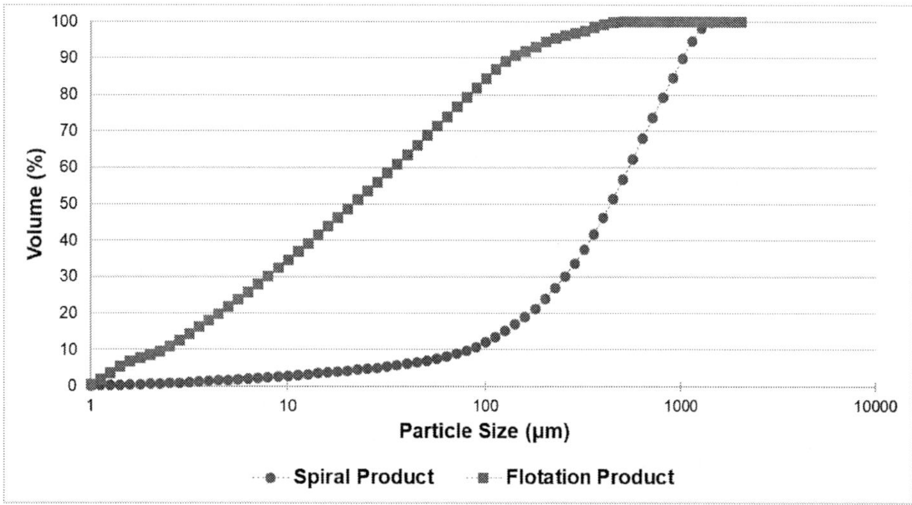

Figure 1: Particle size distribution of coals being investigated

The particle size distribution of the spiral and flotation coal products can be seen in Figure 1. The d_{50} particle sizes of the spiral and flotation coal products are 434µm and 21µm respectively. The total moisture, proximate analysis and calorific value of the coals being investigated are summarised in Table 1.

Table 1: Summary of coals being investigated

Property	Spiral Product	Flotation Product
Total moisture (%) (*ar*)	17.2 (±0.2)	23.5 (±0.4)
Inherent moisture (%)	2.0	4.4
Volatile matter (%)	22.5	22.1
Ash (%)	18.3	18.3
Fixed Carbon (%)	57.2	55.3
Calorific value (MJ/kg) (*ar*)	21.9	19.2

Adsorbents

The adsorbents being used is Porocel Dryocel 848 activated alumina adsorbents which are spherical in shape. 3 mm and 5 mm adsorbent particles are used to dewater the coal in order to investigate the performances of each particle size. The adsorbent being used has a surface area of 320 m^2/g and a pore volume of 0.5 cm^3/g.

The adsorbents were drenched in water for 48-hours to determine maximum water adsorptive capacities. The 3 mm and 5 mm adsorbents reached maximum adsorptive capacities of 46% and 45% respectively within 1 hour of drenching. The adsorbents are able to adsorb approximately 92% of its maximum adsorptive capacity within 10 minutes.

Preliminary experiments showed that the adsorbents are chemically stable and retain its structural integrity when exposed to low pHs. The adsorbents were drenched in $H_2SO_{4(aq)}$ solutions of pH=1, 2 and 3 respectively for 30 minutes while the mass loss in adsorbent was determined. No degradation or mass loss of the adsorbents was recorded after being drenched in the $H_2SO_{4(aq)}$ solutions. Figure 2 shows an intimate view of the dry adsorbents, before and after the drenching. No significant damage can be seen on the adsorbents.

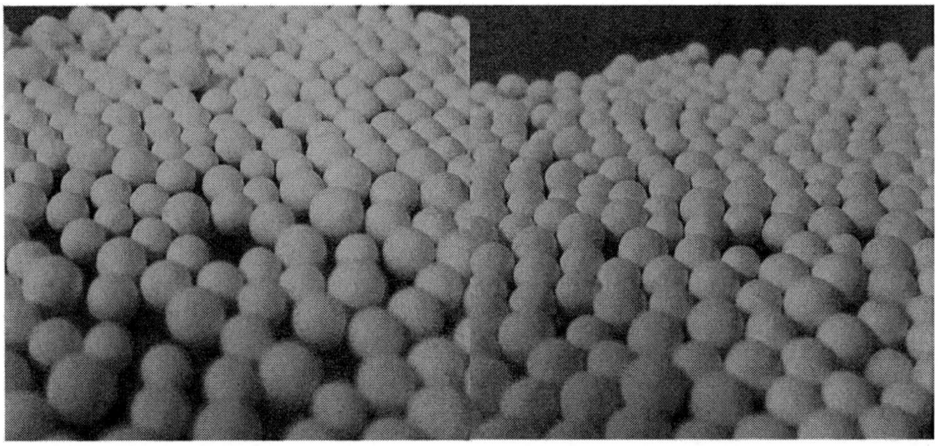

Before **After**

Figure 2: Effect of a low pH on the adsorbents

Apparatus and procedure

A rotating bed with three rollers in tandem and a variable speed motor is used on which cylindrical vessels of 50 mm inner diameter is placed. The cylindrical vessels (Figure 3) have each a 10 mm baffle plate inside to improve mixing and contact between the coal and adsorbents. An illustration of the drying process can be seen in Figure 4.

Figure 3: Side view of cylindrical vessels being used

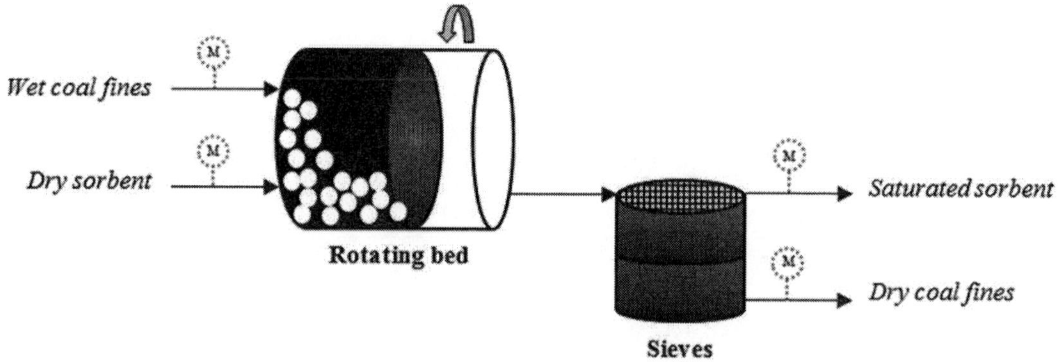

Figure 4: Illustration of the coal drying process, taken from (Van Rensburg, et al., 2018)

For each experimental run, the coal, as listed in Table 1, with its "as received" total moisture, is dried in the cylindrical vessels. 20g of coal is weighed and inserted into the empty cylindrical vessels. The respective adsorbent amount is weighed and added with the wet coal while the vessels are sealed off. The cylindrical vessels are placed on the rotating bed, rotating at 50rpm, and the drying process starts. Moisture analysis samples of the coal and adsorbents are taken in time intervals of 2 minutes until 16 minutes of drying. Calorific values of the coals are determined at feed and time intervals 2 min, 8 min and 16 min. Moisture contents are determined as per SANS 5925:2007 standard and calorific values are determined as per ISO 1928: 2009 standard.

RESULTS AND DISCUSSION

Spiral product vs flotation product drying curves

The adsorbents dewatered both spiral product and flotation product successfully, reaching the target moisture of 8%$_{wt}$ in less than 2 minutes of contact sorption time. Figure 5 shows the drying curves of spiral product and flotation product using new 3 mm adsorbents in a 2:1 adsorbent to coal mass ratio. The initial desorption rate for the first 2 minutes of contact time is found to be 0.076 and 0.085 $g_{moisture}/g_{coal}$.min for spiral product and flotation product respectively. The coal reached final moisture contents of 0.9%$_{wt}$ and 1.3%$_{wt}$ for spiral and flotation product respectively after 16 minutes of drying.

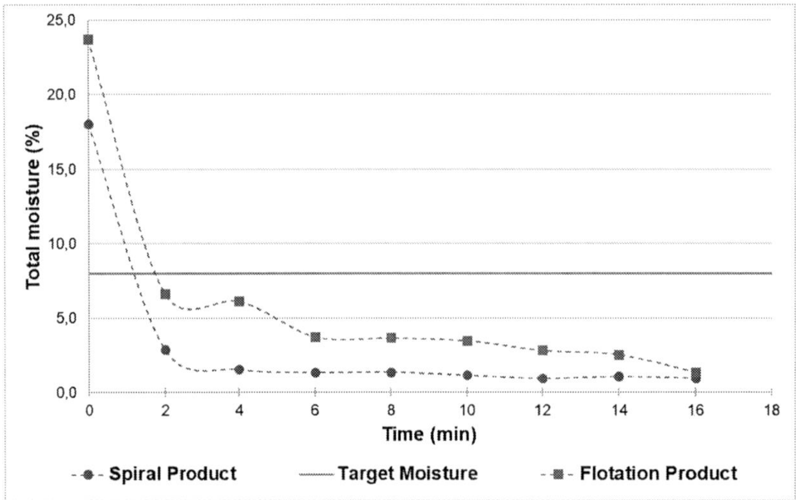

Figure 5: Spiral product vs flotation product drying curves

Effect of drying on the heating value

The drying of the coal had a significant impact on the calorific value (Figure 6), causing a 16% increase in calorific value on the spiral product and a 29% increase on flotation product. Spiral product's calorific value increased from 21.9 MJ/kg to 25.3 MJ/kg within 8 minutes of contact-sorption. Flotation product had an increase in calorific value from 19.2 MJ/kg to 24.7 MJ/kg within 8 minutes of contact-sorption.

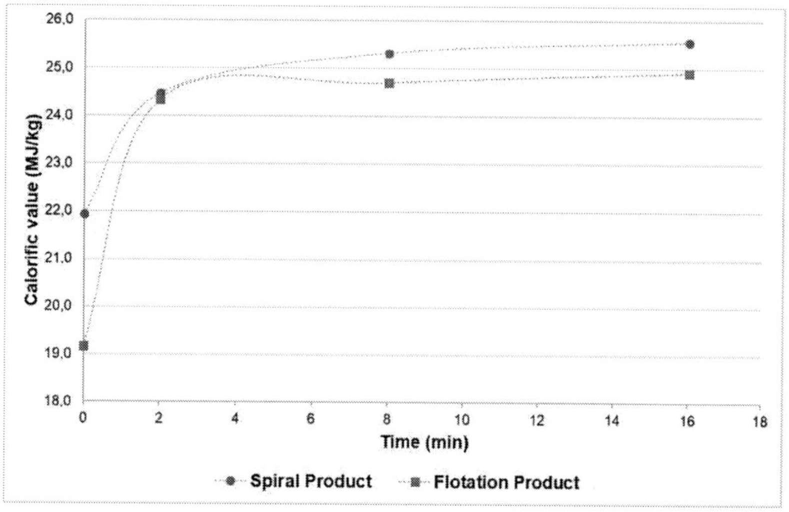

Figure 6: Calorific value of spiral and flotation product

Effect of adsorbent particle size

Both the 3 mm and 5 mm adsorbents showed successful drying capabilities with the latter having slightly slower drying capabilities, but still sufficient. Both 3 mm and 5 mm adsorbents managed to dry the coal to below target moisture of 8%$_{wt}$ within 2 minutes of contact-sorption when using a 2:1 adsorbent to coal mass ratio (Figure 7). From Figure 7, the initial desorption rates of spiral product using new 3 mm and 5 mm adsorbents are recorded as 0.076 and 0.052 $g_{moisture}/g_{coal}$.min respectively.

Figure 8 shows the drying curves of spiral product using new 3 mm and 5 mm adsorbents in a 1:1 adsorbent to coal mass ratio. The initial desorption rates of spiral product using 3 mm and 5 mm adsorbents are 0.052 and 0.031 $g_{moisture}/g_{coal}$.min respectively.

However, using the 5 mm adsorbents in a 1:1 adsorbent to coal mass ratio, required more than 10 minutes of contact-sorption time to dewater sufficiently, opposed to 2 minutes required when using a 2:1 adsorbent to coal mass ratio. In general, 3 mm adsorbents delivered faster drying performances than 5 mm adsorbents, having a higher average initial desorption rate of approximately 0.023 $g_{moisture}/g_{coal}$.min.

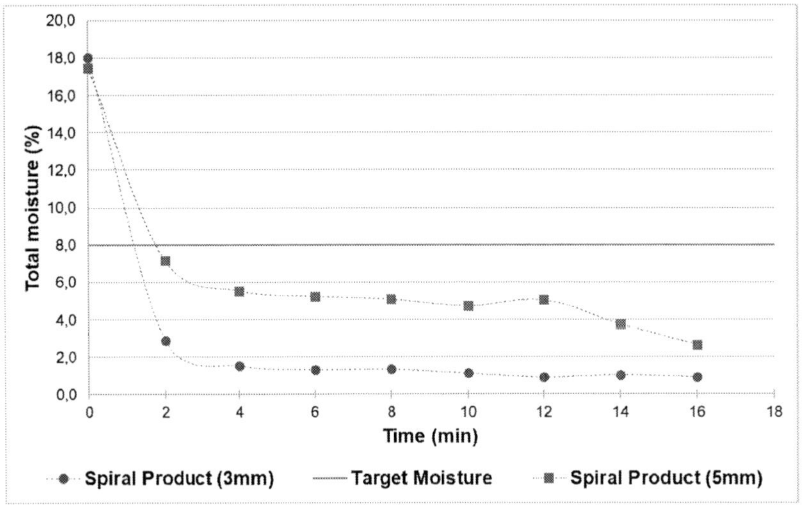

Figure 7: Effect of adsorbent particle size on drying efficiency (2:1)

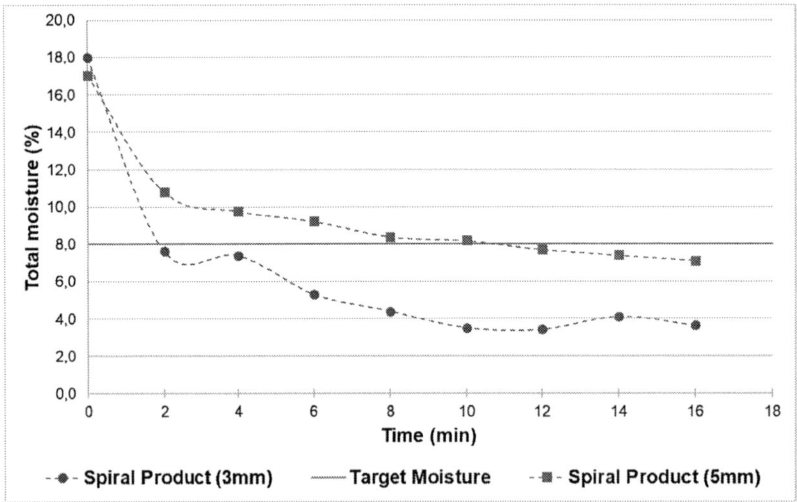

Figure 8: Effect of adsorbent particle size on drying efficiency (1:1)

Effect of adsorbent to coal mass ratio

The 1:1 and 2:1 adsorbent to coal mass ratios showed successful drying capabilities when using 3 mm adsorbents, reducing the coal's moisture content to below target moisture within 4 minutes of contact-sorption time. However, the 0.5:1 adsorbent to coal mass ratio did not dewater the coal to below target moistures within 10 minutes of contact-sorption time. The initial desorption rate of the coal increases with increasing adsorbents to coal mass ratio (Figure 9). The initial desorption rate of the spiral product increases from 0.039 to 0.052 and finally, 0.076 $g_{moisture}/g_{coal}$.min for 0.5:1, 1:1 and 2:1 adsorbent to coal mass ratios respectively. The flotation product had higher initial desorption rates for adsorbent to coal mass ratios of 1:1 and 2:1, due to the smaller particle sizes, which leads to higher contact surface area for drying. Flotation product's initial desorption rate increased from 0.036 to 0.072 and finally 0.085 $g_{moisture}/g_{coal}.min$ for 0.5:1, 1:1 and 2:1 adsorbent to coal mass ratios respectively.

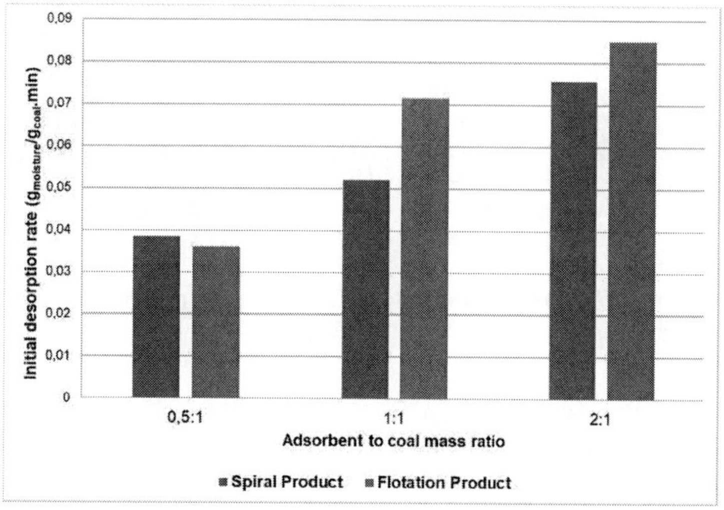

Figure 9: Effect of adsorbent to coal mass ratio on coal drying

Effect of used adsorbents

Figure 10 shows the total moisture contents of the spiral product, flotation product and 3 mm adsorbents after 2 minutes of contact-sorption time using the adsorbents 5 times without regeneration in a 2:1 adsorbent to coal mass ratio. Up to the 3rd time of adsorbent re-use, the spiral product's moisture content was reduced to below 8%$_{wt}$ within 2 minutes of contact-sorption time. Up to the 5th time the adsorbents were re-used, the spiral product's moisture content was reduced to between 8 and 9%$_{wt}$ within 2 minutes of contact-sorption time.

From Figure 10, it can be seen that, although the adsorbent's moisture content increases far above the coal's moisture content, it still has the ability to adsorb moisture from the coal.

For flotation product, dewatering to below 9%$_{wt}$ within 2 minutes of drying, was only possible up to the 2nd use of the adsorbents without regeneration. Furthermore, the flotation product's moisture content could be dialled in, to below 9%$_{wt}$ and 8%$_{wt}$ using the adsorbents up to five times, but required longer contact-sorption periods.

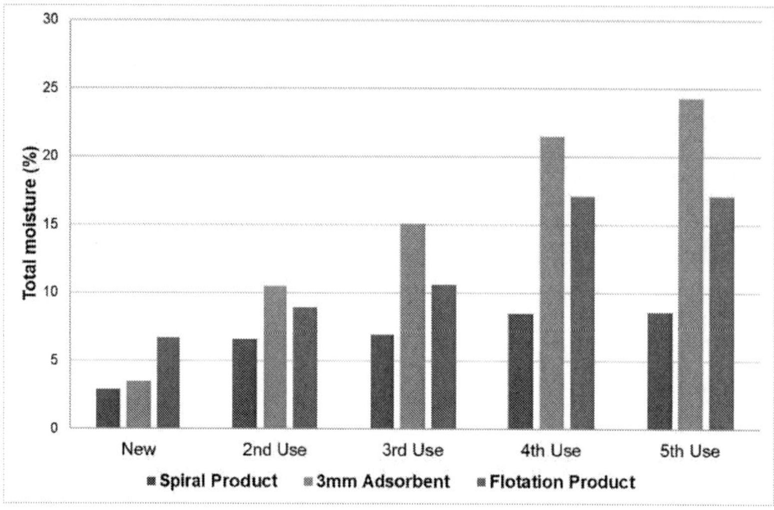

Figure 10: Drying capability of used adsorbents

CONCLUSION

Alumina-based adsorbents showed feasibility in drying spiral and flotation coal products sufficiently up to market specification. The coal products' calorific value increased significantly within 8 minutes of drying, delivering a more marketable product. It is reported that both 3 mm and 5 mm adsorbents dewater the fine coal products sufficiently, with the former yielding the best dewatering performance.

By using 1:1 and 2:1 adsorbent to coal mass ratios, sufficient dewatering could be achieved. The adsorbent to coal mass ratio of 2:1 delivered the best dewatering performance. The adsorbents have the ability to dewater fine coal successfully when re-used up to 5 times without regeneration. Physical contact between the adsorbents and coal is imperative for successful dewatering. An increase in contact surface area between the adsorbents and the coal reduces the contact-sorption time required for sufficient dewatering.

ACKNOWLEDGEMENTS

The authors would like to thank Coaltech 2020 Research Programme, the Southern African Coal Processing Society and the Centre of Excellence in Carbon Based Fuels, North-West University, South Africa for their contribution to this project, financial and/or otherwise.

This work is based on the research supported by the South African Research Chairs Initiative of the Department of Science and Technology and National Research Foundation of South Africa (Chair Grant No: 86880, UID 85643, Grant No: 85632 and Grant UID 72310). Any opinion, finding or conclusion or recommendation expressed in this material is that of the author(s) and the NRF does not accept any liability in this regard.

REFERENCES

Bland, R. W. & McDaniel, B., 2014. Coal Drying Method and System. United States of America, Patent No. US 2014/0144072 A1.

De Korte, J., 2019. The current market specifications and demand for coal fines in South Africa [Interview] (14 February 2019).

Eskom, 2016. *Coal in South Africa.* [Online]
Available at:
http://www.eskom.co.za/AboutElectricity/FactsFigures/Pages/Facts_Figures.aspx
[Accessed 18 April 2018].

Eskom, 2017. *Generation Plant Mix.* [Online]
Available at:
http://www.eskom.co.za/AboutElectricity/FactsFigures/Pages/Facts_Figures.aspx
[Accessed 18 April 2018].

Mohanty, M., Akbari, H. & Luttrell, G., 2012. Fine Coal Drying and Plant Profitability. *Challenges in Fine Coal: Processing, Dewatering and Disposal*, pp. 329-344.

Peters, E., 2016. *Adsorbent assisted drying of fine coal.* Potchefstroom: s.n.

Prat, G., 2012. Dewatering Fine Coal and Tailings with a Filter Press. *Challenges in Fine Coal: Processing, Dewatering and Disposal*, pp. 279-292.

SACPS, 2015. *Coal Preparation in Southern Africa.* 5th ed. s.l.: The Southern African Coal Processing Society.

Tao, D., Groppo, J. & Parekh, B., 2000. Enhanced ultrafine coal dewatering using flocculation filtration processes. *Minerals Engineering,* **13**(2), pp. 163-171.

Van Rensburg, M., Le Roux, M., Campbell, Q. & Peters, E., 2018. Moisture transport during contact sorption drying of coal fines. *International Journal of Coal Preparation and Utilization,* Volume DOI: 10.1080/19392699.2018.1543663.

Venter, P. & Naude, N., 2015. Evaluation of some optimum moisture and binder conditions for coal fines briquetting. *The Journal of The South African Institute of Mining and Metallurgy,* Volume **115**, pp. 329-333.

Assessment of flotation performance of different coal samples by different reagent dosage.

H. Assefa, N. Tshiongo-Makgwe and D. Nyembwe

Corresponding author. Cell +27(0) 74 993 3206 E-mail address: hamamaassefa@gmail.com

Alternative contact Tel.: +27(0) 11 559 6168 cell +27(0) 72 870 0466 E-mail address: ntshiongo@uj.ac.za

University of Johannesburg, Faculty of Engineering and the Built Environment, School of Mining, Metallurgy and Chemical Engineering, Department of Metallurgy, Minerals Processing and Technology Research Centre, South Africa, PO BOX 526, Wits 2050, South Africa

Over the past years, there has been a rise in froth flotation application in South African coal fines cleaning. of the reason for this is the fast depletion of high-grade coals and the complexities in mining, cleaning, and processing low-grade coal, which requires to be treated as fines. Flotation reagents among other parameters have a big influence on the overall flotation process and the quality of coal fines produced. The coal yield and quality of final product obtained depends on types and the amount of reagent dosage, flotation operational parameters and mineral matter composition in coal. In this project, different reagent dosages and other flotation parameters were investigated. The froth flotation test of three types of bituminous coals were carried out by pairing up of different frother and collector dosages. The floatability performance of coal samples where decreases during the addition of frother higher than 150 g/t and collector lower than 2000g/t, but the ash rejection and high calorific value was considerably improved by using latter frother and collector dosage. Coal A, B & C had a yield upgrade of ash content of 15.38%, 15.41% and 13.02% with calorific values of 25.3MJ/Kg, 24.86MJ/Kg and 26.06MJ/Kg respectively for the sample studied.

Keywords:

Froth flotation, reagent dosage, frother, collector, fine coal recovery

1. Introduction

South Africa is the sixth largest coal producer country in the world, supplying approximately 290 million tons of ROM coal per annum (Le Roux & Cambell, 2003). The coal reserve has been extensively mined to satisfy the coal demand for power generation, different industrial requirements and for the export market. Currently, different reports indicating that South African coal reserve will be depleted by 2050. The report of South African mineral bureau demonstrates that the amount of coal which is extracted increases each year and the coal reserve worth mining decreasing from 58.4 billion tons to just 39.1 billion tons in a period of 17 years (from 1983 to 2000) (Jeffrey, 2005). Simultaneously, in South Africa from the total mined coal about 14% reported as a fine (-500μm) and 3% ultra-fine (-100μm), this indicates about 30 million tons of fines being produced annually (Behzad, et al., 2014) (Le Roux , et al., 2005). Coal fines and ultra-fines has been discarded not due to its lack of thermal potential, but it's by nature difficult handling and transport characteristics, which is loss of revenue to the country.

For coal fine, froth flotation is the most effective method of separating ash forming mineral matter from the carbonaceous material, and it is one of the most complex mineral processing operations as it is affected by several variables (Hussin & Jan , 2004). The mineral matter which is present in varying amounts in coal also has a major impact on solid-liquid interaction.

Previously, the cleaning efficiency of two different reagents (Montenol 505 and 800) were investigated to improve the quality of spiral discards. By using Montenol 505 a high combustible recovery and ash rejection was obtained. In this study further upgrading were conducted by agglomerating the flotation product using different binders, however, 15% diesel binder gives a combustible recovery of 25.72 MJ/Kg (Tshiongo-Makgwe , 2017).

Various techniques of fine and ultra-fine coal flotation have been done on improving the floatability of difficult to float coal (Atesok & Celik, 2000), increasing the selectivity by applying different flotation reagents (Zhinqiang , et al., 2015) (Horsfall, et al., 1989), and (Li, et al., 2013) studied the effect of micro-emulsified diesel and conventional diesel collectors, they noted that the performance of micro-emulsified diesel was high at low dosage. Furthermore, the effects of collector (dodecane) and frother (MIBC) dosage were investigated by (Kazadi, et al., 2013) on medium-rank bituminous waste coal. It was found that increasing collector and frother dosage increases the yield of clean coal, however, the combustible recovery was poor.

Due to high demand of coal and extensive mining in the former coalfields (Witbank, Highveld and Ermelo) reserves become depleted. However, South Africa has remaining large unexploited coal reserves in Springbok flats, Free state, and Molteno area. The coal from these areas is relatively low quality compared to the coal from the previous mine reserves (De korte, 2015).

Currently, coal-processing industries facing different problems to process low quality coal to produce good quality (high-grade) products. This work investigating and optimizing the flotation process in low-grade quality coal as fine upgrade. This process removes the noncombustible materials in coal, which produces an upgraded coal quality.

2. Experimental

Material

Three different South African bituminous raw coals from Witbank area were chosen for the study. Diesel oil and MIBC were used as a collector and frother respectively, and CaO (lime) was also used to maintain the pH condition. Equipment used for flotation process, such as advanced flotation cell and all the required materials are acquired from the University of Johannesburg, Doornfontein campus mineral processing laboratory.

Methodology

Approximately 18kg coal sub-sample was obtained from the bulk large sample of 40kg. The coal sample was screened using standard screening procedure to separate into three different size fractions, the size fractions were -5mm (coal A), +5-10mm (coal B) and +10-19mm (coal C). According to Tshiongo-Makgwe (2017), the grindability of coal depends on the mineral content of the specific coal sample and grinding time. The coal sample A, B, and C were ground by controlled dry milling for about 25, 50 and 60 minutes respectively, to obtain P_{80} passing 300 μm. From the grinding test, the cumulative 80% passing 300μm of coal sample A, B, and C were 84, 86 and 83% respectively. Proximate analysis was conducted on the feed samples which includes the determination of ash content, moisture content, volatile matter, calorific value and Sulphur content (table 1). X-ray diffraction (XRD) analysis was employed to reveal the mineral composition of feed samples used in this study. All analysis was carried out at the University of Johannesburg Doornfontein campus extraction metallurgy laboratory.

Table 1. Proximate analysis of feed samples.

Parameters	Coal A	Coal B	Coal C
Calorific value (MJ/Kg)	20.43	21.03	21.08
Ash (%)	24.88	24.37	22.11
Moisture (%)	3.9	4.2	4.8
Sulphur (%)	0.41	0.66	0.61
Volatile matter (%)	23.7	24.77	23.57
Fixed carbon (%)	47.46	46.66	49.52

In this work different flotation operational parameters were investigated, such as % solid, flotation time, impeller speed, collector and frother dosages. After determining the intended flotation operation conditions, the test was conducted by using optimum conditions.

Laboratory flotation experiment was carried out by using advanced flotation machine with the cell volume of approximately 3.4 - liter. The coal slurry was conditioned for 2 minutes until the pH becomes stable calcium oxide (lime) was used to maintain the pH Collector (Diesel) was added and conditioned for 4 minutes, after which the frother (MIBC) was added and the pulp was conditioned for 2 minutes. The aeration was also introduced at 6 L/M and the froth was collected within 10 seconds interval. The product (upgraded) coal was oven dried at 40^0 C overnight, and the quality analyzed using proximate analysis.

3. Result and discussion

3.1 Effects of reagent dosage

To establish the optimum conditions of flotation process different parameters were performed in preliminary experiments. The results of the experiments clearly show that effects of collector and frother dosage in different coal samples. The results of coal flotation were evaluated in terms of ash content, calorific value and %yield.

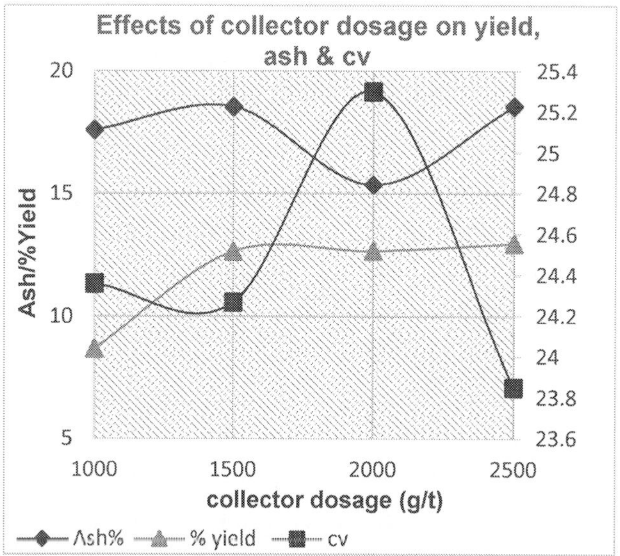

Figure 1 (a) Effects of collector dosage on coal A (-5mm coal milled to minus 300μm) at 150g/t frother, pH of 8.5.

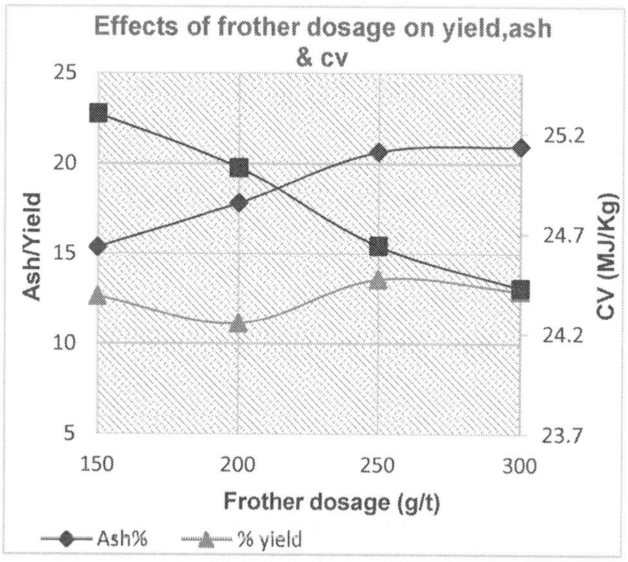

Figure 1 (b) Effects of frother dosage on coal A (-5mm coal milled to minus 300μm) at 2000g/t collector, pH of 8.5.

The figures above illustrate, the flotation results obtained using MIBC and Diesel as flotation reagents on coal A using different dosages. The flotation feed of about 24.88% ash and 20.43 MJ/Kg calorific value obtained, after cleaning by flotation process using collector dosage of 2000g/t and frother dosage 150g/t gives good quality coal. An increase of frother dosage to higher than 150g/t decreases combustibility (cv) from 25.3 to 24.43MJ/Kg and increases ash from 15.38 to 20.67%. The collector dosage also had a significant effect on flotation recovery of coal A. As diesel dosage increases from 2000 to 3000g/t while frother dosage was maintained constant at 150g/t, combustible recovery decreased from 25.3 to 23.85, and ash content was high, ranging from 15.38 to 18.57. This indicating that flotation selectivity was poor while increasing collector dosage.

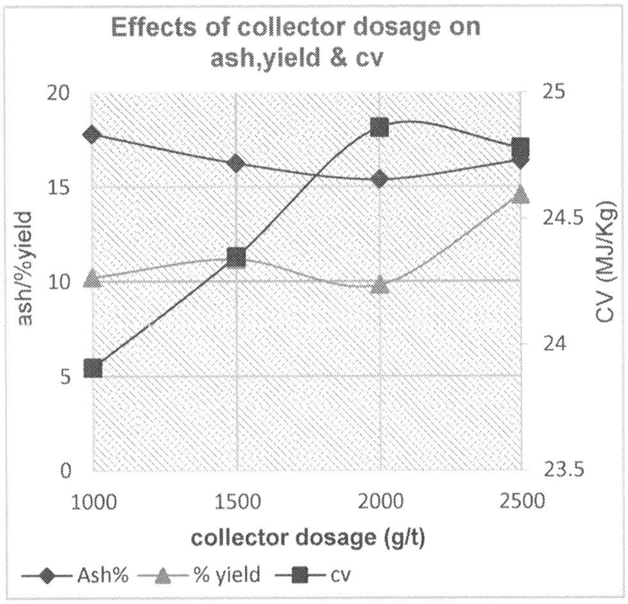

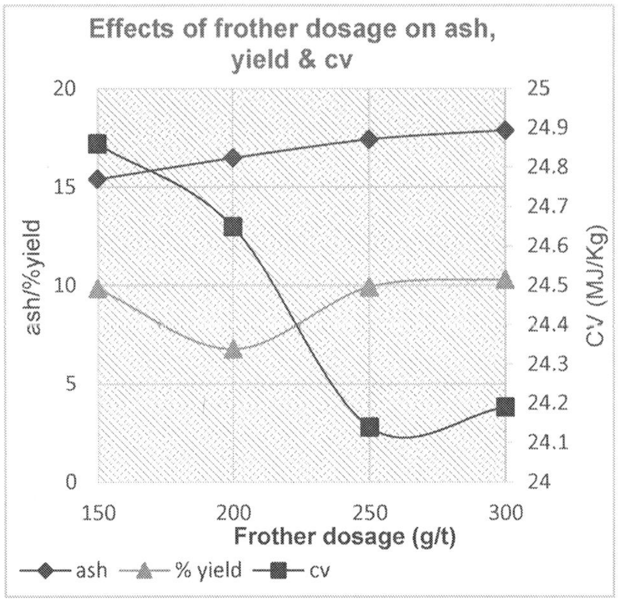

Figure 2 (a) Effects of collector dosage on coal B (+5-10mm coal milled to minus 300μm) at 150g/t frother dosage, PH of 8.5.

Figure 2 (b) Effects of frother dosage on coal B (+5-10mm coal milled to minus 300μm) at 2000g/t collector dosage, pH of 8.5.

As the figure above shows, the effects of frother and collector dosage on ash, calorific value and % yield of coal B. Floatation efficiency was more effective at the collector and frother dosage of 2000 and 150g/t respectively. Increasing collector dosage from 1000 to 2000g/t decreases the ash content by 2.39% and calorific value increases from 23.91 to 24.86 MJ/Kg. While collector dosage is kept constant at 2000g/t and increasing frother dosage from 150 to 300g/t increases the ash content and decreases the calorific value by 2.47% and 0.67MJ/Kg respectively. This indicates increasing the frother dosage decreases the selectivity efficiency. However, the % yield of clean coal is very low (9.83%).

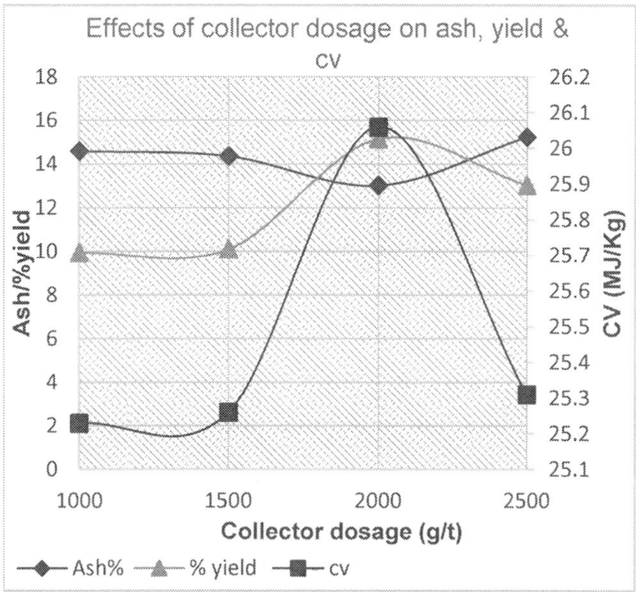

Figure 3 (a) Effects of collector dosage on coal C (+10-19mm coal milled to minus 300μm) at 150g/t frother dosage, pH of 8.5.

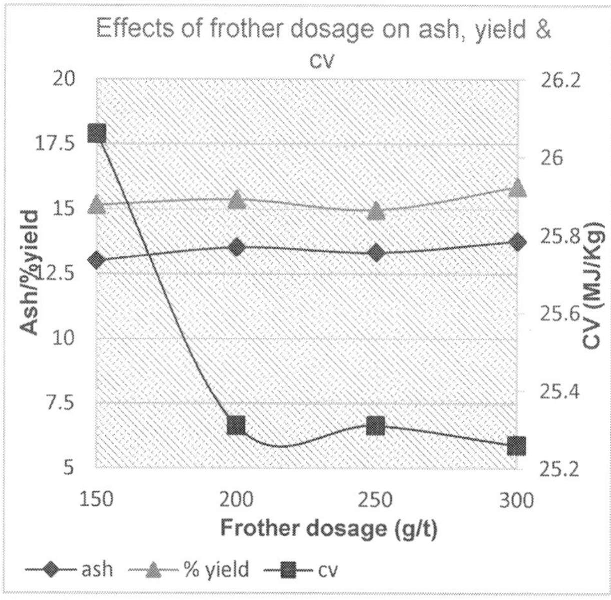

Figure 3 (b) Effects of frother dosage on coal C (+10-19mm coal milled to minus 300μm) at 2000g/t collector dosage, pH of 8.5.

The above figure shows, effects of collector (Diesel) and frother (MIBC) dosage on ash, calorific value and yield of coal C. Increasing collector dosage increases calorific value and % yield from 25.23 to 26.06 and 9.95 to 15.16 respectively, simultaneously the ash content decreases by 1.58%. Increasing frother dosage from 150 to 300g/t decreases calorific value from 26.06 to 25.26. Ash content and % yield decreases by only 0.76% and 0.71% respectively. The previous study by (Tshiongo-Makgwe , 2017) shows, increasing the reagent dosage (Montenol 55) improves the coal quality (low ash and high calorific value), however, in this study further increasing the collector and frother dosage decreases the coal quality.

In each coal sample flotation, the combustible recovery and ash rejection were increased at the optimum frother and collector dosage of 150g/t and 2000g/t respectively. Nevertheless, coal C has a high combustible recovery and low ash content compared to coal A and B. High ash content attributed to high mineral content in the coal. From the mineralogical analysis, coal A and B has high silica minerals compared to coal C, which may increase the ash content (Siddhartha , 2006). The product clean coals can meet the specific requirement of Electricity, Synthetic fuel, and Domestic markets by calorific value and an ash content of 25.3, 24.86, 26.06 and 15.38, 15.41 and 13.02 respectively (Steyn & Minnitt, 2010).

3.2 Flotation time

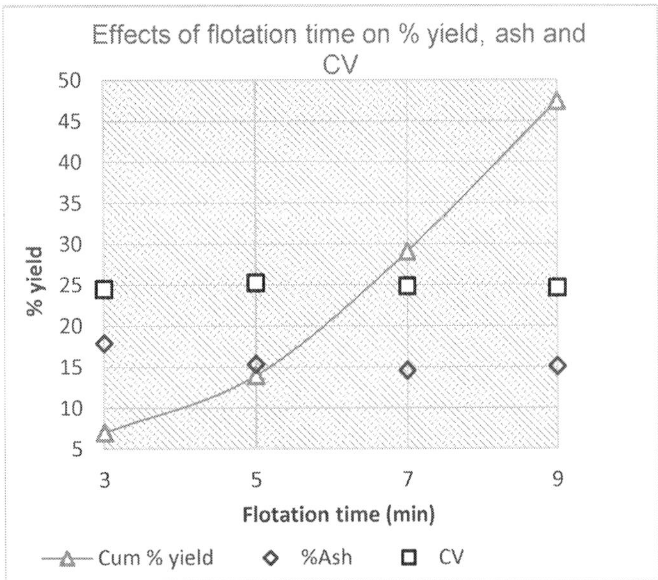

Figure 1 (a) Effects of flotation time on coal A

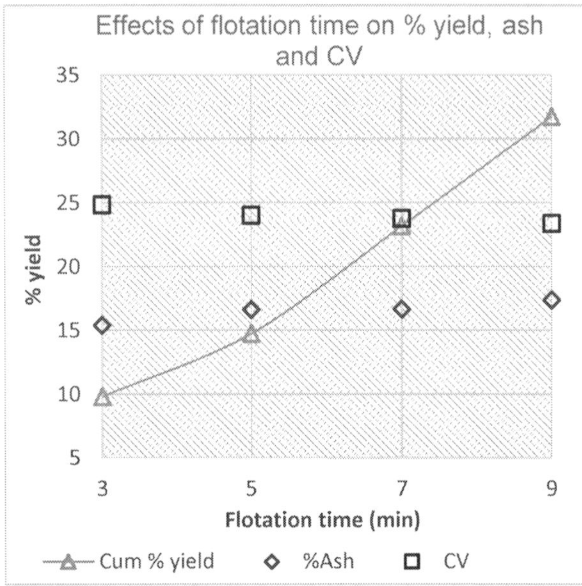

Figure 4 (b) Effect of flotation time on coal B

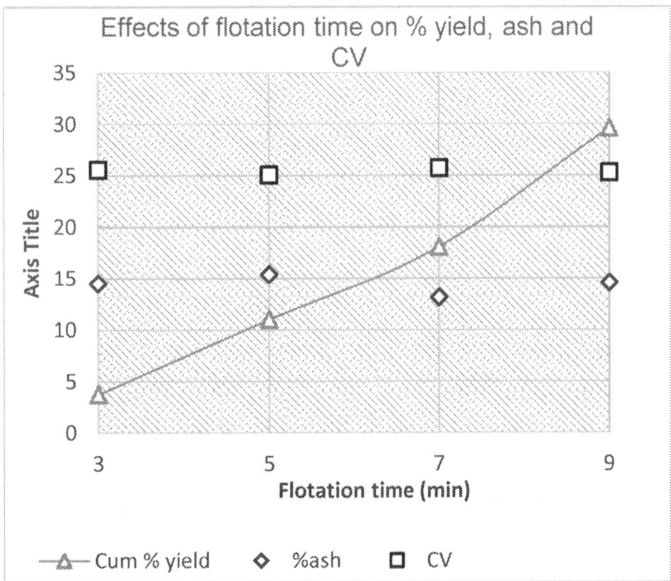

Figure 4 (c) Effects of flotation time C

The graph above illustrates the effects of flotation time on % yield, ash and calorific value. Increasing in flotation time from 3 to 9 minutes increases the % yield by 40.43, 21.98 and 25.85% respectively, however, further increase in flotation time decreases the quality of final product.

4. Conclusion

It should be emphasized that the enhanced flotation performance of South African low-grade coal with collector (diesel) and frother (MIBC) was achieved with good quality coal at much lower reagent dosages. The substantially lower dosage of reagents would considerably reduce the flotation cost and increase the grade of coal in industrial practice. Lower MIBC and Diesel dosage yielded higher combustible recovery and ash rejection, However, % yield was lower, which could not be cost-effective. The current price of diesel and MIBC is 16.12 and 31.68 rand per liter respectively. To clean 10 tons of raw coal, 24 liters of collector (diesel) and 1.85 liters of frother (MIBC) will be needed, which will cost about a total of 445.48 rand. The clean coal product of sample A, B and C from 10 tons of raw coal will be 1.2, 0.98 and 1.5 tons respectively. At the current price of coal in the domestic market of 332.75 rand per ton (Modisane & Ramaano, 2017). this indicates, the process is not economically viable. From the experimental result increasing frother dosage affects the combustible recovery and ash rejection, it is recommended that a further study on MIBC dosage of less than 150g/t be carried out. It is also observed that the % yield of clean coal was very low, it deserves further investigation on another type of collector.

5. Acknowledgment

The authors would like to acknowledge;

- The University of Johannesburg Extraction Metallurgy and Mineral processing laboratories for facilitating the laboratory equipment.
- The Southern Africa Coal Processing Society (SACPS) for financial support.

References

Anderson , C., 2015. *Evaluation of the flotation response of a waterberg coal.* Secunda , The South African coal processing society.

Atesok, G. & Celik, M. S., 2000. A new flotation scheme for a difficult-to-float coal using pitch additive in dry grinding. *FUEL,* Volume 79, pp. 1509-1513.

Behzad, V. H., Fırat , K. & Mehmet, S. C., 2014. Ultrafine coal dewatering: Relationship between hydrophilic lipophilic balance (HLB) of surfactants and coal rank. *International Journal of mineral processing,* pp. 97-104.

de Korte, G. J., 2015. Processing low-grade coal to produce high-grade products. *Journal of the South Africa institute of mining and metallurgy,* pp. 567-572.

Horsfall, D. W., Zitron, Z. & De Korte, G. J., 1989. The treatment of ultrafine coal, especialy by froth flotation. *Journal of the South African institute of mining and metallurgy,* 86(10), pp. 401-407.

Hussin , A. A. & Jan , D., 2004. Effect of flotation procedure and composition of reagents on yield of a difficult-to-float coal. *Physicochemical problems of mineral processing,* Volume 38, pp. 53-63.

Jeffrey, L., 2005. Characterization of the coal resource of South Africa. *Journal of the South Africa institute of mining and metallurgy,* pp. 95-102.

Kazadi, C. M., Franzidis, P. J., Harison, S. T. L. & Broadhurts, J. L., 2013. Flotation of coal and sulphur from South African ultrafine colliery wastes. *The Journal of the South African Institute of Mining and Metallurgy,* Volume 113, pp. 399-405.

Le Roux , M. & Cambell, Q. P., 2003. An investigation in to an improved method of fine coal dewatering. *Mineral engineering,* Volume 16, pp. 999-1003.

Le Roux , M., Campbell , Q. P., Watermayer, M. S. & De Olivera , S., 2005. The optimization of an improved method of fine coal dewatering. *MINERALS ENGINEERING,* pp. 931-934.

Li, L., Lu, X., Qiu, J. & Liu, D., 2013. Effect of microemulsified collector on froth flotation of coal. *The Journal of The South African Institute of Mining and Metallurgy,* Volume 113, pp. 877- 880.

Modisane, M. & Ramaano, N., 2017. *South African Energy Price Report ,* Pretoria : Department of Energy.

North , B., Engelbrecht, A. & Oboirian, B., 2015. Feasibility study of electricity generation from discard coal. *Journal of South African institute of mining and metallurgy ,* Volume 115, pp. 573-580.

Siddhartha , B. M., 2006. Effect of ash in coal on the performance of coal fired thermal power plants.. *Energy source ,* Volume 28, pp. 25-41.

Steyn, M. & Minnitt, R. C. A., 2010. Thermal coal production in South Africa. *The Journal of The South African Mining and Metallurgy ,* Volume 115, pp. 593-599.

Tao, D., Li, B., Johnson, S. & Parekh, B. K., 2002. A flotation study of refuse pond coal slurry. *Fuel processing technology,* Volume 76, pp. 201-210.

Tshiongo-Makgwe , N., 2017. *Quality improvements for South African fines by Flotation with Montenol and Agglomeration.* Johannesburg, The Journal of the South African Institute of Mining and Metallurgy.

Zhinqiang , Z., Coili , W. & Kefeng, Y., 2015. Adsorption of collectors on model surface of wiser bituminous coal ; A molecular dynamics simulation study. *Mineral Engineering,* pp. 31-39.

Introduction to speaker

Mark Cresswell

- Minerals Processing Consultant
- University of London - BSc (Eng) - Royal School of Mines 1974 –77, MSc - 1983
- 1977 - Started with Head Wrightson (UK) who had a DSM (Dutch State Mines) license for designing DMS cyclone plants
- 1985 - No new coal processing plants in the UK after early 1980's, so moved to South Africa and joined VEL in Johannesburg who also had a DSM license. Worked on the design of a large variety of minerals processing plants.
- 2005 Joined DRA when a new phase of large throughput coal plants (max 2 400t/hr) were being planned.
- 2006 – 2014 Lead Process Engineer for a series of plants all using the latest DM Cyclone technology with capacity of 950 – 4 000 t/hr.
- 2014 Minerals Processing Consultant

DRA // 2

About DRA Global

- DRA Global is a diversified global engineering, project delivery and operations management group, with an impressive track record spanning over three decades.
- Known for its collaborative approach and extensive experience in project origination and delivery, as well as turnkey operations and maintenance services, DRA Global delivers optimal solutions that are tailored to meet clients' needs.

DRA

Global Footprint

OVER 4300 STAFF WORLDWIDE

OVER 3 DECADES OF EXPERIENCE

OVER 3000 → PROJECTS → & STUDIES COMPLETED SUCCESSFULLY AROUND THE WORLD

LARGE ANNUITY BASED EARNINGS

17 OFFICES AROUND THE GLOBE

DRA

Coal Technology Changes over the last 40 years

Sector	What has changed
Mining	Massive Open Pits up to 50 Mtpa
Crushing	MMD, Abon Double & Single Roll Crushers, Disc Roller Screens
Large Coal – (300 x 25)mm	Larcodems, LD cyclones
Small Coal (50 x 0.5)mm	LD Cyclones, Banana screens, FGX dry separator, 3 Product Cyclone, Evolution of Flowsheets
Fine Coal (2 x 0.15)mm	LG Spirals, Reflux Classifier
Slimes (0.25 x 0)	Column & Jameson Flotation Cells
Dewatering	Hyperbaric & Steam filtration, Screen Bowl Centrifuges, Bokela Filter, mass production of Filter Presses (China) Thickener Feed well design
Materials Handling	Conveyor design, Stockpile design
Design & Construction	CAD, Ground Floor up construction

DRA // 8

Coal Technology - Mining

- Larger throughput open cast mines, particularly in South Africa, Australia, Western USA, China
- Less selectivity of coal seams in open pit mining
- In-pit crushing option
- More fines generation from mechanized underground mining methods
- Higher raw coal moisture content through use of de-dusting water
- Possible introduction of Ore Sorting in the pit

DRA // 7

Coal Technology - Crushing

Huge increase in Crushing Capacity

- Abon Double Roll Sizer – 2 500 t/hr
- Similar models from MMD

DRA

// 8

Coal Technology - Crushing

High Capacity Disk Roller Screens - 50mm aperture, 2 000 t/hr feed

DRA

// 9

Coal Technology - Innovation In Crushing Circuits

Wood removal after primary crushing

DRA

// 10

Coal Technology - Bradford Breaker

Still a unique rotary secondary crusher for raw coal as well as producing up to 10% of coarse discard removal. Particularly useful for Open Pit mines.

DRA

// 11

Coarse Coal Separation - very little change over the years, except for better automation.

/ Large Baths:
- Drewboy
- Tesla Bath
- Wemco Drum
- Disa Bath
- Peters Bath

/ Jigs:
- Baum Jigs still favoured in India

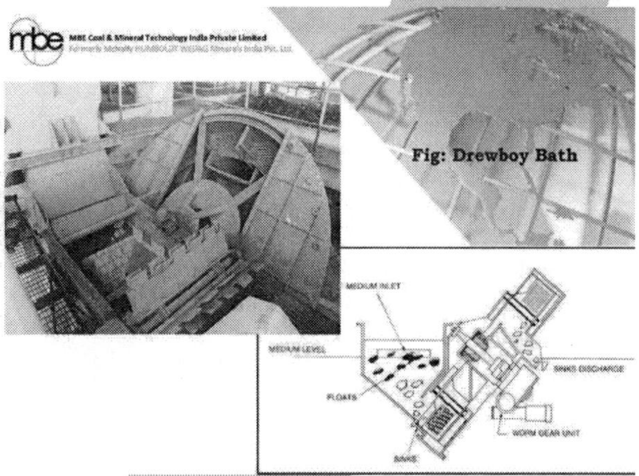

Fig: Drewboy Bath

DRA

Small Coal Technology (50 x 1)mm - Large increase in Cyclone Capacity

Comparison of Cyclone capacity. Krebs, Multotec, Minco all have similar cyclone ranges.

Standard DM Cyclone Capacities					
Multotec cyclone capacities at 9D inlet pressure			DSM Capacities at 9D inlet pressure		
mm dia	m3/hr	coal (t/hr)	mm dia	m3/hr	coal (t/hr)
250	23	9	200	16	6
360	63	25	350	56	22
420	92	37	400	77	31
510	142	57	500	125	50
			560	155	62
610	210	84	600	190	76
660	248	99	660	225	90
710	290	116	700	265	106
			750	300	120
800	372	149			
900	474	190			
1000	589	236			
1150	784	314			
1300	1,006	402			
1450	1,255	502			

Checking out the use of 1150mm dia LD Cyclones in Australia before building them in South Africa.

DRA

LD cyclones can treat up to 90 mm topsize, but normally treat a (50 x 1)mm size fraction.

Extension of Cyclone Capacity

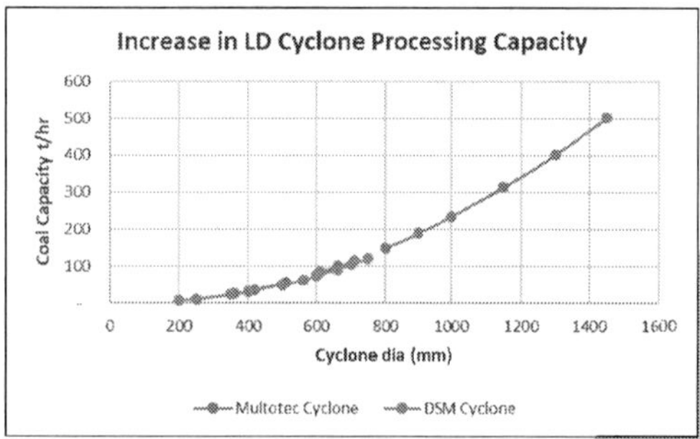

Cyclone Spigot Capacity Increase

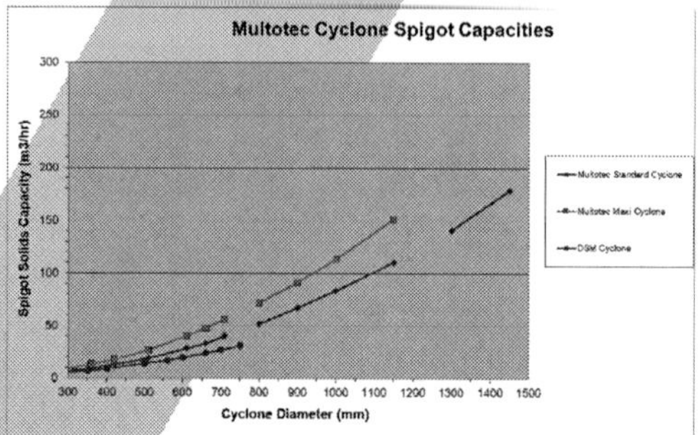

DRA // 15

DM Cyclone Sizing - results from testing ~40 Australian plants

Driven by Chinese coking coal demand, production in Australia increased during the early 2000's with at least 40 new washing plants. The market wanted cheap, simple and reliable plants which drove the development of the LD Cyclone so that one cyclone could replace multiple small ones. Initial worries that LD cyclones would be inefficient compared with smaller ones led to studies of the plants.

Audit Results – Effect of Cutpoint +4 mm

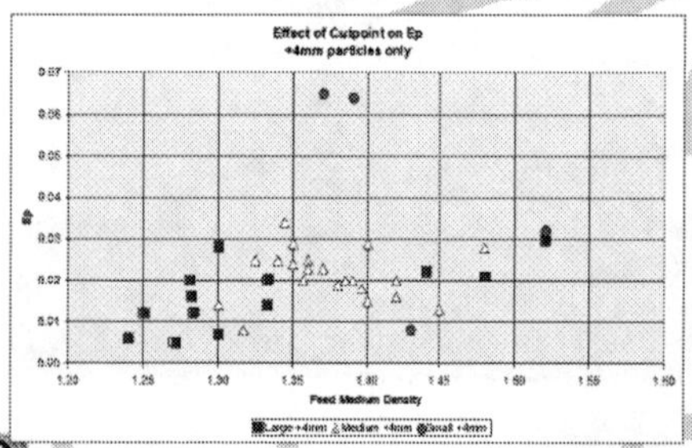

Audit Results – Effect of Cutpoint -4 +1 mm

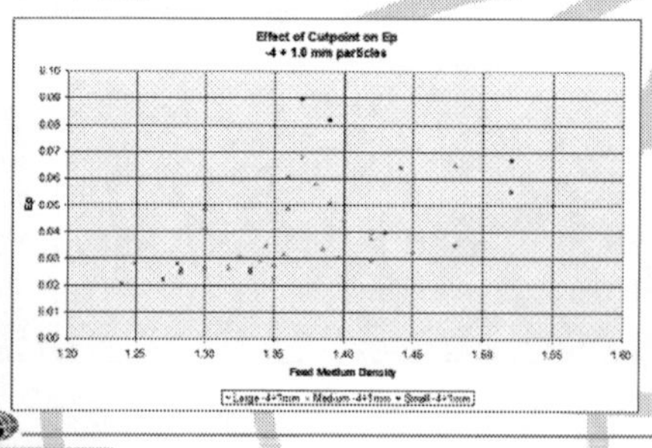

DRA // 16

DM Cyclone Sizing - results from testing ~40 Australian plants

No discernable difference in efficiency using LD Cyclones down to ~2mm.

At lower bottom sizes, the size where the epm increases significantly, the "breakaway epm" does follow diameter differences.

Audit Results – Effect of Diameter All size fractions

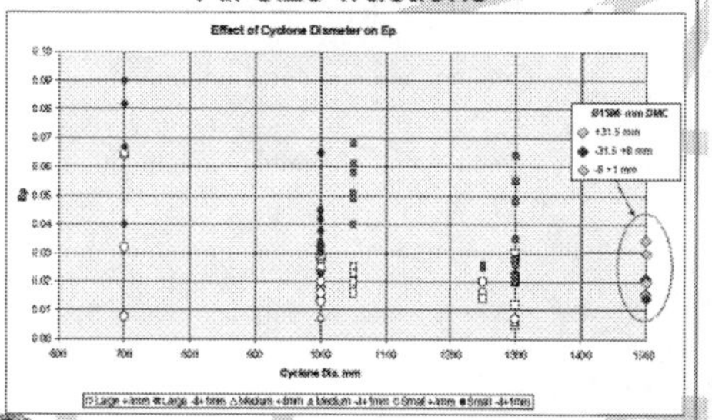

Audit Results – Summary of Findings

- M:C ratio
  - Negligible effect on +4.0 mm down to ratios as low as 2.5:1
  - Does affect efficiency on -4.0 +1.0 mm up to a ratio of 4:1
- Increasing feed pressure improves Ep
- Increasing medium density worsens Ep, but mitigated to some extent by amount % near gravity and magnetite sizing.
- No significant Ep effect using cyclones up to Ø1300 mm

DRA

Banana Screens for Desliming, Drain & Rinse duties

Use of Banana Screens has significantly increased unit capacity

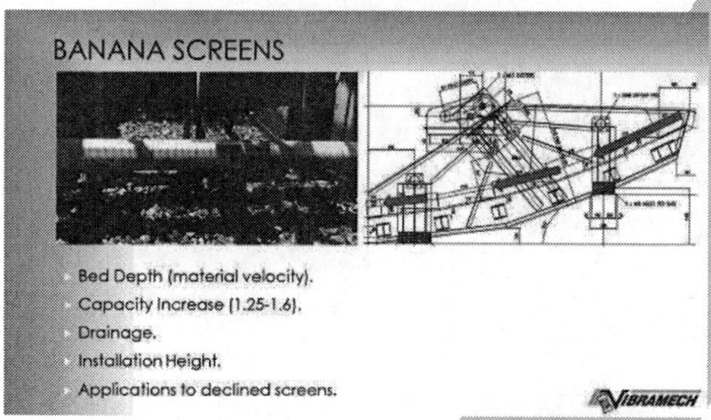

The sloping deck increases the travel velocity of the bed which in turn makes it thinner.

This reduces the load on the screen and enables it to process significantly higher capacities.

DRA

Banana Screens go through extensive design tests during construction

Schenk Screens

DRA // 19

Schenk Banana Screens in a two module 2 000 t/hr Coal Processing Plant

DRA // 20

Evolution of Coal Processing Flowsheets - Lo Density / Hi Density

South African high throughput and efficiency circuit for medium yield thermal coals – 1200 t/hr with high "near-density" material

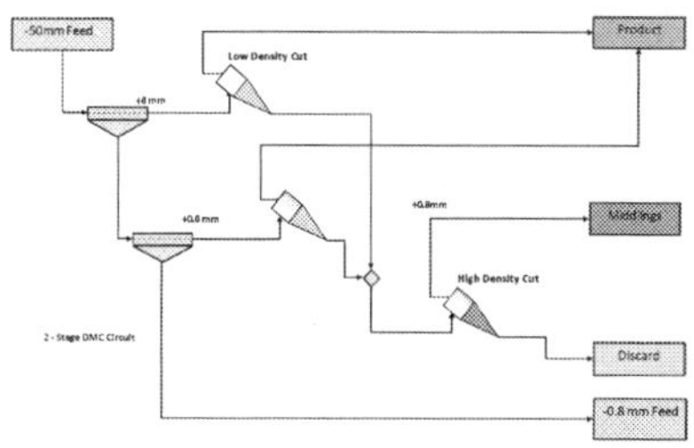

Simpler Australian circuit – 1000 t/hr for high yield coking coals with low "near-density" material.

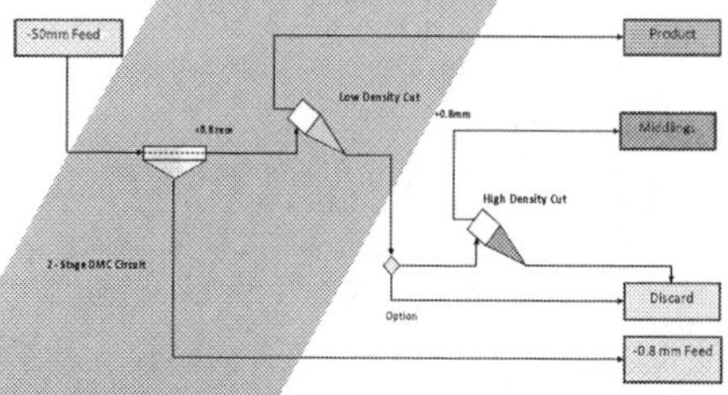

DRA

Evolution of Coal Processing Flowsheets - Lo Density / Hi Density

South African high throughput and efficiency circuit – 1200 t/hr – more cyclones

Simpler Australian circuit – one cyclone per stage

DRA

Evolution of Coal Processing Flowsheets - Hi Density / Lo Density

Coking Coal Circuit for low yield Mozambique coals – 1 000 t/hr

Single DM Cyclone per stage and inter-stage crushing to improve liberation

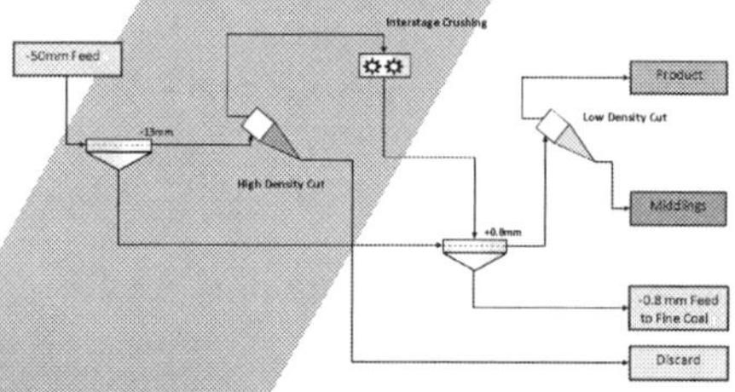

DRA // 23

Evolution of Coal Processing Flowsheets - Fine Coal

Standard Spirals circuit from 1980s

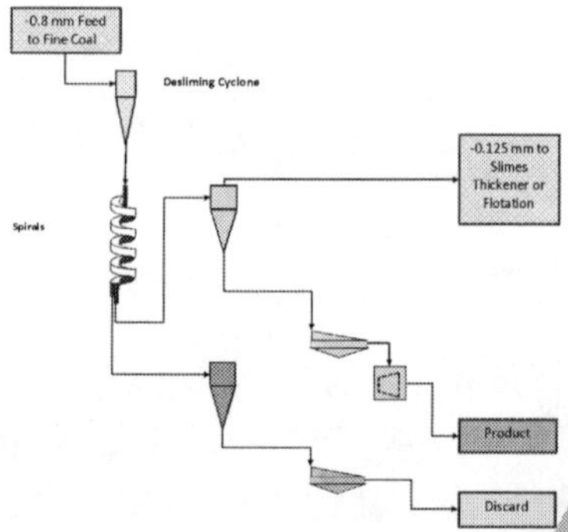

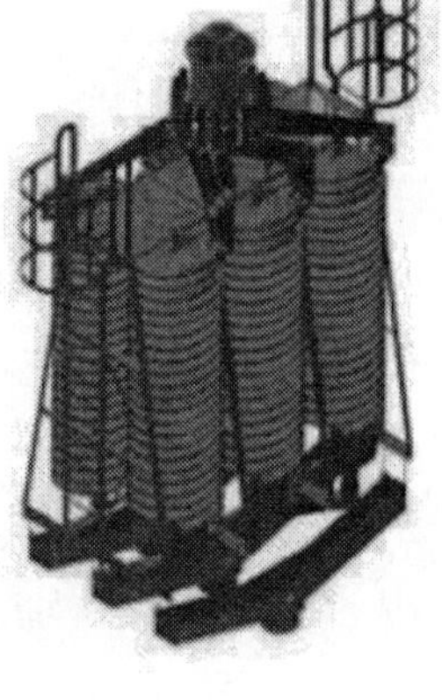

DRA // 24

Fine Coal - New Spirals Development

- Traditional Coal Spirals limited to separating at a density ~ 1.7 SG
- New design "LC3" Spirals by Mineral Technologies, have lowered the cutpoint to 1.4 - 1.5 SG
- Should enable higher overall yields by allowing optimizing of the DMS circuits

DRA

// 25

Evolution of Coal Processing Flowsheets - Fine Coal

Introduction of TBS separator in a Spirals circuit from 2000s

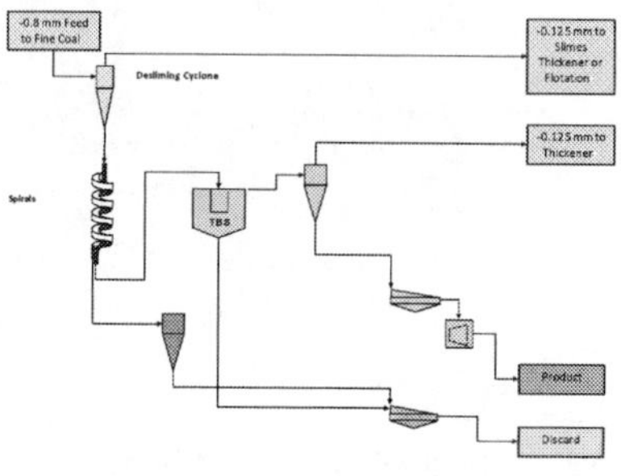

TBS Separator - Mt Arthur North, Australia

DRA

// 26

Evolution of Coal Processing Flowsheets - Fine Coal

Replacement of TBS separator by the Reflux Separator from 2010

A REFLUX™ Classifier (RC™) is a combination of a:

- Lamella Settler
- Autogenous dense medium Separator
- Fluidized Bed Separator

A REFLUX™ Classifier is a volumetric unit.

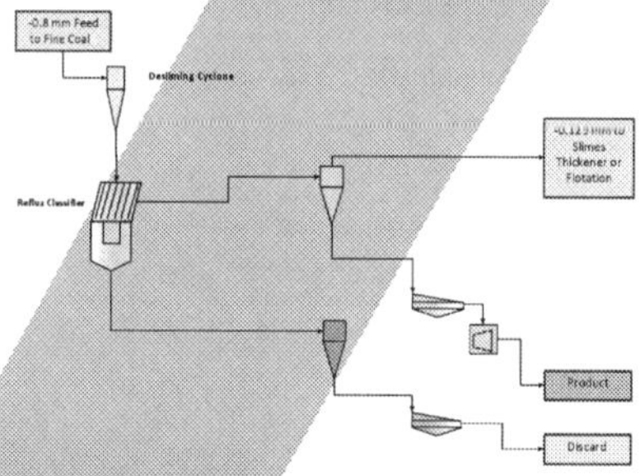

Slimes Treatment - Flotation

Jameson Cell Mk4 - Benga Coking Plant, Mozambique

Slimes Treatment - Flotation

Alternative technology - Eriez Column Cells - Moatise Coking Plant, Mozambique

DRA

// 29

Evolution of Coal Processing Flowsheets

General Trends in Wet Coal Preparation Plants

South Africa, Australia, India - 1970's - 2010

1970's	**Coarse Coal** -(120 x 12)mm *Bath/Wemco Drum/Jig*	**Small Coal** -(25 x 0.5)mm *DM Cyclones / Jig*		**Slimes** -0.5mm *Discard or Flotation*
1980's	**Coarse Coal** -(120 x 12)mm *Bath/Wemco Drum/Jig*	**Small Coal** -(25 x 0.63)mm *DM Cyclones / Jig*	**Fine Coal** -(0.63 x 0.125)mm *Spirals / DM cyclones*	**Slimes** -0.125mm *Flotation or Discard*
1990's	**Coarse + Small Coal** -(90 x 12, 12 x 0.63)mm or -(50 x 0.63)mm *(Bath + DM Cyclones) or DM Cyclones*		**Fine Coal** -(0.63 x 0.125)mm *Spirals*	**Slimes** -0.125mm *Flotation or Discard*
2000's	Coarse + Small Coal *(combined - Austr)* Coarse + Small Coal *(separate - RSA)* -(50 x 0.63)mm *DM Cyclones*		**Fine Coal** -(0.63 x 0.125)mm *Spirals / DM cyclones* *Classifier / DM Cyclones*	**Slimes** -0.125mm *Flotation* *or Discard (RSA only)*

DRA

// 30

Design & Construction - Pipe Clash detection using CAD

- Use of CAD drawing techniques particularly "Clash Detection" when laying out pipes and electrical cable racks, saves significant time during the construction phase.
- Ability to design "around the clock" by using teams situated around the world e.g. main CAD team in South Africa supported by a specialist CAD team in India.

Examples of Pipe clashes with steelwork

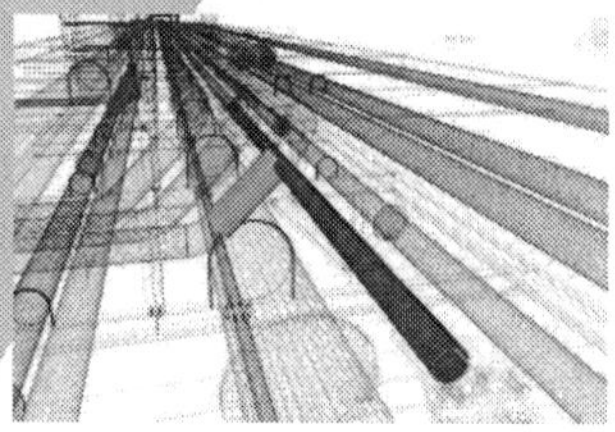

DRA // 31

Use of fully equipped modular Conveyor structures (made in China)

DRA

Construction of 300 000t Stockpiles above ground using pre-cast concrete

Pre-cast Culvert Tunnels

DRA

Plant Washdown - making it easier using round concrete plinths for Columns

Construction

Washing Plant

Primary Crushing

DRA // 35

Importance of Placing the pump tanks correctly

Starting point of DMS design

Placing Pump tanks in a row

DRA // 36

Old Method of Constructing Large Plants was to first erect the Columns

Previous "Stick" Assembly Method

Site Conditions in the summer rainy season

DRA

// 37

Assembly of Plant floor by floor

Modular Assembly on Ground Level

DRA

// 38

Example of how clean a modern DMS coal processing plant can be

Glencore - Mt Owen, Australia

DRA // 39

Coal washabilities: Comparison between traditional methods and the RhoVol analyser

[1]Q.P. Campbell, [1]M. le Roux, [2]M. Fofana

[1]North-West University, Private Bag X6001, Potchefstroom 2520, South Africa;
[2]DebTech, PO Box 82851, Southdale 2135, South Africa

ABSTRACT

The determination of coal washability is a tedious and expensive exercise for finer particle ranges. The recent development of the RhoVol machine by DebTech can bring about a step change in coal washability determinations over the size range 25 mm to 3 mm. It can at the same time measure the density, size and shape of a given population. A project was done to compare RhoVol results to densimetric and size data using traditional zinc chloride float-sink procedures on a Witbank coal sample in the size range 3 mm to 8 mm. The results indicate that the RhoVol can be used in a research environment to facilitate simple, rapid and safe size by washability information for coal samples, provided some pre-calibration is done.

KEYWORDS: Washability determination, RhoVol, densimetric curve

BACKGROUND

The key to coal preparation plant design and performance management is densimetric information of the coal feed or products. In a research environment, such a densimetric information is also important, since any work on novel beneficiation processes, especially for fines, involves a thorough understanding of the density profile of the population of particles involved. It is also often required that a given coal sample must be split into narrow density classes for further experimental work. The determination of washability information is usually done using laboratory float-and-sink techniques, which are often tedious and hazardous. For finer particles especially, the accuracy of float-and-sink data becomes problematic.

The RhoVol densimetric measurement system, developed by DebTech, provides an alternative to laboratory densimetric procedures. This machine can measure a material's density profile using a camera-based imaging system for optical particle size and shape measurement combined with a sensitive balance for mass determination. These measurements are used to determine a density value for each individual particle in a population, which can then be summed over the entire population. The advantages of the RhoVol is a safe and rapid densimetric analysis with additional information on particle size and shape. It is also non-destructive, meaning that the same sample can be used repeatedly.

This study will compare data obtained from traditional laboratory float-and-sink analyses to that of the RhoVol, and provide recommendations on its applicability and accuracy. Particle size comparisons are also made.

LITERATURE SURVEY

Float-and-sink tests

Float-and-sink analysis is a standard laboratory method performed on ores to assess the response to dense medium separation, and to determine some economical predictions (Wills and Napier-Munn, 2006). Dense medium separation is still considered to be the most effective technology to separate ores based on density (Kong *et al.*, 2018). Since coal quality is related to its mineral content, and there is a direct correlation between the ash forming minerals in coal and the density (Sivrikaya, 2014), hence a densimetric analysis is an indication of the quality of a coal. The shape of the densimetric curve also provides information about the difficulty of separation, as well as the yield and expected efficiencies and specified cut points. Float-and-sink analysis are also routinely used on coal preparation plant products and discards to evaluate plant performance (Bhattacharya & Anand, 1998).

Problems associated with conventional float-and-sink analysis include the high cost of medium and the adverse health and environmental effects of especially organic materials (Aktas *et al.*, 1998). Also, for finer coals, the method is very time consuming due to long settling times (Franzidis & Harris, 1986).

The porosity of coal is one factor that influences the accuracy of float-and-sink results. Poor separation occurs when the porosity is too high because $ZnCl_2$ solution can fill the empty macropores of the coal during the process. If there is moisture present in the coal pores, it dilutes the $ZnCl_2$ solution resulting in a decrease in the $ZnCl_2$ solution density. Since the solution diffuses far into the pores, a density gradient occurs from the inside of the pore to the coal particle surface. This leads to a variation of both the liquid density and the apparent density of the coal particle (Aktas *et al.*, 1998). This could result in particles reporting to the incorrect density fraction, compromising the float-and-sink results.

For porous materials like coal different density measurements, like the true skeletal density and the apparent "envelope" density differ (Mahajan, 1982). Hence it can be concluded that the density as determined by float-and-sink is somewhere between the skeletal density and the envelope density, depending on the pore characteristics like pore size distribution, and whether they are blind or open pores.

Density analysis by helium pycnometry

Pycnometry is a well-established method involving the determination of the volume of solid matter in a closed container by comparing the pressure of the surrounding gas before and after expansion into a known volume. These pressure differences can be related to volume differences, and hence the sample volume, by using the ideal gas relationship. Because of its small molecular diameter and inert nature, helium is used as a gas that gives precise measurement of the pore volume, resulting in the determination of the skeletal density (Rodrigues & de Sousa, 2002).

The RhoVol density analyser

Figure 1: RhoVol density analyser (courtesy of DebTech)

The RhoVol was originally developed by DebTech for the density distribution determination of kimberlitic ores. The suitability of the technique to be applied to other materials was recognised, and this particular project was defined. The RhoVol is a densimetric measurement system that determines the density of an ore by weighing and by estimating the volumes of individual particles (Fofana & Steyn, 2018). Apart from the weight and volume parameters, the RhoVol also gives particle information about density, size (specifically an inferred sieve size) and particle shape. Shape descriptors include compactness, flatness and elongation (DebTech, 2018).

The machine operates in an automated batch mode, processing 1 000 particles average per hour. During its operation the sample can be sorted into discrete bins based on the operator's choice of measured parameters; density, volume or shape. There are currently two models available; a fine particle version that can handle particles of sizes between 8 mm and 3 mm (the one used in this project), and a coarse particle unit for 25 mm to 8 mm particles. The weighing of a particle occurs on a high-speed weighing cell, and the limits of the scale used the present project is between 10 mg (limit set in RhoVol software) and 150 g on the 8mm to 3mm unit (DebTech, 2018). The Coarse unit can measure up to 250g maximum.

The volume parameter is determined from a 3D reconstruction of the particle generated by multiple silhouette images captured by seven cameras situated in the machine at different angles. A 3D model, called the visual hull, is then generated (Mangera *et al.*, 2016). By definition, a visual hull is the single, largest object that is consistent with a finite set of available silhouettes (Forbes *et al.*, 2003). This defines the envelope volume of the particle.

It has been shown in previous studies that any visual envelope-based volume reconstruction methods overestimate the particle volume (Forbes *et al.*, 2003), especially for non-regular shaped particles. This is also true for the RhoVol, where mean volume estimation errors of up to 20% have been found by Mangera *et al.* (2016). Figure 2 shows the volume discrepancy between the true volume (ground truth) and the visual envelope particle volume for three different shape particles, round (BB1), elongated (EL3) and flat (FL1).

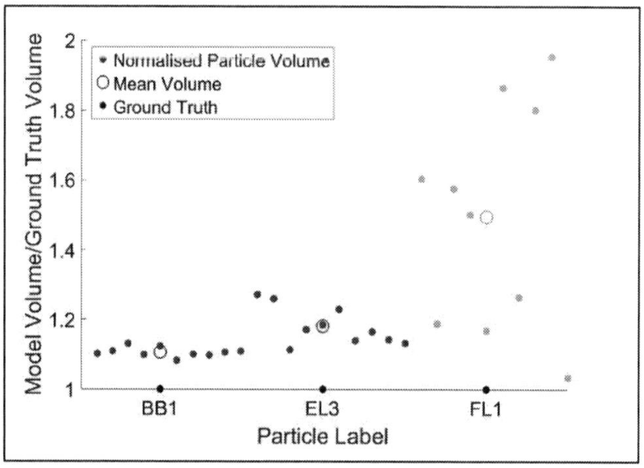

Figure 2: Normalized volume distribution of 3 particles captured 10 times in the RhoVol system compared to the true volume (Mangera et al., 2016).

From this, it is clear that the RhoVol consistently overestimates the volume of the particles. The particle shape plays a significant role in the degree of overestimation of volume by the RhoVol; it is easier to accurately determine the volumes of rounded or elongated particles than those of flat particles. The effect of shape on the volume estimation is mainly due to particle orientation during the image capture process. According to Mangera *et al.* (2016) and Forbes *et al.* (2003), varying orientations of particles allow for different silhouettes to be used to reconstruct more accurate visual hulls. For example, a silhouette of the side of a flat particle may generate a very different hull than that generated from the silhouette of the face of the same particle.

To improve on the 3D re-constructed volumes of particles, a multi-pass mode, which allows for the capturing of particles multiple times (DebTech, 2018) is used. When doing this, the RhoVol has the capability to recognise an individual particle based on a stored database of silhouettes and then tags it. When it recognises a particle again during subsequent passes, the silhouettes from the first pass are combined with the silhouettes of the second pass to construct a higher fidelity 3D model of the particle, thereby increasing volume estimation accuracy. RhoVol however will not be able to determine the ultimate volume of a particle due to the fact that porosity and surface texture of the material being measured cannot be determined with this method of volume determination.

Thus, a volume correction factor will always be required to align the particles' true volumes with the RhoVol estimated volumes. Two methods of generating volume correction factors are possible. The first one is termed "bulk correction factor", which is achieved by comparing a sample's total bulk volume obtained from the RhoVol, to the same sample's total bulk volume measured using techniques such as pycnometry or Archimedes. In this instance, a common volume correction factor is assigned to every particle as the sample is treated through the RhoVol. This correction factor compensates for porosity and surface texture issues as well and produces highly repeatable results.

The second method is the so called "dynamic correction factor". Since the error of visual hull is mainly as a result of shape, RhoVol utilizes a shape dependant volume correction factor that allows for more accurate results. The unique correction factor function is determined experimentally for a particular material by comparing the raw RhoVol volumes, after sorting in ten shapes categories, to the volumes of the same sub-samples measured using gas (helium) pycnometry. Ultimately, a relationship between shape factor and volume correction factor is established. In this scenario, every particle with a shape factor, in the sample, is assigned its unique volume correction factor from the "dynamic correction factor" plot.

EXPERIMENTAL

Coal samples

Three coal samples from different sources were used in the study, but only one is described in this paper, as described in table 1. The samples were screened to exclude any +8 mm and -3 mm material, which are the limits of the particular RhoVol machine that was used. Sufficient material was prepared for all the analyses to be performed.

Table 1: Origin and properties of the sample used.

Sample	Properties	Comments
Witbank blend	Size 3 mm – 8 mm Moisture (AD) 3.5% Volatile matter (AD) 19.2% Ash yield (AD) 38.7%	This sample consists of a mixture of leftovers from the Witbank area used in previous experiments; it is not representative of any particular coal source.

The RhoVol BDD127SM density analyser situated at the DebTech campus in Crown Mines, South Africa, was used. While it was possible to use the separation capability of the RhoVol to exclude extremely flat or round particles for improved volume estimation, the entire population including all shapes was used for this project.[1] For each type of coal, a sample of about 3 000 particles, equivalent to about 400 g - 500 g, was put through the machine using a historical dynamic correction factor for a previously tested coal from Leeuwpan. It was assumed to be similar to the Witbank sample, at least for this purpose.

While the sample was analysed for size, shape, mass and consequently density, on a particle-by-particle basis, the machine was set to simultaneously physically

[1] It is was reported by Mangera *et al.* (2016) and Forbes *et al.* (2003) that these particles would add to inaccuracies of the optically determined volume measurements, but for this study, the calibration procedure was assumed to compensate for this.

separate the sample into 10 density bins. The separation criteria for each bin were determined by dividing the entire density range (1.2 – 2.2 g cm^{-3} in this case) by 10, hence the exact values of the interval density limits for each bin was rather arbitrary. In the interest of time, only one RhoVol pass was used in this project, but it is acknowledged that the accuracy of the volume determinations improves significantly with each additional pass. Each pass of 3 000 particles takes about 3 hours to complete.

To quantify the correlation between the RhoVol envelope density and the skeletal density, the material in each of the bins was subjected to standard helium pycnometry. A correction function was developed by assuming the difference in densities were linear across the density range. This correction function was applied to the actual individual RhoVol generated densities as well, to back-calculate a skeletal density for each particle in the population.

Standard zinc chloride float-and-sink analysis (SANS 7936:2010/ISO 7936:1992) was used to determine the densimetric curve of the same sample that went through the RhoVol. The sample size was smaller (235 g), but this was considered acceptable since it was the entire population and not a sub-set of a larger sample.

Each sample was subjected to standard sieve analysis prior to introduction to the RhoVol to compare the particle size analysis as determined by the RhoVol with laboratory sieve data. This would confirm the accuracy of the visual shape and size determination.

RESULTS AND DISCUSSION

Figure 3 shows the calculated average RhoVol density, compared to the pycnometer determination for that particular bin for the Witbank blend sample.

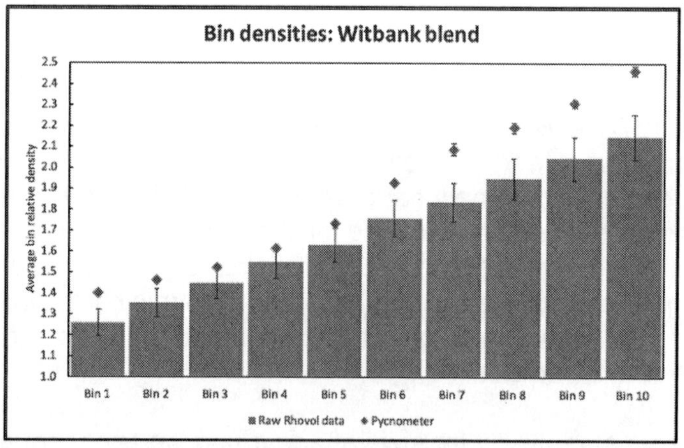

Figure 3: Comparison of RhoVol calculated bin densities with that determined by pycnometry. The error bars on the histograms represent a 5% error for reference.

From the difference observed a linear correction function was applied to the RhoVol densities. For this particular coal, the correction was:

$$RD_{corrected} = 1.252\ RD_{RhoVol} - 0.254 \qquad \text{(Equation 1)}$$

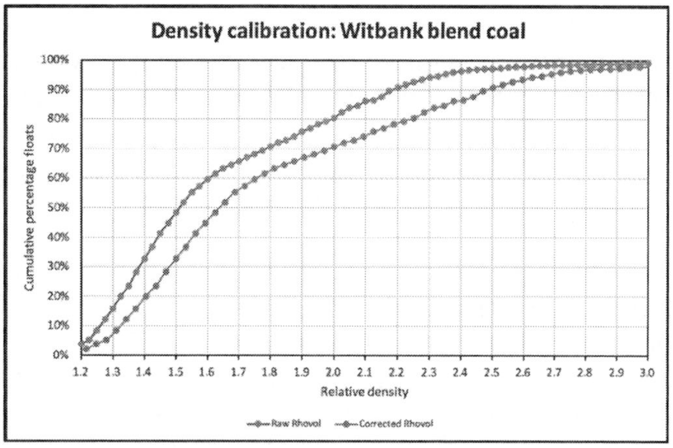

Figure 4: Data showing the effect of the correction function on the RhoVol generated densimetric curve.

Figure 4 compares the calculated cumulative skeletal densimetric curve as determined by RhoVol measurements with the raw envelope densities after application of the correction above. While it is not expected that skeletal density will correspond to the laboratory float-and-sink results (if the assumption is made that $ZnCl_2$ does not fully penetrate most of the pores), it may still be useful in future for pseudo-porosity calculations.

The results from the laboratory float-and-sink procedure is plotted with the RhoVol envelope densimetric curve to show the correlation between the two determination methods (Figure 5).

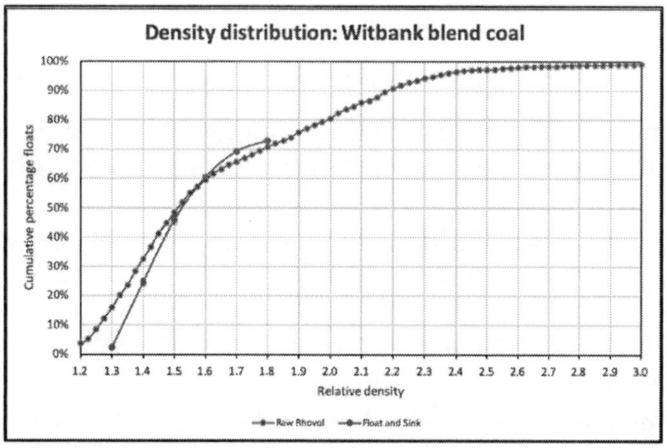

Figure 5: Comparison of the RhoVol densimetric curve with the laboratory generated curve.

Good correspondence at higher densities can be observed, but an offset occurred below about RD 1.4 where the experimental yield was lower than that predicted by the RhoVol densimetric data. There may be a number of reasons for this deviation, like an inaccurate dynamic correction factor; limited RhoVol passes resulting in poor volume estimation; particle shape and size irregularity over the density range resulting in inconsistent volume estimation; and differences in porosity, and hence the degree of $ZnCl_2$ intrusion over the density range. These explanations will be discussed in more detail at the end of the paper. The relatively good overall fit of float-and-sink data to the RhoVol envelope density distribution supports the suggestion that float-and-sink analyses measures an envelope density rather than a skeletal density. This is, of course, dependant upon the pore characteristics of the coal.

Despite the errors at lower densities, the RhoVol shows great potential in unlocking additional information about the population of particles. Of particular significance is that data can be obtained beyond the traditional upper density limit of float-and-sink tests (between RD 1.3 and RD 1.8 in most cases). It is also very easy to differentiate the cumulative density curve to give more information about the shape of the true density distribution (Figure 6).

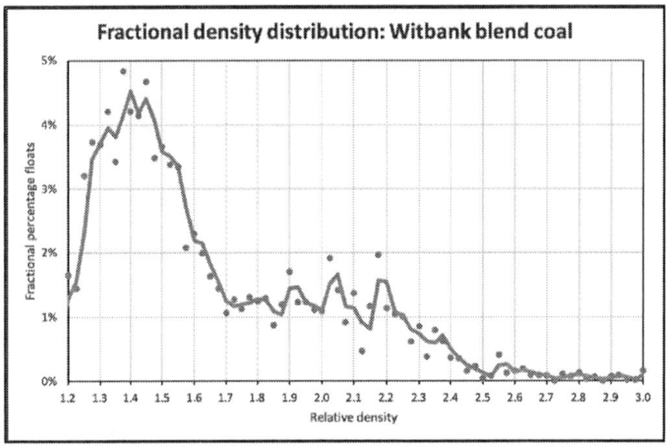

Figure 6: Fractional distribution of envelope densities in the Witbank blend sample. A 5-point moving average was used to highlight the trends of the data.

This type of curve can potentially be used to help identify macro-components, both organic and inorganic, in the sample, and to study coal liberation.

Finally, the particle size shape distribution can easily be extracted from the RhoVol results. The RhoVol particle size distribution obtained for the Witbank blend coal is compared to a standard sieve analysis in Figure 7. Size data can be expressed in a number of ways, and the imitation of a "minimum square aperture size" was chosen here to compare with the sieve data. It can be seen that a very accurate correspondence was achieved, indicating that the RhoVol's volume estimation, and hence derived particle sieve size, was very good.

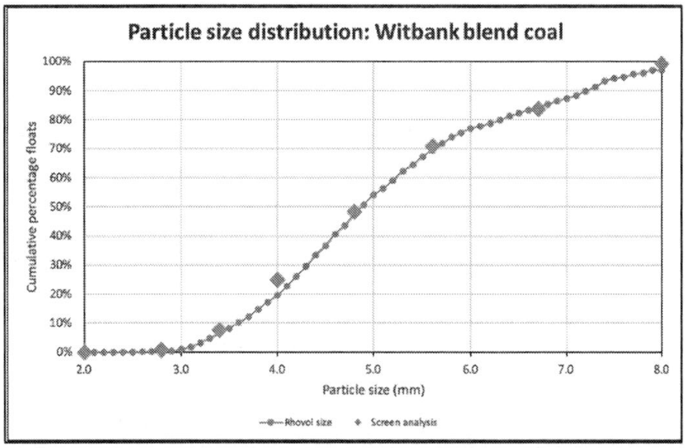

Figure 7: Comparison of the RhoVol particle size analysis curve with a sieve analysis curve.

It may often be of interest to obtain information about particle shapes, for example to understand pneumatic or hydraulic processes involving the motion of coal particles through a gas or a liquid. This includes dry processing and most wet gravity processes. The distribution of the so-called compactness factor - one of the metrics used by the RhoVol describing how the particle shape deviates from a perfect sphere – is shown on Figure 8.

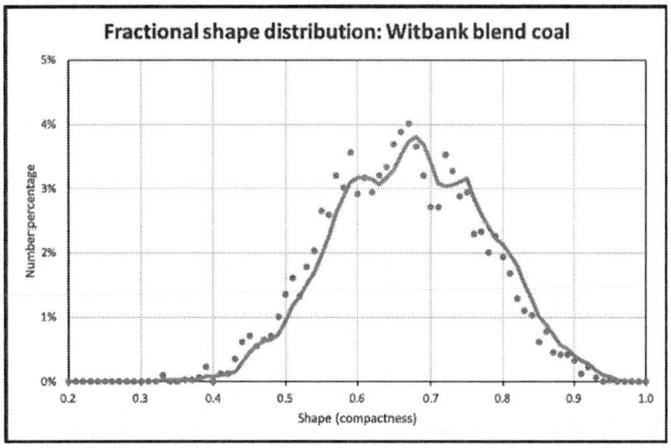

Figure 8: Fractional distribution of particle shape (expressed as compactness) in the Witbank blend sample. A 5-point moving average was used to highlight the trends of the data.

The information on Figure 8 shows that the majority of particles have a compactness of between 0.6 and 0.8, indicating fairly regular shapes. There are also a significant number of flat and elongated particles below 0.5 compactness. This paper will not explore this further, suffice to mention that there is still a wealth of information to be considered in future studies.

CONCLUSIONS

It was shown the envelope density, as determined by the RhoVol, and not the back-calculated skeletal density, matched the float-and sink data the best. Evaluating the possible reasons for the discrepancy between float-and-sink data and the RhoVol envelope density at lower densities (Figure 5), the following can be stated:

Reason 1: Limited RhoVol passes results in poor volume estimation. It was reported by the operators that volume errors of up to 20% per sample can occur for a single pass, but this will improve to less than 5% after 5 passes (Forbes et al., 2003). While this is a likely cause of errors, the fact that the PSD fits the RhoVol data almost exactly demonstrates that the volume estimation was accurate despite the single pass. It depends on the particle shape and will differ from coal source to coal source. This effect should be investigated in a next phase of the project.

Reason 2: The dynamic correction factor used by the machine was inaccurate. The correction factor used was for a previously tested coal, similar but not the same as the Witbank coal used in this project. This factor can be influenced by the shape distribution of particles as a function of the density profile. Ideally, every new source of coal should be calibrated to increase accuracy, but again, the good PSD correspondence obtained indicated an accurate volume estimation. However, this should be confirmed when testing widely different coal types in future.

Reason 3: Particle shape and size irregularity over the density range resulting in inconsistent volume estimation. Analysis of the distribution of shape and size over the full density range (not shown in this paper) indicated that there was no relation between the density and either shape, or size, therefore this did not have an effect.

Reason 4: Differences in porosity due to the degree of $ZnCl_2$ intrusion. This was the most likely explanation, even though it could not yet be confirmed. The theory is that particles of lower density are more likely to contain less mineral matter, and possibly be more vitrinite rich. This causes higher porosity at the low end of the density spectrum. The behaviour of particles in a $ZnCl_2$ medium is highly dependent on how much of the medium can penetrate what pore size, and how this affects the density in float-and-sink tests. Additionally, the difficulty of washing off the $ZnCl_2$ after the tests increases with an increase in porosity, hence affecting the mass of the particles. It is expected that with the variability in porosity as a function of the density range (as deduced from Figure 3) will be different between low density porous, and high density, less porous particles. For this reason, it should be stated that although float-and-sink analysis is the accepted industry standard, it's measurement of density may not be fundamentally correct.

In summary, the RhoVol is an exciting new analytical tool that has the potential to safely and easily generate accurate densimetric, size and shape distributions of coal samples. One area that needs further attention is the effect of the coal dependant dynamic correction factor on the accuracy of the volume, and consequently the density measurement, as well as the effect of pore size and shape.

ACKNOWLEDGEMENTS

The authors wish the acknowledge the contribution of the staff at DebTech while using the RhoVol, the work of our students, Lee-Roy Morgan and Dyllan Stone, in generating the information. The assistance of Mr Thinus Steyn (DebTech) in training and assisting the students on the RhoVol is hereby gratefully acknowledged. This work is based on the research supported in part by the National Research Foundation of South Africa (Coal Research Chair Grant Numbers: 86880; and UID85643, UID85632). Any opinions, findings and conclusions or recommendations expressed in any publication generated by the NRF supported research is that of the author(s) alone, and that the NRF accepts no liability whatsoever in this regard.

REFERENCES

Aktas, Z., Karacan, F., & Olcay, A. 1998. Centrifugal float – sink separation of fine Turkish coals in dense media. 235–250.

Bhattacharya, S. & Anand, V. 1998. Estimation of grindability from sink-float test data for two different coals. *Int. J. Miner. Process.* 53:99–106.

DebTech, 2018. Product brochure on the RhoVol. http://www.DebTech.com/prod12.htm. Date accessed 17 July 2019.

Fofana, M. & Steyn, T. 2018. Monitoring the performance of DMS circuits using RhoVol technology. *J. Southern African Inst. Min. Metall.* 119(2):133–138.

Forbes, K., Voigt, A., & Bodika, N. 2003. Using Silhouette Consistency Constraints to Build 3D Models. *Proc. Fourteenth Annu. Symp. Pattern Recognit. Assoc. South Africa.* (PRASA 2003):33-38.

Franzidis, J. & Harris, M.C. 1986. A new method for the rapid float-sink analysis of coal fines. *J. South African Inst. Min. Metall.* 86(10):409–414.

Kong, L., Bai, J., Li, H., Chen, X., Wang, J., & Bai, Z. 2018. The mineral evolution during coal washing and its effect on ash fusion characteristics of Shanxi high ash coals. *Fuel.* 212 (October 2017):268–273.

Mahajan, O. P. 1982. Coal Porosity. (In Meyers, R.A., ed. Coal Structure. New York: Academic Press. p. 51-84).

Mangera, R., Morrison, G., & Voigt, A. 2016. Particle Volume Correction using Shape Features. *2016 Pattern Recognit. Assoc. South Africa Robot. Mechatronics Int. Conf.* IEEE. C:1–6.

Rodrigues, C.F. & de Sousa, M.J.L. 2002. The measurement of coal porosity with different gases. *Int. J. Coal Geol.* 48(2002):245–251.

Sivrikaya, O. 2014. Cleaning study of a low-rank lignite with DMS , Reichert spiral and flotation. *Fuel.* Elsevier Ltd. 119:252–258.

B.A. Wills & T.J. Napier-Munn. 2006. Wills' Mineral Processing Technology. 7th ed. Elsevier.

DEFINING THE TRANSPORT MECHANISM DURING CONTACT SORPTION DRYING OF COAL FINES

M le Roux[*+], **QP Campbell*** **and J Hoffman****
**School of Chemical and Minerals Engineering, North-West University, South Africa.*
*** South African Nuclear Energy Corporation (Necsa), South Africa.*
[+]*Corresponding author: Marco.leroux@nwu.ac.za.*

ABSTRACT

Coal fines, defined here as +0.15 mm -0.5 mm, are notorious for being difficult to dewater and can contain large amounts of moisture, even after filtration. In an effort to minimize waste, these wet fines are blended with coarser product until a maximum contractual moisture content is reached.

An emerging technology that proofed to be successful is contact sorption drying. It involves the mixing of a sorbent, known for having a large surface area, with wet coal particles in an attempt to aid moisture transport from the coal to the sorbent. It was found that commercially available sorbents have the ability to dry a fine coal filter cake, typically starting at about 0.30 g(moisture)/g(coal and moisture), to well-below 0.08 g(moisture)/g(coal and moisture) within 10 minutes of operation.

The aim of this paper is to quantify the moisture transportation mechanism during contact sorption drying of fine coal.

***Keywords*:** Fine coal dewatering; Contact sorption drying; Moisture transport, Activated alumina

1 BACKGROUND AND INTRODUCTION

An increase in the use of mechanize mining has led to an increase in the production of fine coal (-500µm), which in turn has had a detrimental effect on achieving set product moisture targets. Blending of moist fines with dryer coarser particles is common practice, but does not eliminate the problem. The only viable option is to find cost effective drying methods that will address this issue. One such an option is the use of contact sorption drying.

During contact sorption drying, porous sorbents like ceramics, zeolites or some polymers are mixed with the wet coal fines, after which moisture is transferred from the coal particles to the sorbent. Separating the coal from the sorbents then results in a dry fine coal product and moisture loaded sorbent. The sorbents are hereafter dried and recycled to be used again. Van Rensburg et.al. (2018) tested the validity of this process on fine South African coals and proofed that it is possible to reduce the moisture of a typical coal filter cake to below 8 $\%_{wt}$ in less than 10 min. In addition, it was further proven that moisture removal from sorbents are possible by subjecting the sorbents to high ambient airflow in a packed bed (Van Rensburg et.al, 2016). The subsequent energy balance showed the process to be energy positive when comparing the gain in coal calorific value with the use of energy for the total drying route.

To be able to implement this process on a larger scale, it is necessary to understand that mechanism of moisture movement from the coal to the sorbent. The aim of this paper is to determine this moisture transfer mechanism by making use of a) a series of experimental setups and b) X-ray computed tomography.

2 PREVIOUS WORK

Work done by Van Rensburg et.al. (2016) proof the validity of the process. Figure 1 shows the moisture transfer measured between +0.5 mm -1 mm coal particles and alumina rich ceramic spheres over the course of 30 minutes. It shows a rapid initial reduction in coal moisture, followed by a more gradual transition towards the final equilibrium moisture content. Also indicated on Figure 1 is the adsorption of moisture onto the ceramic beads, clearly indicating the transfer of moisture from one material to the other.

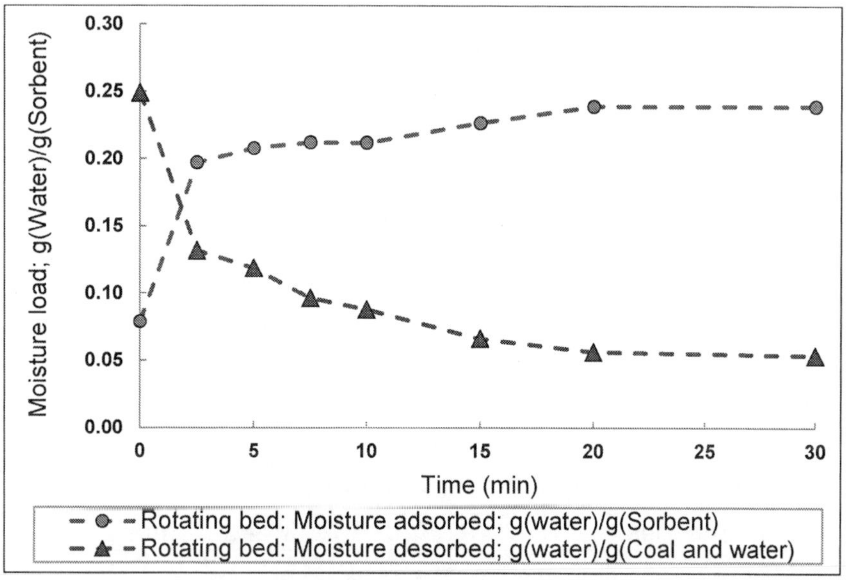

Figure 1: Moisture transfer between +0.5 mm -1 mm coal and ceramic spheres (Van Rensburg et.al., 2018).

To address the question of feasibility, Le Roux et.al (2018) showed that alumina-rich ceramic sorbents can be reused up to six times before regeneration is necessary. To add to the feasibility of the process, it was further shown that, unlike current patents such as Bland et.al. (2011), no heat or microwaves are needed to regenerate the sorbents. Using ambient airflow through a packed bed will disturb the vapour-liquid equilibrium between the moisture in the sorbent and the surrounding air, hence removing the moisture from the sorbent to the air. The sorbent exits the packed bed at moisture levels acceptable for reuse in the process.

3 EXPERIMENTAL

3.1 Materials used.

Medium Rank C Bituminous coal from the Highveld coal fields in Mpumalanga, South Africa, was used in this investigation. Table 1 gives the proximate analysis of the coal.

Table 1: Proximate analysis of coal used.

Measured quantity	Value	Standard
Inherent moisture content (%wt)	3.6	ACT-TPM-010 rich on ISO11722: 1999
Volatile matter (%wt)	29.6	ACT-TPM-012 rich on ISO 562: 2010
Ash yield (%wt)	14.9	ACT-TPM-011 rich on ISO 1171: 2010
Fixed carbon (%wt)	51.9	By difference
Total sulphur	0.97	ACT-TPM-013 rich on ISO 19579: 2006

As received coal was stored under ambient conditions and later split into three size ranges namely -150 μm, +150 μm – 250 μm and +250 μm – 500 μm. Prior to drying, the coal was drenched for 24 hours in water, filtered, sealed and then later used as feed to the drying vessels.

The sorbents used for this series of tests were both 3 mm and 5 mm Alumina rich ceramic spheres that are currently commercially available. The composition of the ceramic can be seen in Table 2 while Table 3 gives the physical properties of the beads.

Table 2: Composition of the ceramic sorbent.

Content	Composition (weight percentage)
Aluminium oxide (Al_2O_3)	92.7
Silicon oxide (SiO_2)	0.02
Iron (III) oxide (Fe_2O_3)	0.02
Sodium oxide (Na_2O)	0.3

Table 3: Physical properties of the ceramic beads.

Properties	Measurement
Surface area (m^2/g)	350
Pore volume (cm^3/g)	0.5
Bulk density (kg/m^3)	769
Inherent moisture (%)	1.8

The ceramic beads were dusted and used as received to form the feed, in specific ratios with wet fine coal, to the drying vessels.

3.2 Experimental setups

Two experimental setups were used in an attempt to quantify the phase of moisture movement from the coal particles to the ceramics.

3.2.1 Laboratory setups

A series of eight small scale setups were designed with the purpose to isolate either the sole possibility for moisture liquid transfer, or for moisture vapour transfer. The loss in mass of the coal and/or the gain in mass of the sorbents were then measured to ascertain the amount of moisture desorbed, adsorbed or transferred due to the isolated mechanism. A schematic of the eight setups are shown in Figure 2.

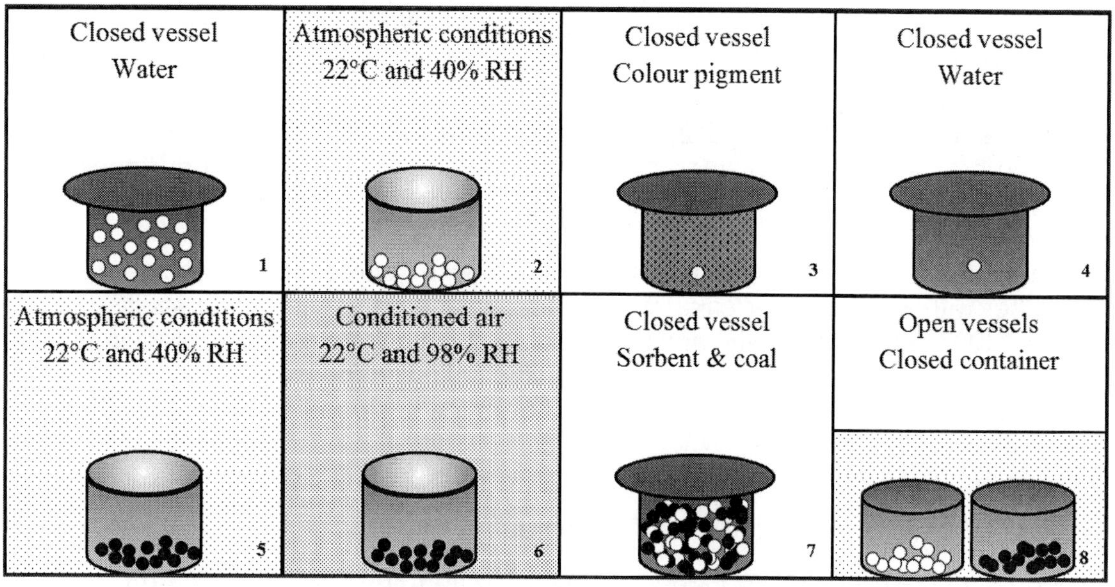

Figure 2: Schematic of the eight experimental setups.

The sorbent and coal samples in each of the experimental setups shown in Figure 2 were weighed at different time intervals to determine the moisture loss/gain. This was confirmed by doing moisture analyses on either the coal particles or the sorbents, based on the ISO 11722:1999 standard. A description of each of the setups in Figure 2 is given below:

(1) In the first setup, a container was filled to the top with pure liquid water, to which sorbents with known moisture content were added. At predetermined time intervals, some beads were removed and the moisture thereof analysed in an attempt to quantify the maximum liquid adsorption rate as well as the maximum final liquid moisture holding capacity of the beads.
(2) The second setup was similar to the first, but instead of liquid moisture in the container, the container was placed in a climate chamber of which the conditions were kept constant at 22 °C and 40% relative humidity. This allowed for the measurement of water vapour adsorption under these conditions. In addition it was used to determine moisture load capacity of the sorbent material at equilibrium.
(3) In the third setup a single sorbent particle was placed in liquid water containing gentian violent colouring. Upon removal, the particle was cut open and examined under a Light Emitting Microscope (LEM) to visually see the diffusion paths of liquid into the beads.
(4) Setup four repeated setup one, but instead of a population particles, a single sorbent particle was added to the water. This allowed for the elimination of sorbent-sorbent interaction and isolated the adsorption capabilities of a single bead exposed to liquid water.
(5) In setup five, a wet coal filter cake was placed in the climate chamber of which the conditions was again controlled at 22 °C and 40% relative humidity. By definition, coal left under the condition for a long enough period will lose moisture to finally achieve inherent moisture conditions. Through this setup, the moisture desorption

rate from a fine coal filter cake at said conditions can be measured. If this information is used together with setup two, a vapour transfer balance can be concluded.
(6) The previous setup was repeated, but this time the air as conditioned to be at 22 °C and 98% relative humidity. These conditions simulated saturated air and allowed to determine how it would influence desorption of water vapour from a wet coal filter cake, typically during rainy days.
(7) In the seventh setup, wet coal filter cake was mixed with conditioned sorbents and compressed into a container to eliminate as much as possible air surrounding the system. This allowed for the measurement of predominantly liquid transfer from the coal to the sorbents.
(8) In the last setup two open vessels, one filled with conditioned sorbent beads and the other with wet coal filter cake, were placed inside a third container starting at 22 °C and 40% relative humidity, and then sealed close. Direct contact between the coal and sorbents were hereby prohibited which allowed for solely vapour transfer from the coal fines to the surroundings and from the surroundings to the sorbents to be studied.

3.2.2 Micro focus X-Ray computed tomography

Viljoen et.al (2015) described x-ray computed tomography as a non-destructive method used to map and draw three-dimensional images of the internal structures of objects by utilising, for example, the difference in x-ray scatter due to the density of the objects. Figure 3 shows a schematic illustration of the facility located at the Nuclear Corporation of South Africa's (NECSA) Pelindaba campus.

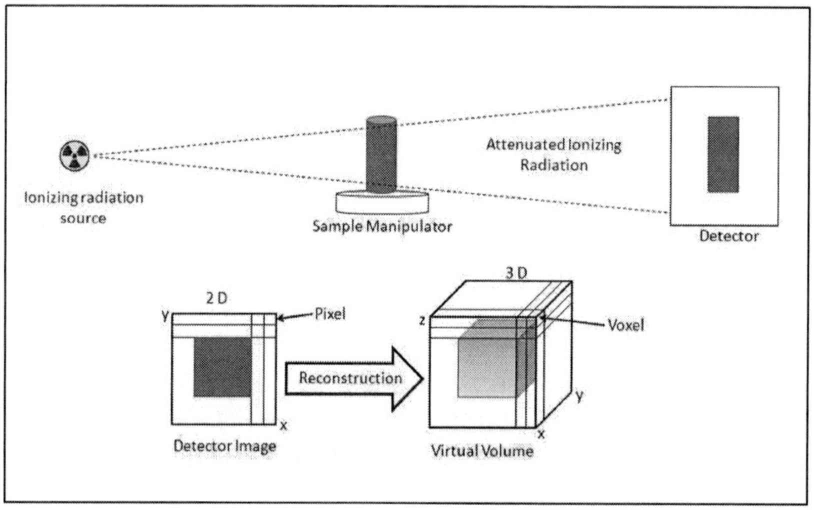

Figure 3: Schematic illustration of the SANRAD facility at NECSA.

Using a similar setup than the one in Figure 3, Van der Merwe et.al. (2006) found that it was possible to trace liquid water, containing barium chloride ($BaCl_2$), inside activated alumina spheres. The addition of $BaCl_2$ was used to enhance the contrast between the solid and liquid phases. Such a setup opens up the possibility to image a container, encompassing a mixture of sorbents and wet coal filter cake, over an extended period to visually determine the transfer of liquid from the coal to the sorbents.

4 RESULTS AND DISCUSSION

4.1 Laboratory experiments

The information obtained from the eight setups depicted in Figure 2 were grouped together and used to explain the transfer mechanisms. This was done for both moisture adsorption and desorption on/from either the coal or the ceramic beads.

4.1.1 Adsorption of liquid and vapour onto the ceramic

Setup one and two from Figure 2 was used to determine the rate and total moisture adsorption onto a 3 mm ceramic bead. This is shown in Figure 4.

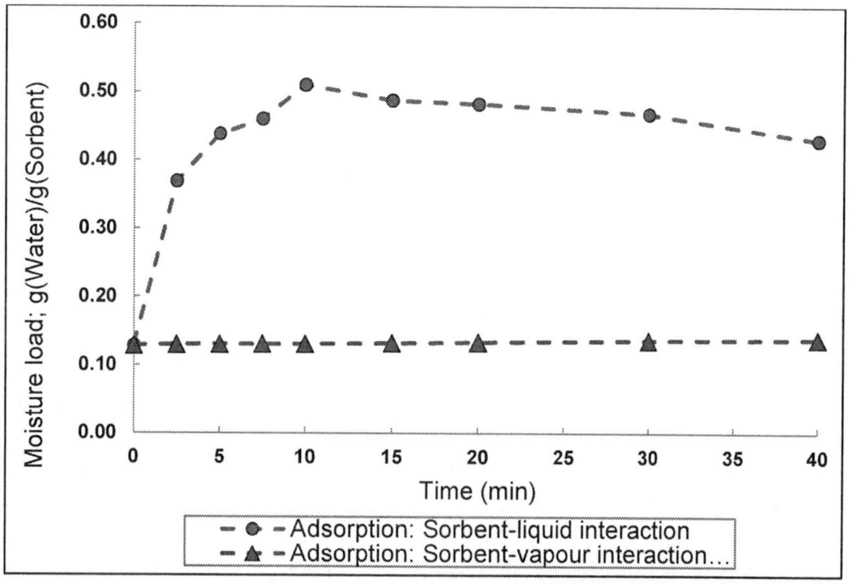

Figure 4: Adsorption: Sorbent-liquid vs. sorbent-vapour interaction.

The graphs from Figure 4 shows a significant difference between liquid moisture adsorption and vapour moisture adsorption onto the ceramic beads over a 40 minute time period. The rate of liquid adsorption is magnitudes faster than that of vapour adsorption, and reached an equilibrium around 40%$_{mass}$ increase over the allowed time period. This denotes an additional moisture load 30%$_{mass}$ over the 40 minutes. The vapour adsorption in contrast was only able to adsorb an additional 3%$_{mass}$ moisture over the same time. The linear increase in moisture content for the sorbents exposed to water vapour indicates that the sorbents are still far from capacity and hence would have continued to adsorb moisture if the exposure time was continued until equilibrium was reached.

4.1.2 Desorption of liquid and vapour from coal

The fifth and sixth setup from Figure 2 were used to study the desorption of moisture from a coal filter cake if subjected to different atmospheric conditions. These two setups therefore isolated the transfer of vapour from the coal into a saturated and non-saturated environment. This was compared to the results obtained from setup seven in Figure 2, where moisture was mainly transferred between the coal filter cake and the adjacent sorbents. The comparison between these three setups are shown in Figure 5 and gives an indication between the difference in vapour and liquid desorption from the coal filter cakes if subjected to different environments.

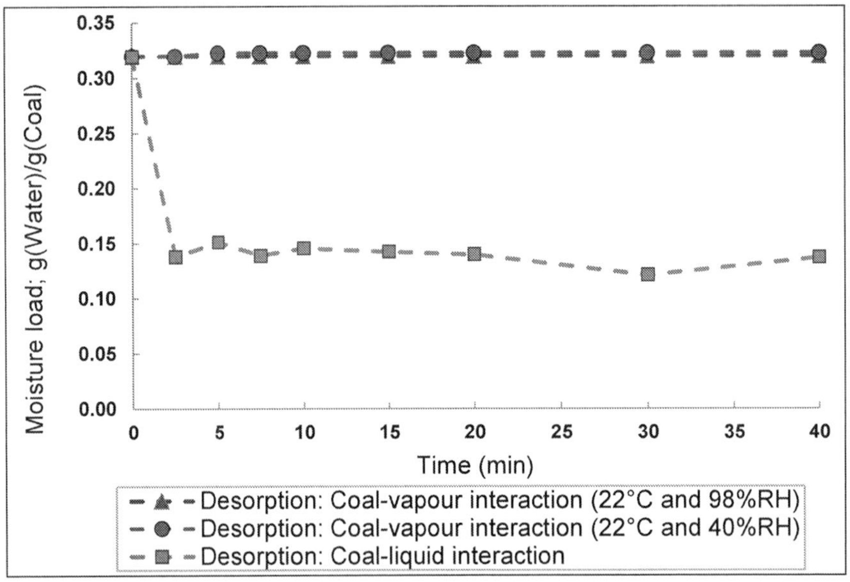

Figure 5: Desorption: Coal-liquid vs. coal-vapour interaction.

It is easy to see from Figure 5 that the amount and rate of moisture desorption was significantly more for the seventh setup than for the fifth and/or sixth. As a matter of fact, very little moisture loss was recorded for the two setups that was exposed to atmospheric conditions, leading to the conclusion that moisture vapour loss from a coal filter cake to the surroundings is negligible over a 40 minute time period, irrespective of the conditions thereof. However, when in contact with a sorbent, liquid transfer from the coal to the sorbent is rapid, resulting in a new equilibrium state in less than 5 minutes.

In support of this finding, the change in moisture levels measured for the final setup of Figure 2, was also negligible as it showed virtually no change over the 40 minute exposure period. This was a clear indication that vapour only transport did not contribute towards the transfer mechanism of contact sorption drying.

4.1.3 Description of liquid moisture transport

Single particle behaviour was used to confirm that liquid moisture transfer was the predominant mechanism for contact sorption drying of fine coal. The results obtained from setup one and three of Figure 2 were plotted against actual drying data reported by Le Roux et.al. (2018) for a 50% per mass coal/sorbent feed. This is shown in Figure 6, which indicates significant correlations in the initial adsorption rate between all three setups. The single particle reached a maximum moisture holding capacity of 0.39 g(Water)/g(Sorbent) within 10 minutes, where the bulk sorbents continued to adsorb water until an equilibrium moisture load of 0.43 g(Water)/g(Sorbent) was reached. This scenario indicates that the bulk sorbent is able to adsorb and retain more water, known as inter-particulate moisture. It usually reports as thin films on the surfaces of the single particles, or inside the contact angles formed between two adjacent particles. For the real 1:1 ratio dewatering tests, an equilibrium moisture of 0.27 g(Water)/g(Sorbent) was reached. This is an indication of the lack of available moisture for adsorption in the system, and confirms the conclusion by Van Rensburg et.al. (2016) that a single load sorbents can be used more than once prior to regeneration. This was also observed from the photos taken during the third setup in Figure 2 and showed in Figure 7 below which clearly indicates open white areas that can still absorb moisture.

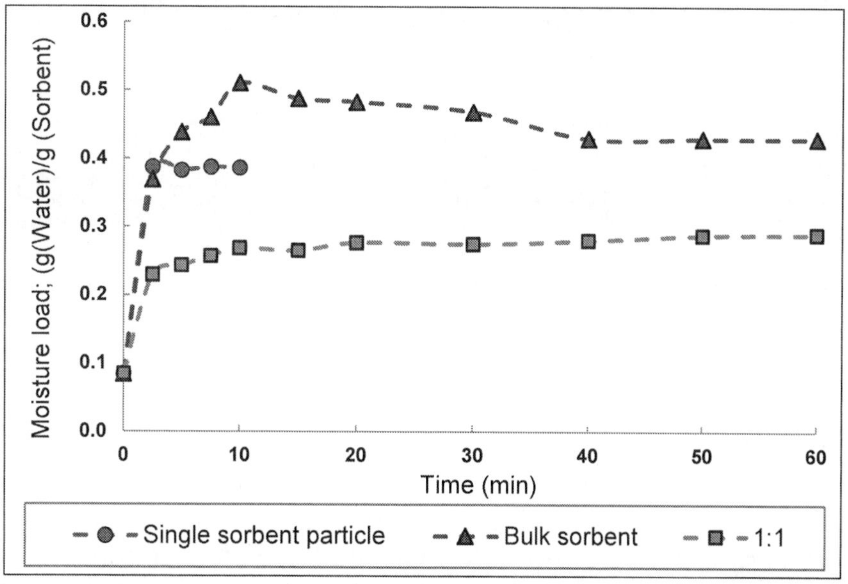

Figure 6: Adsorption: Sorbent-liquid interaction vs. sorbent-liquid-coal interaction.

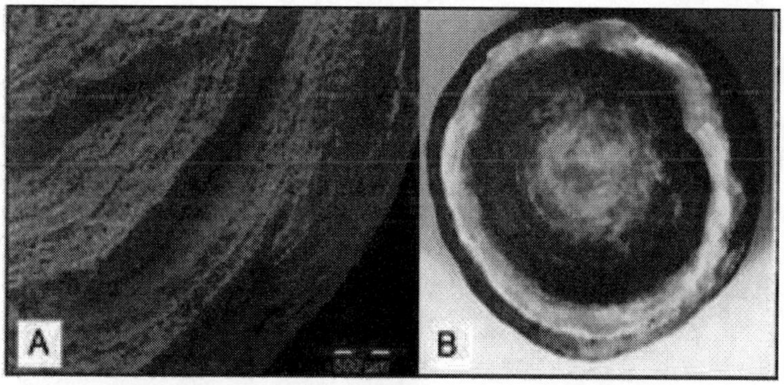

Figure 7: (A) Alumina rich sorbent: SEM scan; (B) Alumina rich sorbent: LEM scan.

4.2 X-ray computer tomography

In a final attempt to quantify the transport mechanism, Micro-focus X-Ray Tomography scans were done on a coal filter cake, containing barium enriched water. The filter cake were mixed with sorbent beads in such a way that a single bead was completely covered by wet coal. Figure 8 shows the progression of moisture into the sorbent as if diffuses to the centre of the bead over time. Each one of the tomopraphs are of the same bead, showing the moisture progression into the sorbent, which are mapped out by the blue lines. Using software, the volume moisture adsorbed by the bead was calculated over time, to give an estimate of the adsorption rate. It was calculated as 0.22 mm^3/min, which correlated well with work done by Le Roux et al. (2014) on static drying of fine coal using ceramic drying agents.

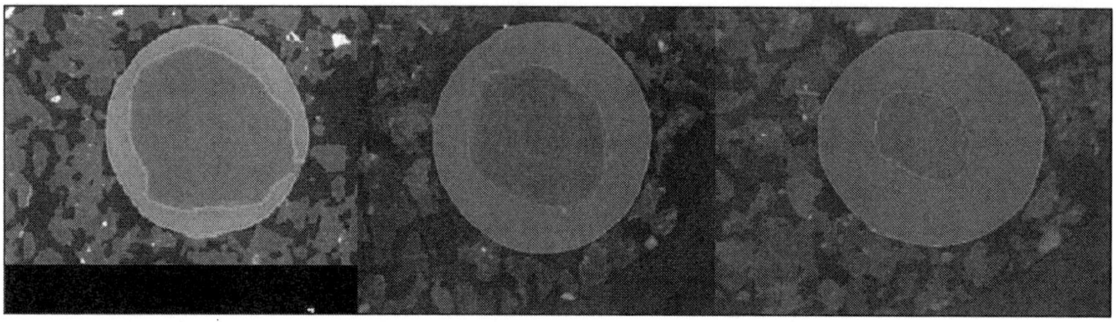

Figure 8: Microfocus X-Ray Tomograph show moisture adsorption onto a ceramic bead.

5 CONCLUSIONS

It can be concluded from this work that contact sorption drying of fine coal is a possible drying alternative during fine coal beneficiation. By isolating the possible transfer mechanisms, it was shown that liquid moisture transfer from the wet coal filter cake to the ceramic beads dominates vapour transfer to such an extent that the latter can be deemed negligible. This was confirmed using x-ray tomography which clearly shows liquid diffusion from the adjacent coal cake into a bead. Therefore, when designing a large scale operation, the drying vessel should be built in such a way that optimal coal/sorbent contact is established in an effort to maximise the transport of liquid from the coal onto the ceramic.

6 ACKNOWLEDGEMENTS

The authors would like to thank Coaltech, the Southern African Coal Processing Society, the Centre of Excellence in Carbon Based Fuels, North-West University, South Africa and NECSA, for their contribution to this project, financial and/or otherwise.

This work is based on the research supported by the South African Research Chairs Initiative of the Department of Science and Technology and National Research Foundation of South Africa (Chair Grant No: 86880, UID 85643, Grant No: 85632 and Grant UID 72310). Any opinion, finding or conclusion or recommendation expressed in this material is that of the author(s) and the NRF does not accept any liability in this regard.

7 REFERENCES

1. Bland, R.W., Harsh, P., Hurley, M., Jones, A.K. & Sikka, V.K. 2011 Coal fine drying method and system. United States patent application publication. Pub. No US 2011/0078917.

2. Le Roux, M., Campbell, Q.P., Van Rensburg, M.J. & Peters, E.S. 2018 Moisture transport during contact sorption drying of fine coal. In: Vince A. (eds) Seventeenth Australian Coal Preparation Conference. Brisbane. ACPS Ltd.
3. Le Roux, M., Campbell, Q.P. & Van Rensburg, M.J. 2014 Fine coal dewatering using high airflow. International Journal of Coal Preparation and Utilization, 34:220-227.
4. Van Rensburg, M.J., Le Roux, M., Campbell, Q.P. & Peters, E.S. 2018 Moisture transport during contact sorption drying of coal fines, International Journal of Coal Preparation and Utilization, DOI: 10.1080/19392699.2018.1543663.
5. Van Rensburg M.J., Le Roux M., Campbell Q.P. 2016 Drying of coal fines assisted by ceramic sorbents. In: Litvinenko V. (eds) XVIII International Coal Preparation Congress. Springer.
6. Van der Merwe, W., Nicol, W. & De Beer, F. 2006 Internal wetting dynamics of alpha- and gamma-alumina catalyst spheres using X-ray computed tomography. South African Journal of Science, 102:585-588.
7. Viljoen, J., Campbell, Q.P., Le Roux, M. and Hoffman, J. 2015 The qualification of coal degradation with the aid of micro-focus computed tomography. South African Journal of Science, 111:9-10.

DENSE MEDIUM SEPARATION USING A MAGNETIC CYCLONE

J.J. Rust
R&D Engineer, Multotec Process Equipment (Pty) Ltd Technology Division

Q.P. Campbell
Professor and Director - School of Chemical and Minerals Engineering, North-West University

M. Le Roux
Director: Center for Engineering Education, North-West University

J. Eves
Director, Eco-nomic Innovations Ltd

R. Fourie
Engineer in Training, Multotec Process Equipment (Pty) Ltd Technology Division

ABSTRACT

It is a well-known fact that good quality coal is being depleted due to the high demand of this product. What is left is coal of less favourable quality, and it is necessary to make efficient use of such coals. Since the majority of washed coals are produced by dense medium cyclones world-wide, there is as clear motivation to research any improvements in this technology. In this paper, the application of a magnetic field around a dense medium cyclone has proven to increase the sharpness of separation within such a cyclone, and has also made it possible to make use of coarser media during the process. Thus, the coal cleaning process can be made more efficient with higher yields and less misplaced material.

INTRODUCTION

The ever-decreasing run-of-mine coal qualities have rendered it necessary for coal beneficiation processes to be continuously optimised in order to make efficient use of the value of the coal. The most popular coal beneficiation method is dense medium separation, a process that exploits the differences in density between the coal and the gangue in order to separate these materials. Dense Medium Separation (DMS) cyclones form part of this dense medium beneficiation process and, as the name indicates, makes use of a medium (a suspension of magnetite in water) in conjunction with the forces acting within such a cyclone to efficiently clean the coal that is being treated.

Traditionally, fine magnetite is used in DMS cyclones, as magnetite which is too coarse tends to settle out within the cyclone, causing instability (England *et al*, 2002). The challenge with fine magnetite, is that it can become viscous, also causing instability. It is more expensive than coarse magnetite, and its losses are also prevalent within the medium recovery process.

The application of a magnetic field around a DMS cyclone influences the medium stability (Svoboda *et al*, 1998). If the medium is unstable, it would influence the separation within a DMS cyclone and consequently render the process inefficient and costly (Myburgh, 2001). Due to an increased stability of the medium with the application of a magnetic field, the possibility to use a coarser media under such circumstances will reduce magnetite costs as well as increase the recovery efficiencies thereof. Coarse magnetite is more readily washed from DMS cyclone products and recovered in magnetic separators.

The solenoid unit used for this application, the SpecSep™, requires a wide Particle Size Distribution (PSD) range in order for magnetite to be stable within the DMS cyclone. SpecSep™ allows for coarser magnetite to be used in conjunction with fine magnetite for effective commissioning. This study aims to prove that coarser, cheaper magnetite can be used in a coal dense medium process by applying a magnetic field around a DMS cyclone, which consequently stabilises the coarse medium.

EXPERIMENTAL PROCEDURE

The test work process focused on the application of a weak magnetic field around a dense medium cyclone. The tests were done using medium only.

Magnetite Samples

For the test work, both commercial (fine) and coarse magnetite were used. Refer to Table 1 for information on the used magnetite.

Table 1: Magnetite Information

	Fine Magnetite	Coarse Magnetite
Supplier	Martin and Robson	Kimony (Pty) Ltd
Material D_{95}	75 μm	150 μm

Magnetic Dense Medium Separation Cyclone Rig

The DMS cyclone rig can be seen in the following figures:

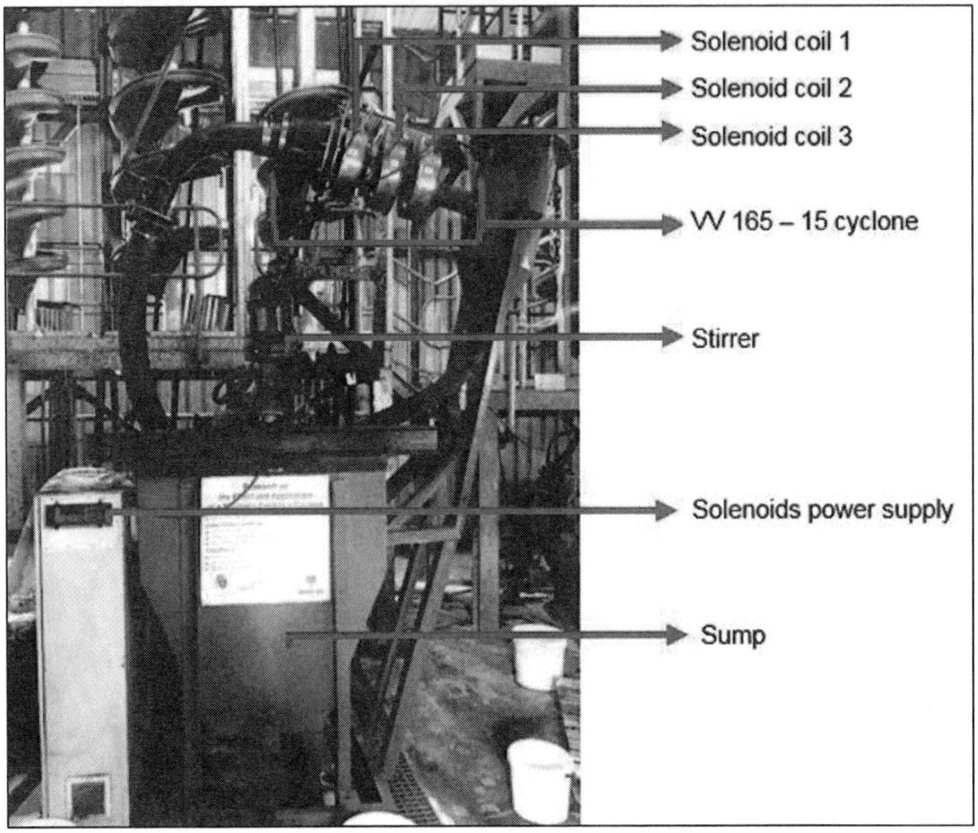

Figure 1: Magnetic DMS Cyclone Rig (Front View)

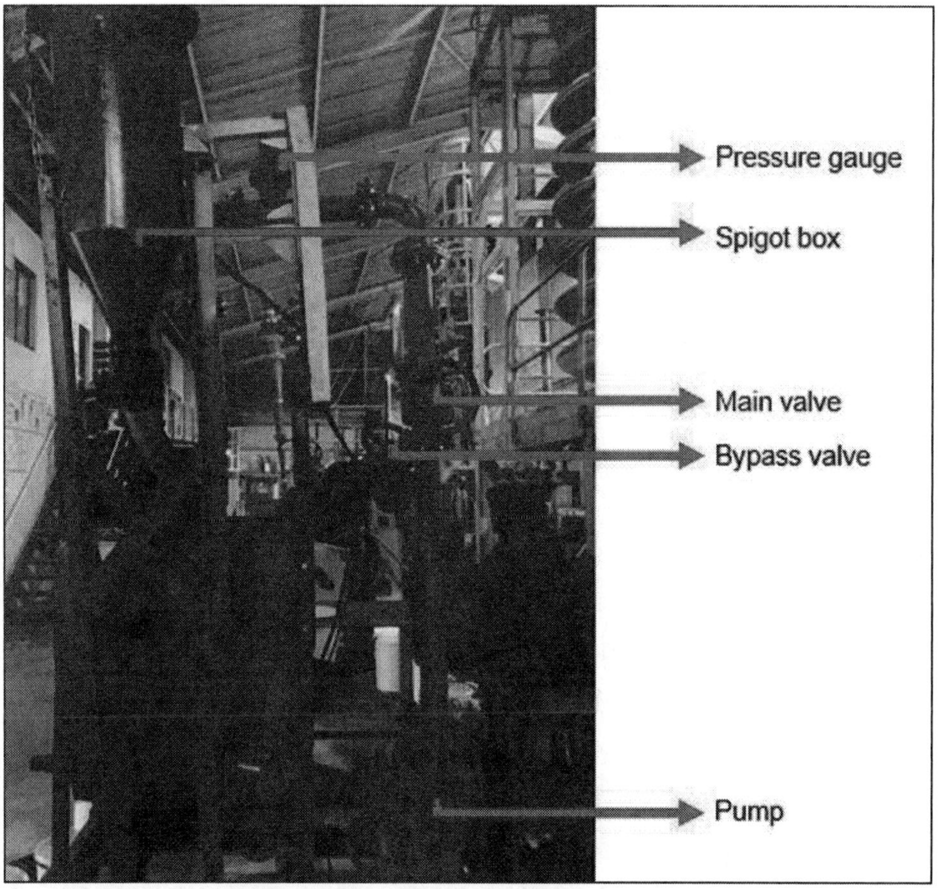

Figure 2: Magnetic DMS Cyclone Rig (Side View)

Test work was performed using the cyclone rig as shown above. The rig includes a polyurethane VV-165-15 cyclone installed with a 35 mm spigot. A polyurethane cyclone was chosen as it contains no metal parts, which is preferred due to the application of a magnetic field to the system. The cyclone has a diameter of 165 mm and is inclined to 15°. The front and side views of the DMS cyclone rig can be seen in Figure 1 and Figure 2 respectively:

The overflow and underflow streams of the cyclone reported to the sump where it was recirculated back through the cyclone. During operation, the main valve was closed and by opening the bypass valve the feed density could thus be measured using a volumetric cylinder. The density differential between the underflow and overflow streams was similarly measured

The cyclone rig is equipped with a pressure gauge, which was controlled at a constant head of 9D (Honaker, 2011), using a variable speed drive to control the flow through the cyclone.

SpecSep™ Solenoids

From Figure 1 it can be seen that the solenoids installed on the VV-165 cyclone rig is a three unit system, with the solenoids spaced equal lengths from each other. This was done to enable testing the effects of the position and strength of the three magnetic fields simultaneously. Test work was done by varying the current through the solenoids so that a suitable density differential would be obtained for the coarse and fine magnetite, respectively, as well as a mixture of both. The density differential is the difference between the cyclone overflow and underflow stream densities, and is an indicator of cyclone separation efficiency. Ideally, this differential should be between 0.2 and 0.5 g/cm^3 (Campbell & Coetzee, 1997:6). When the differential is too high, the separation efficiency decreases due to a longer retention time of near-density material.

The SpecSep™ solenoids were supplied by Eco-nomic Innovations Ltd and was designed as a three-unit system for this specific cyclone. The triple output DC power supply used to power the solenoids is an Aim and Thurlby Thandar Instruments MX100T.

Test method

A suitable mixture of fine and coarse magnetite, together with water, was added to the sump of the cyclone rig in a ratio that would result in a feed Relative Density (RD) of 1.74 g/cm^3. Non-magnetic tracers with densities ranging from 1.5 to 2.2 g/cm^3 at 0.1 g/cm^3 increments, was used to determine the partition curves. The tracers were 2 mm, colour coded cubes for ease of sorting.

Initially, a known amount of fine magnetite was added to the cyclone rig's sump, where coarse magnetite was added incrementally with density tests conducted after each addition of coarse magnetite. It was found that settling problems occurred as the density measurements were inconsistent. A stirrer was then installed to keep the coarse magnetite in suspension. Such a stirrer would not be necessary where the bottom of the sump is conical to prevent coarse media settling.

An amount of fine (conventional sized) magnetite, 95% passing - 45 µm, was introduced to the cyclone rig sump where tracer tests were done without the application of a magnetic field. Incremental amounts of coarse magnetite (Eves, 2015) were added to obtain the suitable coarse medium, where tracer tests followed, with the application of a magnetic field. A triple output DC power supply allowed for the solenoid coils to be operated independently from each other.

Medium density measurements were taken for the overflow and underflow streams and the feed density was measured and maintained throughout. Partition curves were plotted for the fine magnetite test work to obtain a base case to compare results with the case where coarse media was added to the fine media.

RESULTS AND DISCUSSION

During the solenoid calibration phase, it was concluded that by using the two solenoids furthest from the spigot (Solenoids 1 and 2 in Figure 1), better medium stability could be obtained compared to the case where all three solenoids were used together.

The addition of coarse magnetite was halted when approximately half of the total medium was coarse. At this ratio and a feed RD of 2.29 g/cm^3, the measured density differential between the overflow and the underflow was 0.9. This density differential was obtained with no magnetic field being induced over the cyclone. A current of 0.8 amps was then applied to both Solenoid 1 and Solenoid 2. It was evident that, at this current, the best differential was obtained with a value of 0.5. The results showed that by increasing the coarse medium resulted in an increase in the differential regardless of the strength of the current applied to the magnets.

A further test was done where a known amount of coarse magnetite was added to the cyclone rig's sump, and fine magnetite was added incrementally. The differential indicated that stability was obtained when 43 % of the total media was coarse. With no magnetic field exerted onto the cyclone, this differential was found to be 1.14, whereas applying a current of 1.35 A resulted in a reduction of the differential to 0.48. The difference in results between the two approaches (incrementally adding course magnetite versus adding fine magnetite) can be due the fines generated by the attrition of coarse media since the first test was conducted over approximately a week. Less fines were generated during the reverse test when the test unit was operating smoothly.

Better separation efficiencies were evident while using coarse media, since the coarse media promotes lower viscosity. To validate the improvement, fine magnetite was added to the cyclone rig to obtain a baseline. The media was diluted to deliver a typical cut point found in the coal industry. The Ep value was then determined by using the 2 mm non-magnetic tracers. To compare results, the density differential for the test without the magnetic field was kept between 0.2 and 0.5, which is the acceptable range obtained from literature (Scholtz, 2015).

The average of three tracer tests is shown in Figure 3 as follows:

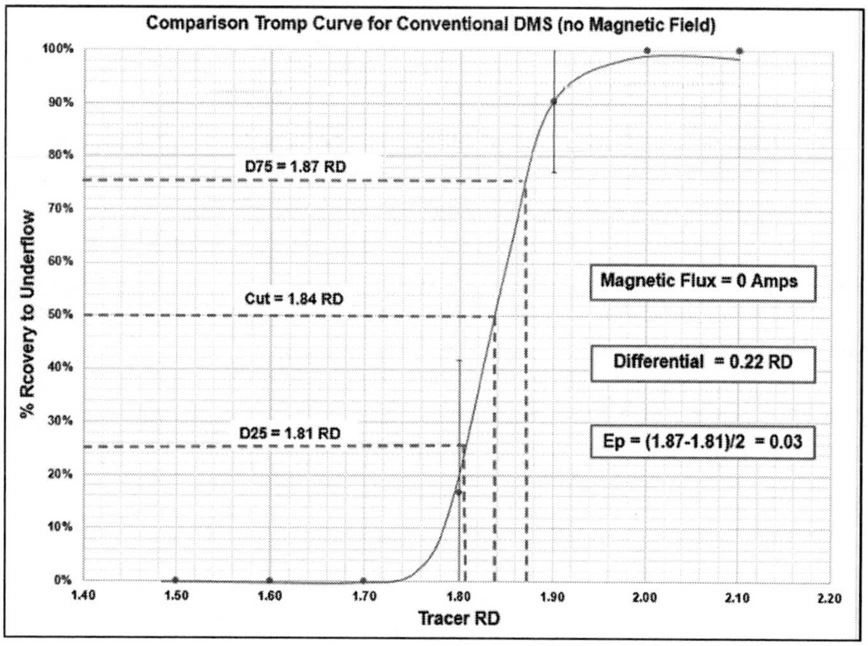

Figure 3: Partition Curve for Fine Magnetite Test Work without Magnetic Field

As calculated from the graph above it was obtained that the Ep value was 0.03 at a cut point density of 1.84 g/cm^3. The cyclone was set to run at 9D, thus giving a pressure of 25 kPa and a feed density of 1.71 g/cm^3.

For the test work with the application of a magnetic field, a suitable coarse medium mixture with feed density of 1.74 g/cm^3 was prepared. Test work was carried out to determine the optimum magnetic field from which to conduct tracer tests on. By varying the current of both solenoid coils from 0 to 1.35 A, the differential was recorded as can be seen in Figure 4:

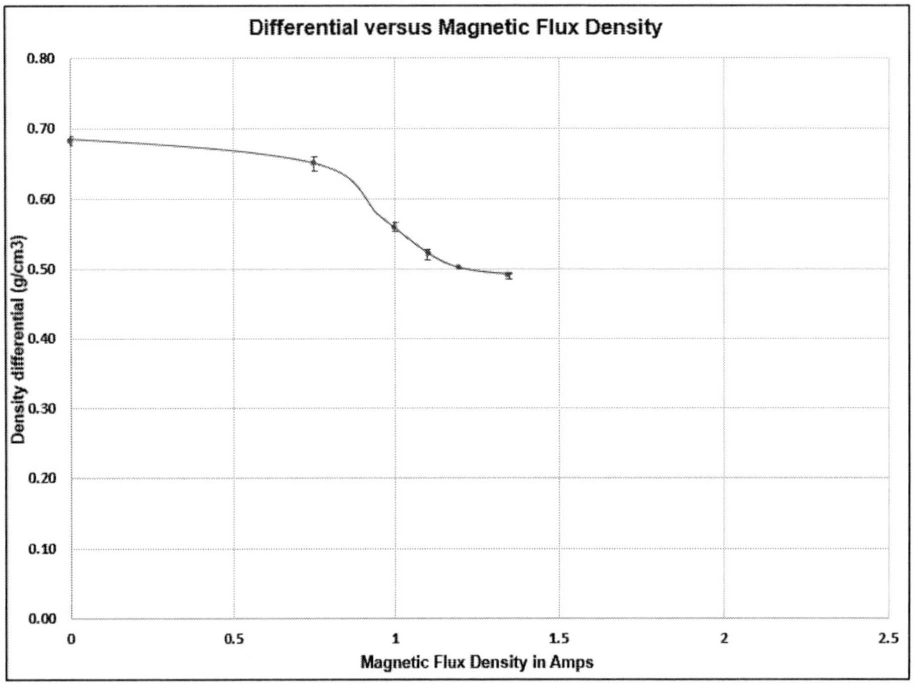

Figure 4: Density Differential versus Magnetic Flux Density

As evident from the figure above, the lowest differential obtained was 0.48 but it is suggested to run the test work at a differential of 0.52 to ensure there is no magnetic flocculation occurring in the cyclone. This chosen value was slightly beyond the upper limit as the magnetic field results in increased stability thus allowing operation at a slightly higher differential.

A tracer test was done by inducing a current of 1.1 A through the solenoid coils. The partition curve for this test is shown in Figure 5:

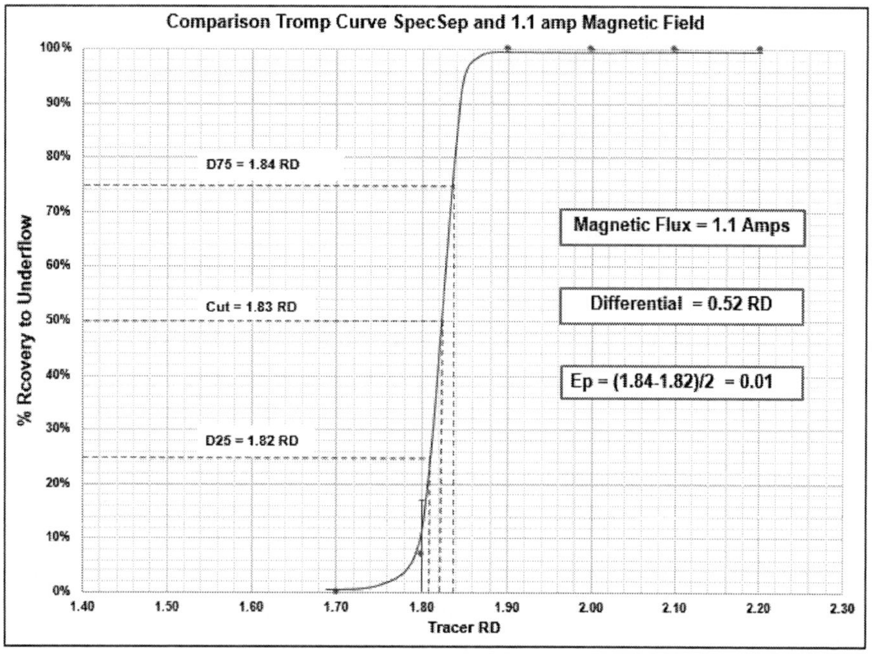

Figure 5: Partition Curve for Coarse Magnetite and a 1.1 A Magnetic Field

Comparing the base case partition curve with the partition curve with the added coarse magnetite, it can be noted that the Ep value decreased from 0.03 to 0.01 at similar cut points of 1.84 g/cm^3 and 1.83 g/cm^3. Table 2 below is a summation of the test work results obtained for the fine and coarse magnetite:

Table 2: Results for fine- and coarse magnetite

	100 wt% Fine Magnetite	43 wt% Coarse Magnetite
Solenoid current (A)	0	1.1
Cut-point (g/cm^3)	1.84	1.83
Differential	0.22	0.52
Ep	0.03	0.01

CONCLUSIONS

To conclude, by using the SpecSep™ Technology (Eves, 2015), the separation efficiency of the cyclone increased as a result of the decrease in the Ep value from 0.03 to 0.01. This is a significant decrease in Ep, and proves that the use of this technology enables the stabilisation of cheaper, coarse media within a dense medium cyclone resulting in improved separation efficiency, cost saving in the processing of coal due to reduced magnetite consumption. It is thus recommended that further work be done with the addition of coal and media to the cyclone, to determine what the actual cost saving is for a large scale application and more importantly, the value of the additional coal recovered.

ACKNOWLEDGEMENTS

The authors would like to thank Multotec Process Equipment (Pty) Ltd Technology Division for the use of their facilities, as well as the assistance of their dedicated staff. Furthermore, special thanks are in order for Eco-nomic Innovations Ltd. and the North-West University for planting the seed many years ago and whose support has made this project come to fruition.

REFERENCES

Campbell, Q.P. & Coetzee, C. 1997. The application of a magnetic cyclone for dense medium separation. (Paper submitted to the De Beers Diamond Research Laboratories on 26 November 1997). Johannesburg. 40 p. (Unpublished).

The South African Coal Preparation Society (SACPS). 2002. Coal preparation in South Africa. 4th ed. Pietermaritzburg: Natal Witness Commercial Printers. 298p.

Eves, J. C., 2015. South Africa Patent No. 2016/06592, US Patent No. US9901932B2, China Patent No. Application CN106061615. Europe Patent No. Application 15710734.3

Honaker, R., 2011. Classification and dense medium cyclone workshop notes, University of Kentucky.

Myburgh, I.K. 2001. Experimental investigation into the application of a magnetic dense medium cyclone in a production environment. Potchefstroom: NWU. (Thesis – MEng).

Scholtz, J. 2015. Dense medium separation. In: Coal Preparation in Southern Africa. s.l.:Coal Preparation in Southern Africa. pp. 187-216.

Svoboda, J. & Campbell, Q.P. Magnetic cyclone and method of operating it, South African Patent No.97/3440, (1996), Australian Patent 731513 (2001), Canadian Patent CA 2205339 (2001).

Svoboda, J., Coetzee, C. & Campbell, Q.P. 1998. Experimental investigation into the application of a magnetic cyclone for dense medium separation. *Minerals Engineering*, 11(6):501-509.

Presentation 2019 Southern African Coal Processing Society Conference.

Designing a Coal Tailings filter press plant.

M. Craddock Mineral Processing Technologies (MPTEQ) (Pty) Ltd

ABSTRACT

This paper describes the design parameters to be considered when designing a Filter Press plant together with the operational issues to consider from the thickener underflow rheology to the stockpiling of filter cake.

INTRODUCTION

Dewatering fine coal tailings is now accepted as an environmentally friendly alternative to tailings impoundments in Southern Africa and are becoming standard practice. This is not a new phenomenon, Peter Hand installed one at Middelburg Mine in the mid 1990's. The first Coal plant built with Larcodems and no tailings dam.

Designing a filter press plant begins with the thickener and thickener underflow pumping system. Thickeners can provide a perfect feed for a filter press. The design of the pumping system from the thickener underflow to the filter press feed or buffer tank is critical.

A high rate thickener with a standard sidewall height, typically 2.4 meters, can produce an underflow that has a viscosity that is difficult or impossible to pump with a standard centrifugal pump. The underflow often has to be diluted to get the centrifugal pump to transfer the slurry. In order to minimize this issue, place the underflow pump as close as possible to the thickener underflow cone.

An effective flushing system needs to be designed, the problem with flushing systems is that pumps prefer pumping water to slurry so the piping, valves and the inlet angle of the flush water piping need to be carefully designed.

At least three flush pipes with the associated valves are advised. The first water addition/flush pipe should be angled at 30 degrees to the underflow suction pipe in the direction the underflow cone. The objective of this pipe it to hopefully remove any foreign objects that may have fallen into the thickener, blocking the outlet. It is also used to assist with starting up the rake under stalled conditions.

A second flush pipe is at the same angle but directed toward the pump suction, the objective of this flush line is to dilute the thickener underflow, thus reducing the viscosity allowing the centrifugal pump to operate. The third is located on the pump delivery, also angled at 30 degrees toward the direction of flow.

The higher the underflow density, the better the filter plant operates, shorter filtration times are achieved.

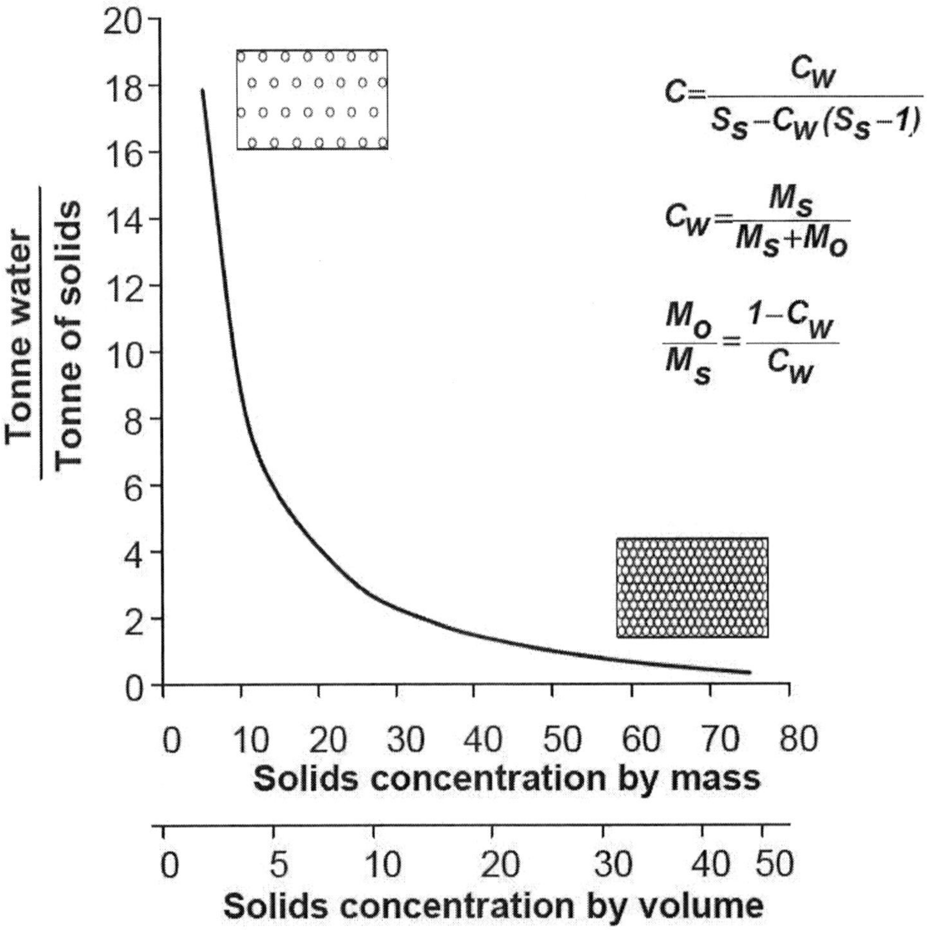

Figure 1. The relationship between water content per ton solids and solids concentration or the volume metric concentration, $S_s = 2.9$. M_s= mass flow rate of dry solids, M_O= mass flow rate of water. C_w is concentration by mass.

It is therefore advisable to operate a system that does not reduce/dilute the thickener underflow density. A shear thinning loop can achieve this objective.

Design and operating conditions require a thorough understanding of the rheological characteristics of the material, both in shear and compression. The system is selected and designed to minimise the pumping energy required whilst delivering a material at the velocity, yield stress and viscosity needed for the transfer to the filter plant agitated buffer tank.

The viscosity of the material will vary with shear rate. As the shear rate is increased, pseudoplastic or shear thinning materials exhibit a decrease in the viscosity. The shear thinning nature of tailings is attributed to the alignment of particles and/or flocs in the flow field. An increase in the shear rate from rest, results in instantaneous alignment of particles in the direction of shear, thus providing a lower resistance to flow. As such, the suspension will show a decreasing viscosity with increasing shear rate.

Non-Newtonian behavior of coal tailings underflow is time-dependent behavior. Here, the shear stress is a function of both shear rate and time of shear. Basically, this is the case where the alignment of the particles and/or flocs occurs on a timescale which can be observed in a rheometer.

Some typical data for coal tailings is listed below:

Variable	Coal tailings
Specific gravity (kg/m^3)	1450
Solids concentration (% w/w)	36
Slurry density (kg/m$^{3)}$	1120
Slump height (mm)	203
Calculated yield stress (Pa)	160
Predicted pressure drop (kPa/m)*	5.07

*Pressure drop prediction assumes:

- Bingham material
- Bingham viscosity – 1 Pa.s
- Horizontal pipeline
- Pipeline internal diameter = 200 mm
- Pipeline velocity – 1 m/s

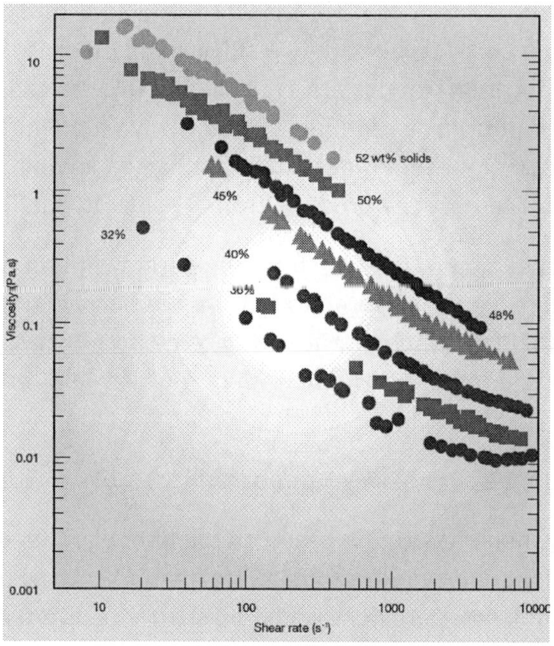

Figure 2. Viscosity shear rate data for a coal tailings thickener underflow. Suspensions can be pumped up to a yield stress of about 200 Pa with a centrifugal pump.

Thickener underflow pumps optimally have a VSD fitted, allowing the transfer of slurry to the filter plant to controlled, slowing the flow when the tank is at about 80% of capacity and increasing the speed when it drops below 60% capacity.

This leads to the obvious, locate the thickener as close as is practical to the filter plant in order to minimize the pumping pipeline length.

The agitated feed tank is the start of the filter plant. When designing a uniformly agitated feed tank the ideal dimensionless numbers are Height H = D Diameter (Add freeboard to H), baffles are critical and 3 or 4 are best. The baffle width is D/12 and the off-wall distance is D/32. The impeller off floor distance is typically 1 Impeller diameter. The impeller diameter is typically 1/3 D. Recommended tip speed is 6 meters per second. Gearbox output speed can be calculated from Tip speed (m/sec) = Pi x D_{imp} (m)x N (rpm)/60.

An ultrasonic level probe is located on top of the feed tank, this gives the tank level and sends this information to the plants SCADA to control the thickener underflow pump speed.

The capacity of the filter press feed tank should ideally have a minimum of twice the filter press fill volume of solids. Remembering that one third of the tank volume is below the agitator and you do not want the slurry level to drop below the agitator or you will get settling.

Filter press feed pumping.

MPTEQ prefer filling a filter press as quickly as possible. This reduces cycle time and ensures a competent cake. The higher the pump final pressure the dryer the final cake moisture at a given feed PSD. A dry conveyable cake with a moisture content of lower than 23% total moisture can be achieved with a feed pressure of 8 to 9 bar. This can be achieved with two stage standard centrifugal slurry pumps. More sophisticated integrated centrifugal pumps are available but this can cause spares holding increments and increased maintenance staff training requirements so we avoid them. This is often an adequate cake moisture, if however, a lower moisture is desired a three-stage pumping system with a third centrifugal or by using a PD pump to increase the pressure from 9 bar to 12 bar.

Automatic valves are fitted to the feed line to allow the core blow to operate without pushing air into the feed tank through the pumps. The core blow is required to remove the slurry in the internal feed port of the filter. This is a core of slurry that cannot be dewatered once the filter i=has reached its terminal pressure. An air compressor is used to push this slurry out of the filter and return it to the feed tank.

An automatic valve isolates this line when the core blow is not operating.

Filter presses are sized on volume. The volume required is based on the filter plant feed tonnage. This can be an interesting number; often junior miners underestimate the fines when they design a plant. This is exacerbated if no large diameter drill cores are done and drop shatter tests are thus not conducted. Often assumptions are made based on similar plants in the area. However these can be different based on what seams are mined and how the seams are blended prior to feeding the plant. For example, it would be considered normal to assume 15% fines from a typical Witbank coal mine. If fines treatment is utilized then it is often assumed that the thickener will see 5% of the plant feed. So five tons per hour per one hundred ton plant feed. This would give twenty tons per hour to a filter press plant on a four hundred ton per hour plant. This is correct, most of the time, however when mining 1 seam or 5 seam, this number could increase dramatically. We typically double the design volume for this reason and for redundancy. So, a 400 tph plant feed will have a filter plant with a capacity of 30 to 40 tons per hour.

There are a number of factors that have to be considered when sizing a filter press. These include:

Cycle time. An automatic filter has, typically the following cycle operations:

- Close bomb-bay doors (Drip trays).
- Close the filter press.
- Open the feed valve.
- Start the pump sequence.
- Open core blow valves.
- Open the bomb-bay doors.
- Start the transfer conveyor(s) (if fitted).
- Open the filter.
- Discharge the filter cakes.

Typical cycle times for a 100 chamber 1500mm x 1500mm x 38 mm thick cake coal tailings filter is in the order of 20 minutes. So theoretically almost three cycles can be achieved in an hour.

If a final cake moisture of 25 % is targeted, the volume of slurry is calculated, this is then divided by the number of cycles per hour. This volume is then divided by the volume of one chamber in the desired filter press size. This depends on the cake thickness, obviously, so this is selected first. On coal tailings we have found cake thicknesses of between 35 to 40 mm work well.

PLANT PEFORMANCE AND DISCUSSION

A filter plant is a complex system, all components must work together to achieve the required production. However, in its simplest form it is intrinsically a big G-Clamp able top withstand the pressure of the feed pumps.

So, if the filter cloths are in good condition and the closing pressure of the hydraulics are greater than the pump final pressure the system is good to go. The balance of the system is proximity sensors, conveyors, plate shifters and drip trays.

The key to successful filter press operation is, in my opinion, cleanliness and maintenance. If the plant is clean you can easily see if there are any issues, such as leaks, damaged cloths, leaking hydraulics immediately and fix them.

The filter press was initially fitted with filter cloths from China, these worked well however they had a plastic filter feed port protection sleeve that protruded into filter chamber. This held up the filter cake necessitating an operator to nudge the cake off. When we fitted local cloths with a rubber or cloth feed port protector this problem disappeared.

One of the concerns we had with filter presses imported from China is local spare parts and maintenance. In order to negate this risk we replaced all the hydraulic power pack components with local, easy to source components and modified the hydraulic pump from its location on the top of the power pack to a flooded suction located on the side of the power pack.

The other important thing we did was to replace the MMI PLC interface with our own Siemens unit with our own code. I was very worried about getting a call in the middle of the night and having no one that could program the PLC or understand the logic.

These modifications have proven to be very successful and removed any uncertainty on maintenance.

Development of a high capacity, low cut spiral for the coal industry

HS Baloyi, MC Tjabadi and MC Diketane

Traditional coal spirals typically treat about 2.5 to 3.5 tons per hour (tph) feed solids per start, with a high relative density (RD) cut point of ±1.6–1.8. In recent years, low cut point spirals were introduced into the coal market. Typical throughputs of about 1.0-2.5 tph solids per start, to prove low RD capabilities, are reported in literature and audits of mines with low RD spirals have shown similar ranges.

Many coal spiral plants have been designed to treat fines contributing 6-10 % of the total plant feed. Audits have shown a drastic increase over the years in the amount of fines reporting to the spirals, with some mines reporting up to 30 % increases. This adds significant pressure onto spiral banks leading to inefficiencies. Replacing the traditional spiral with a low cut spiral, without increasing the number of starts, adds even more pressure to the circuit. Also, given the need for a low cut, and that this comes with significant yield losses, bulk processing of coal with spirals at a low cut point would require the use of significantly more spirals, increasing the required plant footprint. Industry players approached Multotec to look at the design of a high capacity spiral with low RD cut capabilities.

This paper evaluates a new approach: a 3-stage, 4-product, 10-turn spiral with 2 intermediate discard removal points. This arrangement allows for low cuts at high solids percentages and high feed rates, thereby reducing the required plant footprint. Intermediate discard removal results in the ideal percent solids for a low cut at the final stage. A cut point of 1.53 was achieved with good yields at a high capacity. This spiral configuration, with intermediate discard removal, has the added benefit of being able to efficiently wash low quality coal or discard.

INTRODUCTION

Spiral concentrators are widely used in fine coal beneficiation as they offer several advantages that include low operating and capital costs, being simple to operate, low maintenance and high tolerance to variability in feed conditions (Erasmus, et al., 2018).

Fine coal slurry (75 – 3000 µm) is fed to the top of a spiral and gravitates downwards. Particles separate radially across the trough as described by the primary and secondary flow (Matthews *et al.*, 1997). The primary flow is the slurry descending along the inclined portion of the trough. The secondary flow occurs radially across the trough (Holtham, 1990). During their passage through a spiral, particles are subjected to hydrodynamic drag, lift, friction, gravity and centrifugal forces.

In general, the motion of particles depends on the particle properties (size, density and shape), the operating conditions as well as the slurry density and flow rate (Atasoy & Spottiswood, 1995). Spiral design (trough shape, inclination, number of turns) also plays a significant role in particle separation (Holland-Batt, *et al.*, 1984).

Relative Density (RD) Cut

Subasinghe & Kelly (1992) mentioned the RD cut as the relative density of that particle which has an equal chance of reporting to either the product or discard stream. Subasinghe & Kelly (1992) concluded that for small particles, the RD cut is affected by changes in the flow rate and particle size. For large particles, the cut is affected by flow rate and solids concentration and to a lesser extent by the particle size itself.

South African coal mines largely use spirals to clean coal of 2 mm x 0.15 mm in size. In some cases spirals are used to treat feed with a top size of 3 mm. The MX7 model has been used for decades as a spiral of choice for fine coal cleaning, especially for the difficult to clean coal (Klima & Benusa, 2006). The MX7 spiral has a recommended capacity of 2.5-3.5 tph solids. It is mainly in heavy minerals where manufacturers are known to have high capacity spirals that can take up to 9 tph solids per start.

In recent years, low cut spirals were introduced into the coal processing market. It is well known that reducing the flow rate and percentage solids on coal spirals, decreases the RD cut (Arnold *et al.*, 2018). Arnold *et al.* (2018) reported that typical throughputs of these low RD cut spirals are about 1.0-2.5 tph solids per start. Site test-work at a South African coal mine by De Korte (2015), achieved a low RD cut on these spirals at a solids flow rate of 1.3-1.5 tph and it was mentioned that a recommended flow on these spirals would be 1.8-2.5 tph solids. Ramsaywok & Mathumbu (2017) achieved a low RD cut at a solids feed rate of 1.5-2.5 tph. It was also reported that an increase in the RD cut occurs as the feed rate increases.

High Capacity Requirements

Honaker *et al.*, (2007), reported a worldwide survey which showed, that coal preparation plants at that time, 6 % of coal treated was feed to the spiral plant. Enslin (2019), in an audit of South African coal mines, reported a drastic increase over the years in the amount of fines reporting to some spiral banks. In some audits, spiral plants which have been designed to handle 7 % of the total feed to the plant, are now receiving up to 30 % of the total feed (Enslin, 2019). These spirals plants were not designed for this and are constantly under pressure. Replacing the MX7 spirals with low RD cut spirals can place even more pressure on the spiral circuit. In almost all instances the MX7 had to be replaced with the same number of starts as on the low cut spiral. It has been observed that some coal washing plants had to cut on the feed rate to accommodate the additional fine coal and also to achieve a low cut on the spirals. Some mines resort to changes in the desliming cyclone setup, increasing the fines to the cyclone overflow, with the subsequent result of constraining the thickener.

SX10 high capacity, low cut spiral

Aiming for a low RD cut results in significant yield losses. Given that coal is a bulk commodity, whenever there is need for a low cut point, then this requires the use of significantly more spirals, increasing the required plant footprint (Arnold, *et al.*, 2018). The Northern Hemisphere coal mines and plant designers realized the need for low cut spirals and requested Multotec to supply a low RD cut spiral. The plant footprint constraints and capacity issues in some mines led Multotec to develop the high capacity, low RD cut SX10 spiral.

It has been claimed in literature that the motion of slurry reaches steady state after two turns and hence effective separation ceases at this point (Subasinghe & Kelly, 1992). Further studies by Holland-Batt (1995) showed that if minerals are liberated, the mineral recovery does not cease, but the rate of recovery gradually decreases with an increase in the number of turns on the spiral. Holland-Batt (1995) and Klima & Benusa (2006) demonstrated that the addition of repulpers on coal spirals has the added benefit of revitalising separation. For many years, the lack of mathematical models that can accurately predict separation on spirals, has made spiral development more costly due to trial and error. A better understanding of particle transport in recent years has helped manufacturers to design more suitable spirals for specific ores. Due to this, coupled with advances in material of construction, spirals are making it possible to configure the desired pitch, descend angle and number of turns that will abide by the desired separation criteria.

The SX10 is a 10-turn spiral allowing the near dense particles longer retention time for separation. Denser particles (quick to separate from the discard), are removed into the center column after the first four turns, with the remaining clean coal and middlings are subjected to repulping. This is followed by three more turns, another discard removal step, and another repulping stage prior to the final distribution of coal on three more turns before the final splits into clean coal, middlings and discard. Following the removal of the discard from the first two discard draws, the feed rate and percent solids feeding the final three turns is reduced, approximating the ideal feed solids percentage and feed rates for typical low RD spirals. Collecting the discard into the center column means the working surface diameter of the spiral is not affected by the removed discard and hence allowing for extra capacity. Also, the discard removal into the center column has an added benefit, especially when treating low quality coal or discard. It has large volumes of refuse to be collected, without interfering with the separation of the middlings and the clean coal stream.

EXPERIMENTAL PROCEDURE

Sample preparation

Three different types of samples were sourced from two different collieries in South Africa to be referred to as region A and region B. Approximately 600 kg of spiral feed from region A, 800 kg from region B and 600 kg of discard from region B were made available to Multotec.

Sub-samples were taken for specific gravity of solids (SG) and Particle Size Distribution (PSD) analyses. A pycnometer was used to measure the SG and a sieve shaker was used for PSD analysis. The SG for region A and region B spiral feed samples was measured and found to be 1.52 t/m^3 and 1.55 t/m^3, respectively. The SG of the Discard sample was 1.67 t/m^3.

Particle Size Distribution

The PSD of the material as received is shown in Figure 1.

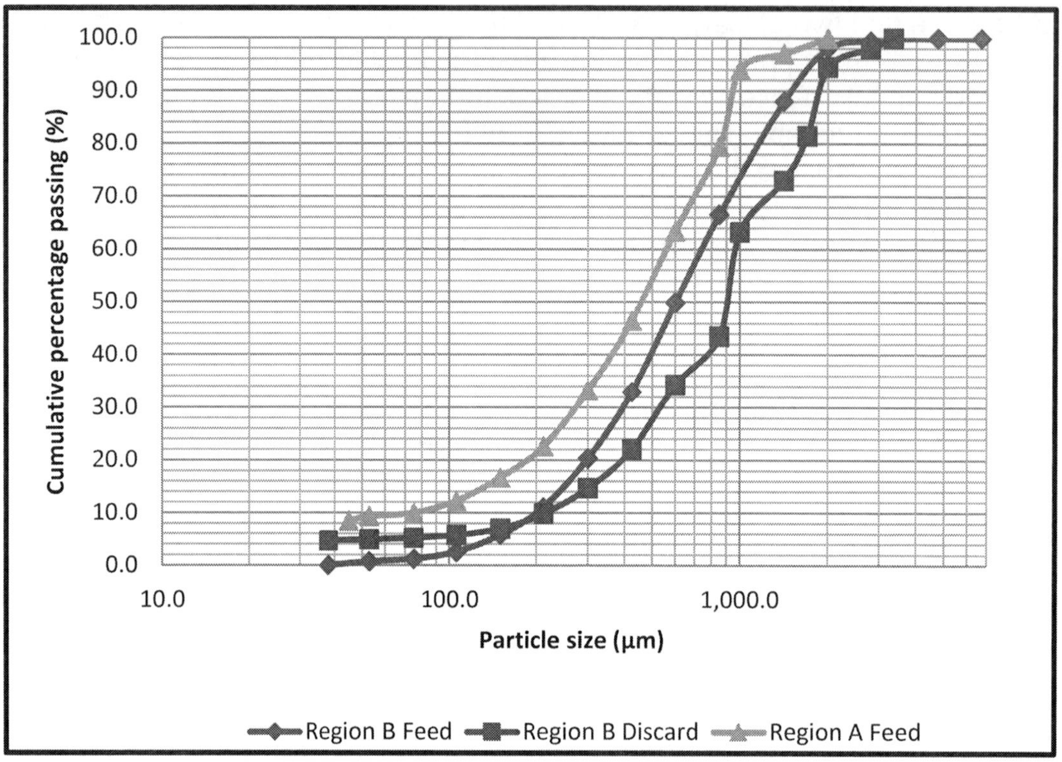

Figure 1: Particle Size Distribution

Figure 1 shows the PSD for region A and B samples. Region A feed appeared to be finer than the region B samples. The region A sample showed a P_{80} of 600 µm while region B samples showed a P_{80} of 1700 and 1200µm for both discard and the feed samples, respectively. The discard sample demonstrates a coarser PSD as compared to both feed samples.

Spiral Concentrators

The test work was conducted using the Multotec MX7 and SX10 spiral models, which were each fitted with a mouth organ product box as shown in Figure 2.

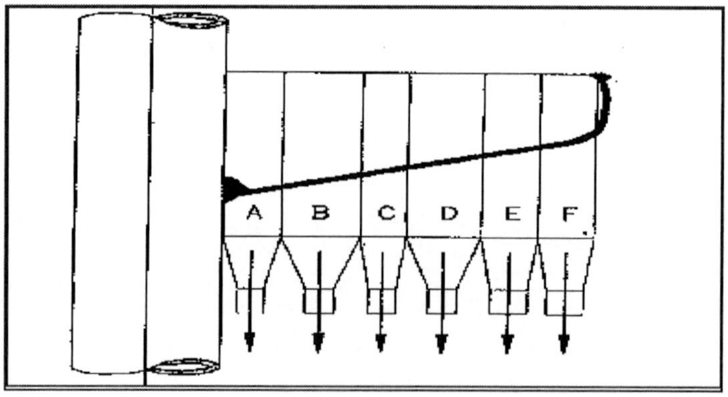

Figure 2: Mouth Organ Fraction

Sample A represents the sample collected from the inside of the trough. It is the heaviest sample and should contain the material with the highest ash content and SG. The product fraction for coal is from the F fraction towards the A fraction, to a chosen point that is still within the product specification. The SX10 has an additional outlet called the auto-reject which is not shown in the figure above. This allows for the immediate removal of high ash content material (discard). The auto-reject outlet discharges into the center column.

Representative samples were mixed with water in order to produce feed solids concentrations as shown in Table 1.

Table 1: Spiral Test Work Parameters for Region A Coal

Spiral Stage	Spiral Model	% Solids by weight (Target)	% Solids by weight (Actual)	Dry solids (tph)	Water (tph)
Secondary	MX7	30	28.5	2.3	5.8
Tertiary	SX10	30	32.9	2.6	5.4

Table 2: Spiral Test Work Parameters for Region B Coal

Spiral Stage	Spiral Model	% Solids by weight (Target)	% Solids by weight (Actual)	Dry solids (tph)	Water (tph)
Secondary	MX7	35	36.6	2.86	4.94
Tertiary	SX10	35	34.7	2.86	5.40
Tertiary	SX10	35	34.0	4.63	8.98
Tertiary	SX10	40	38.4	4.33	6.94
Tertiary	SX10	40	40.5	3.38	4.95
Tertiary	SX10	40	39.9	3.45	5.14

These slurries were fed to the spiral feed box via a distributor with an overflow in order to maintain constant feed conditions during the tests. Figure 3 shows the spiral test rig configuration used for the test work.

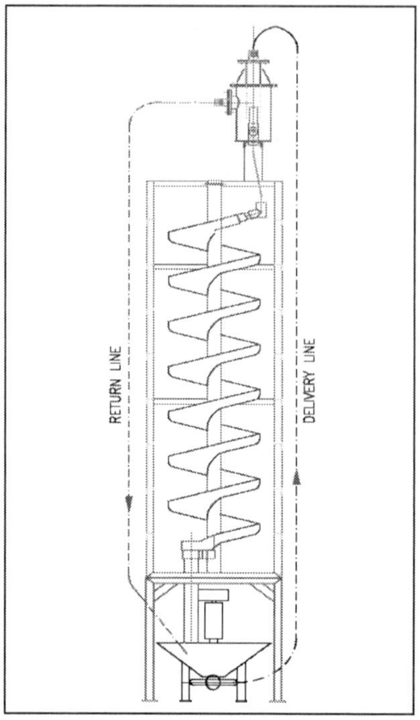

Figure 3: Spiral Test Rig Setup

Each spiral stage was run until steady state conditions were reached before taking the RD measurements at the top of the spiral rig. Three consistent RD measurements and three consistent flow rate measurements were recorded before taking the simultaneous timed samples (Fractions A to F) from the mouth organ product box.

RESULTS AND DISCUSSION

Low cut application: Region A coal

Figure 4 below shows the mass yield comparison between the MX7 and SX10 for Region A.

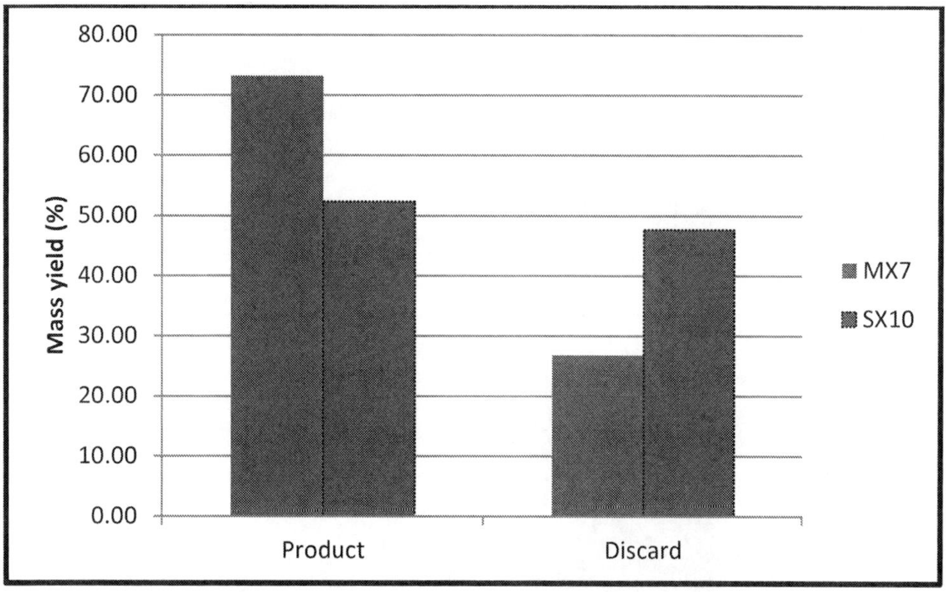

Figure 4: Mass Yield Comparison for Region A Coal

The comparison between the mass yield for the MX7 and SX10 shows the MX7 has a higher mass yield in the product as compared to the SX10. This can be attributed to the intermediate discard removal points, which rejects the material prior to the next stage. The lower mass yield in the product results in a high imperfection (EPM).

Figure 5 shows the percentage ash comparison between the MX7 and SX10 for region A.

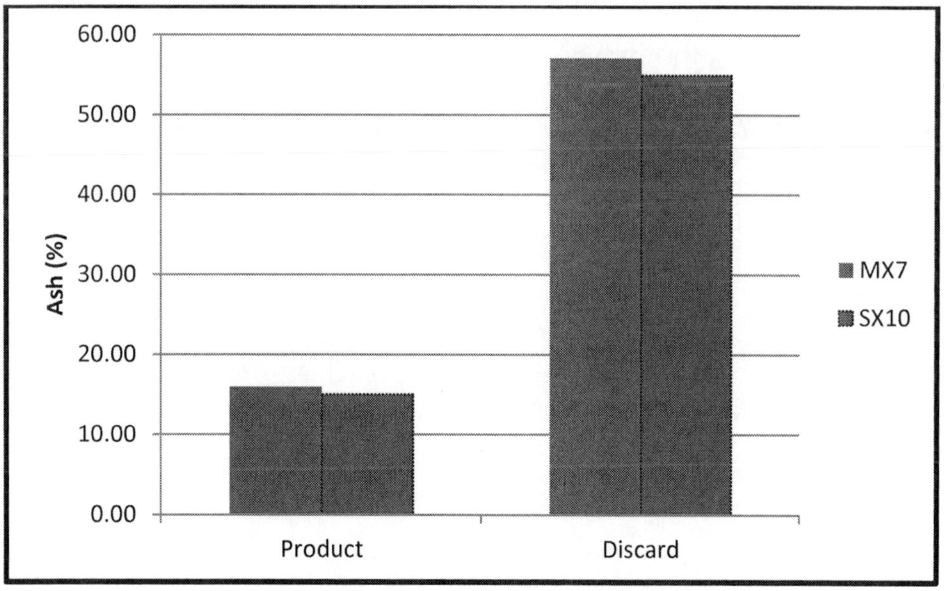

Figure 5: Percentage Ash Comparison Region A Coal

The percentage ash comparison for both spirals show that the MX7 has a higher ash content in the product and discard as compared to the SX10 spiral. The lower ash content in the SX10 can be attributed to the lower mass yield associated with the low cut. Cutting low results in the loss of good quality coal to discard.

Figure 6 shows a Calorific Value (CV) comparison between MX7 and SX10 for region A.

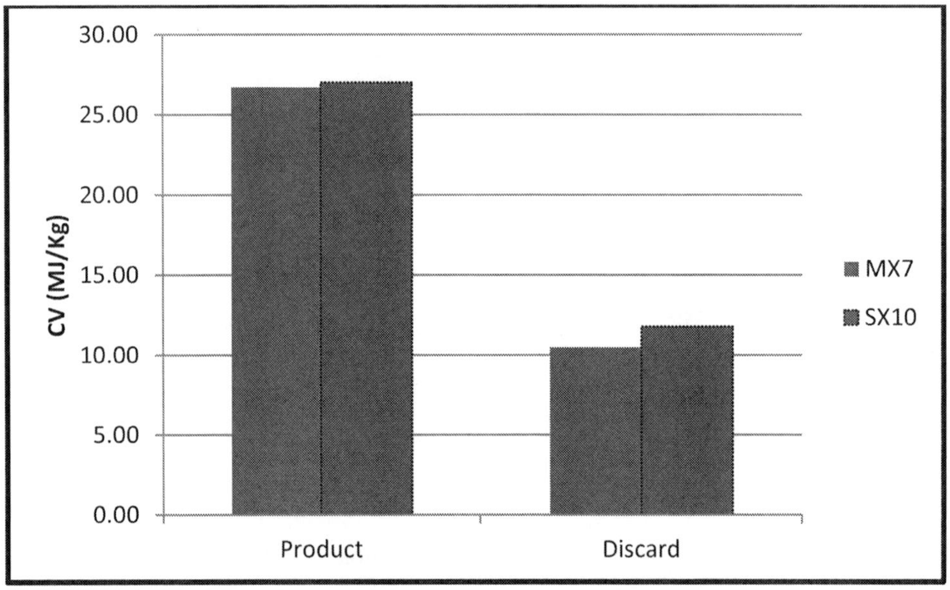

Figure 6: Calorific Value Comparison for Region A Coal

The CV comparison for both spirals shows an additional 1 MJ/kg in the product when using the SX10 in comparison to the MX7 spiral. The CV comparison in the discard shows that there is a higher CV for the SX10 discard as opposed to the MX7.

Figure 7 shows the partition curve comparison for MX7 and SX10 in Region A.

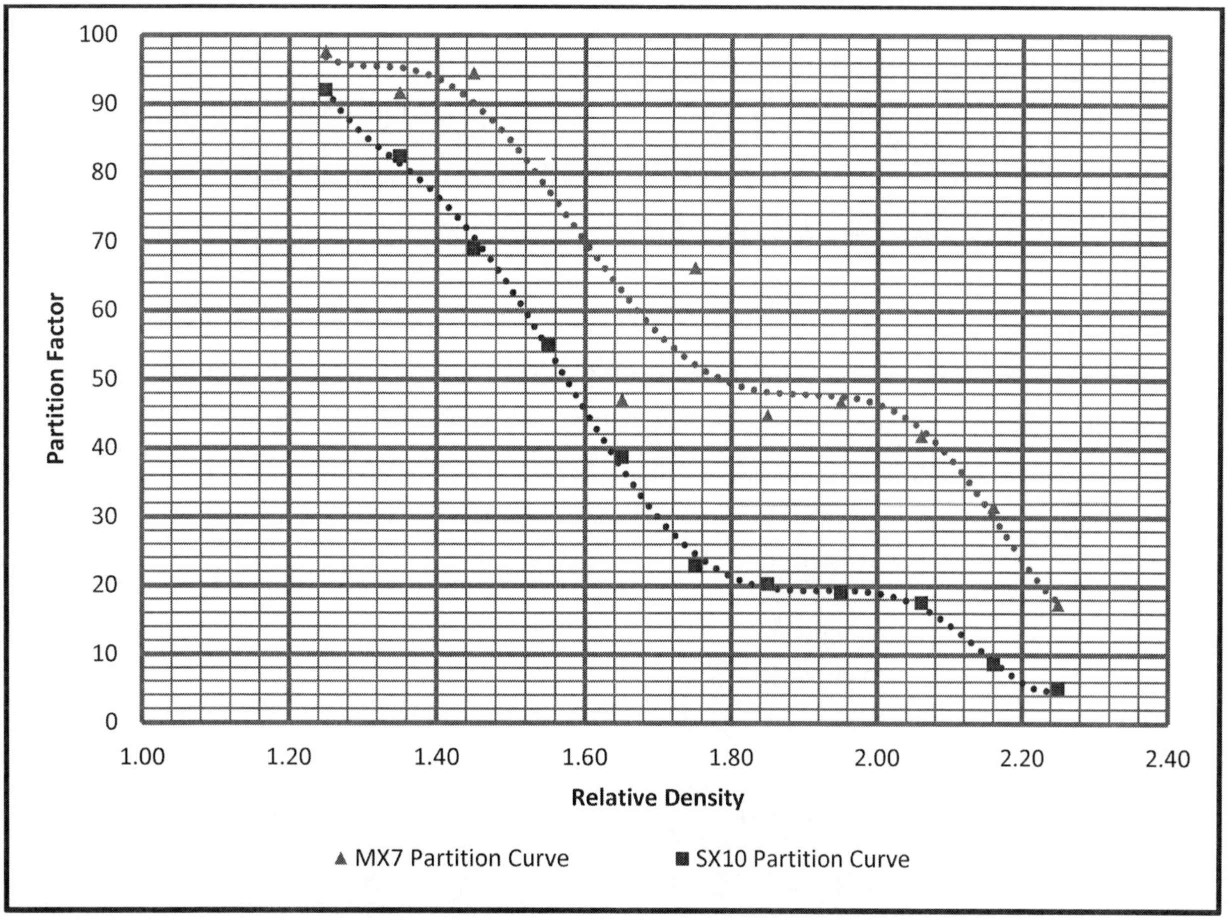

Figure 7: Partition Curves Comparison for Region A Coal

By using curve fitting the SX10 spiral partition curve shows a cut point (D_{50}) of 1.56, whereas the MX7 shows a cut point of 1.78 .There is a 0.22 density differential between the cut point for SX10 and MX7 spirals. Both partition curves show that the gradient flattens out between a density of 1.75 and 2.0. This can be due to near density particles which makes separation difficult.

Table 3 shows the efficiency result comparison for region A coal washing on the MX7 and SX10.

Table 3: Efficiency for Region A Coal

Efficiency Test Results		
	MX7	SX10
Feed % Ash	22.23	21.88
Product % Ash	15.96	15.02
Discard % Ash	57.06	54.93
Product Yield	73.21	52.34
D50	1.78	1.56
Ep	0.26	0.165

The MX7 shows that it has a higher Ep value than the SX10 spiral. It is expected that when you cut at a lower density, the Ep value increases which is a result of some misplacement of the product to the discard. The results suggest that a large number of near density material were recovered into the product by the MX7 spiral whereas the SX10 was able to reject the near density material, thus improving the Ep.

Low cut application Region B Coal

Figure 8 shows the mass yield comparison between the MX7 and SX10 for region B.

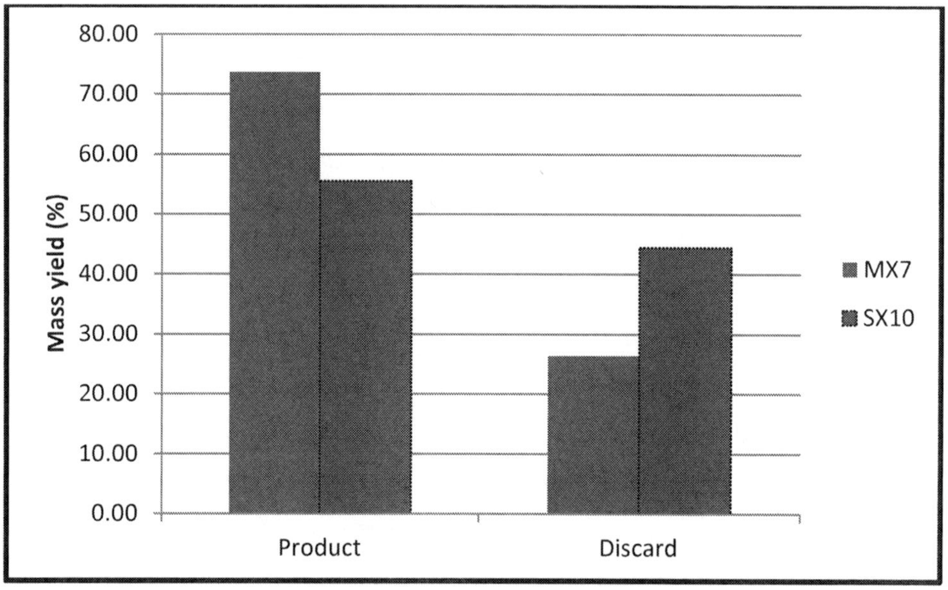

Figure 8: Mass Yield Comparison for Region B Coal

The comparison between the mass yields for both spirals in region B coal shows that the MX7 has a higher mass yield in the product as compared to the SX10. The results also indicate that the SX10 has a higher mass yield in the discard. This behavior was also observed in the region A sample.

Figure 9 shows the percentage ash comparison between the MX7 and SX10 for region B coal.

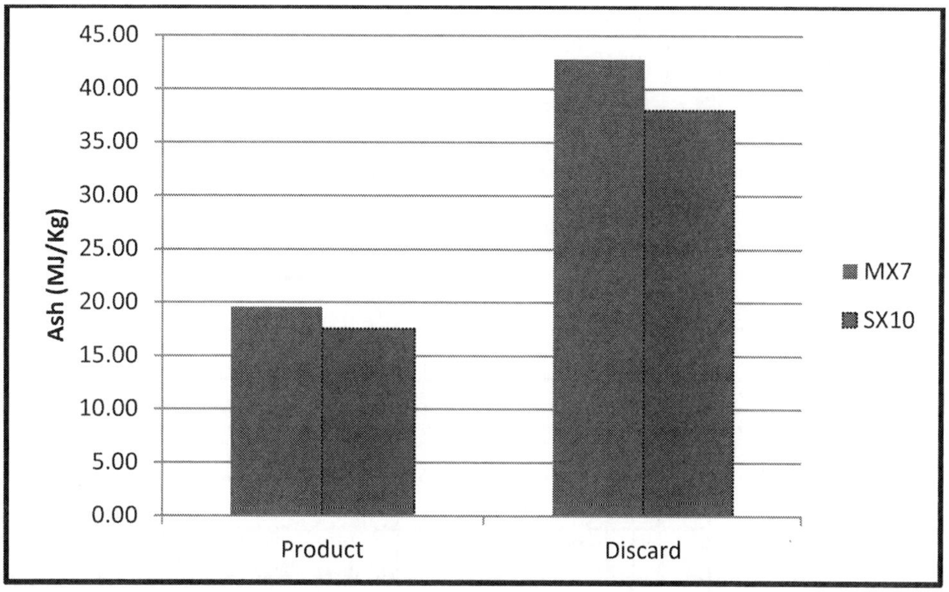

Figure 9: Percentage Ash Comparison for Region B Coal

The ash comparison for both spirals for region B shows that the SX10 yields a 2 % less ash as compared to the MX7 spiral. The SX10 also shows a significant reduction of the ash content in the discard as compared to the MX7. The reduction in the ash content was also observed for the region A sample, but it was not as significant as compared to the region B sample.

Figure 10 below demonstrates the CV comparison between the two spirals for region B.

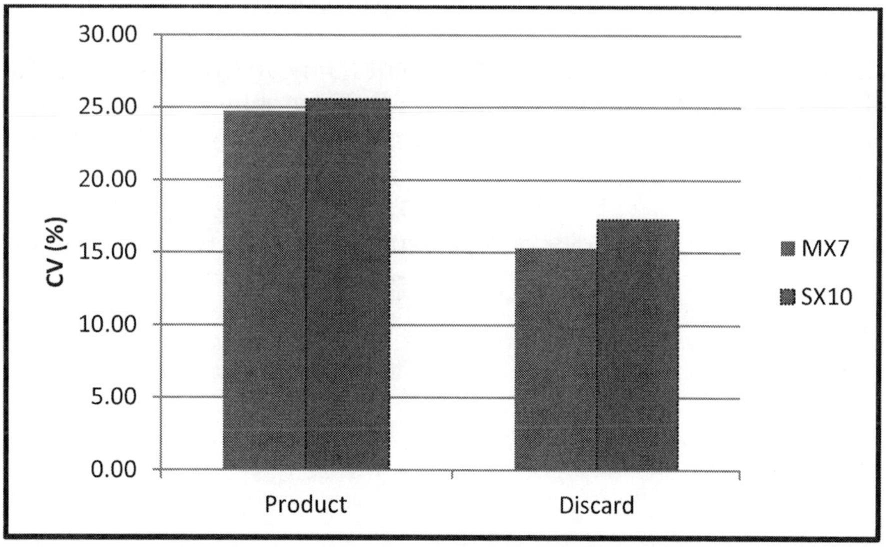

Figure 10: Calorific Value Comparison for Region B Coal

There is an additional 2 MJ/kg CV in the product when using the SX10 on region B coal as compared to the MX7 spiral. There is also a sweeter discard for the SX10 as compared to the MX7.

Figure 11 illustrate the partition curve comparison between the MX7 and SX10 for region B coal.

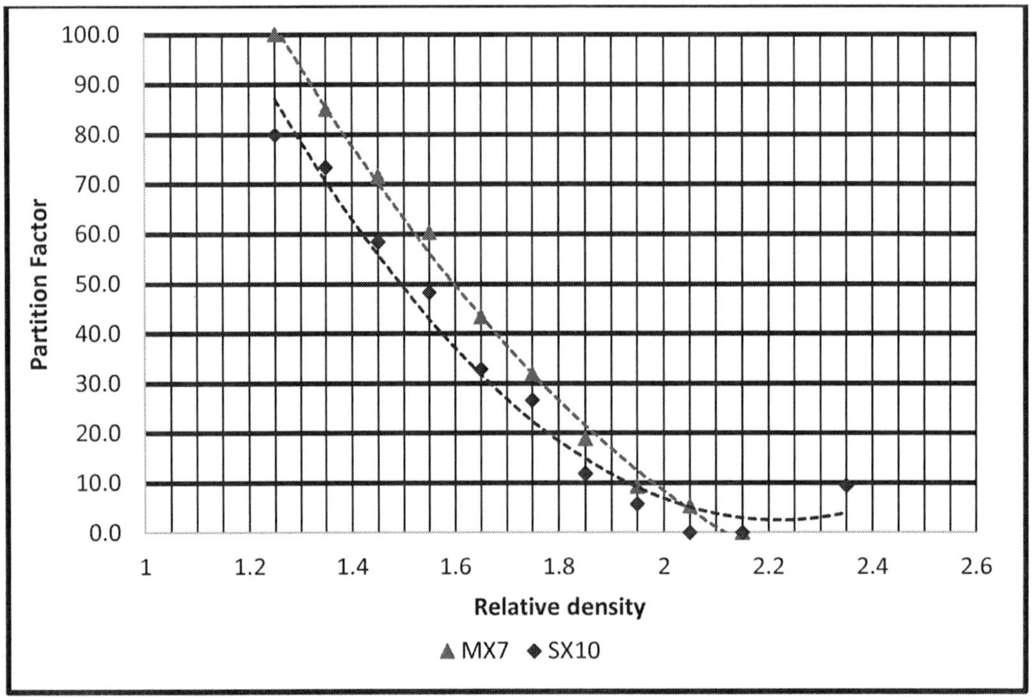

Figure 11: Partition Curves Comparison for Region B

The SX10 spiral partition curve shows a cut point (D_{50}) of 1.53, whereas the MX7 shows a cut point of 1.61 for region B coal. There is a 0.08 density differential for the cut point. This might be due to less near density particles which makes it difficult to clean most of the existing coal ores.

Table 4 shows the efficiency results comparison for region B coal washing on the MX7 and SX10.

Table 4: Efficiency for Region B

Efficiency Test Results		
	MX7	SX10
Feed % Ash	21.74	19.76
Product % Ash	19.02	17.56
Discard % Ash	42.76	38.03
Product Yield	73.2	52.30
D50	1.61	1.53
Ep	0.19	0.22

The MX7 shows a lower Ep value than the SX10 spiral. It is expected that when you cut at lower density the Ep value increases which is a result of a lot of misplacement of good ash material in the discard. This is visible on the SX10 which has a slightly lower cut point.

Throughput evaluation on the SX10 for typical spiral feed treatment

The SX10 spiral was further tested at different throughputs to demonstrate optimum operating parameters.

Figure 12 demonstrates the relationship between the throughput and product mass yield on the SX10 spiral using region B coal.

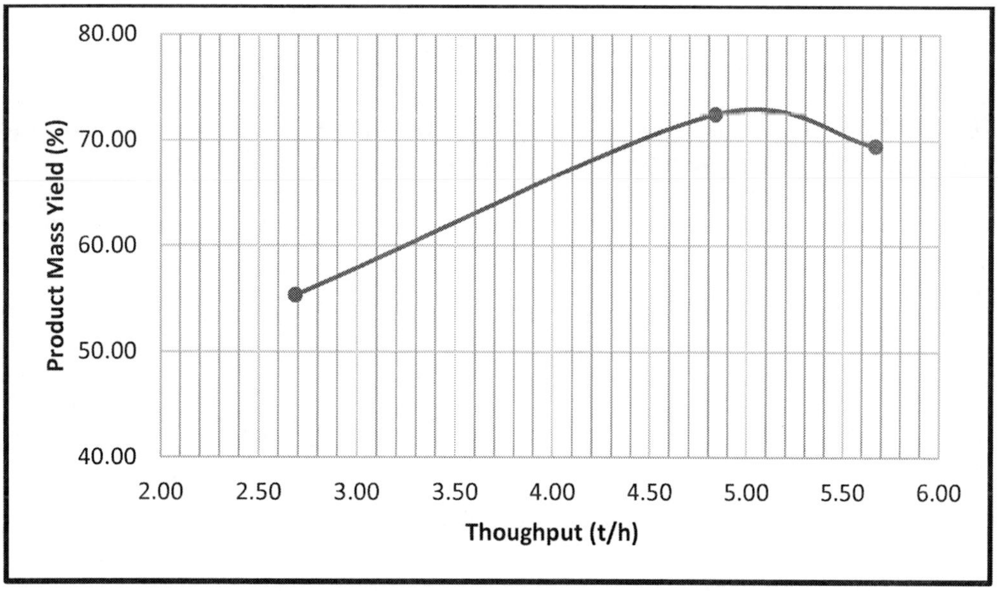

Figure 12: Throughput versus Product Mass Yield

Figure 12 shows that lower throughputs results in lower mass yield. The curve also suggests that the increase in throughput results in the increase in the mass yield to the product up to a throughput of 4.8 t/h. By increasing the throughput further, the mass yield starts to decrease, therefore 4.8 t/h is the optimum throughput for the SX10 spiral.

Figure 13 shows the relationship between the product quality and throughput on the SX10 spiral.

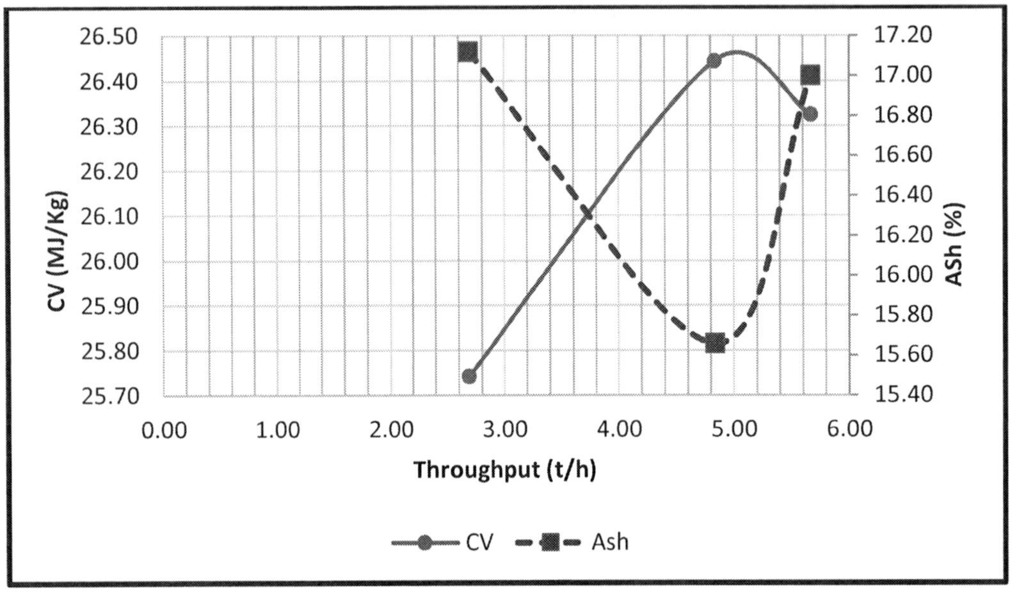

Figure 13: Relationship between Throughput and Product Quality

Figure 13 indicates that increasing the throughput between 2.6 and 5 t/h results in an increased CV and decreased ash value in the product. If the throughput is increased further than 5t/h, the product quality is also affected. The curve suggests that the ash is more sensitive at higher throughput rates, compared to the CV.

Throughput Evaluation on the SX10 for Discard Retreatment

The SX10 was further evaluated for retreatment of a tailings dump. The results obtained are summarized in Figures 14 and 15.

Figure 14 shows the relationship between the product mass yield and throughput on the discard sample.

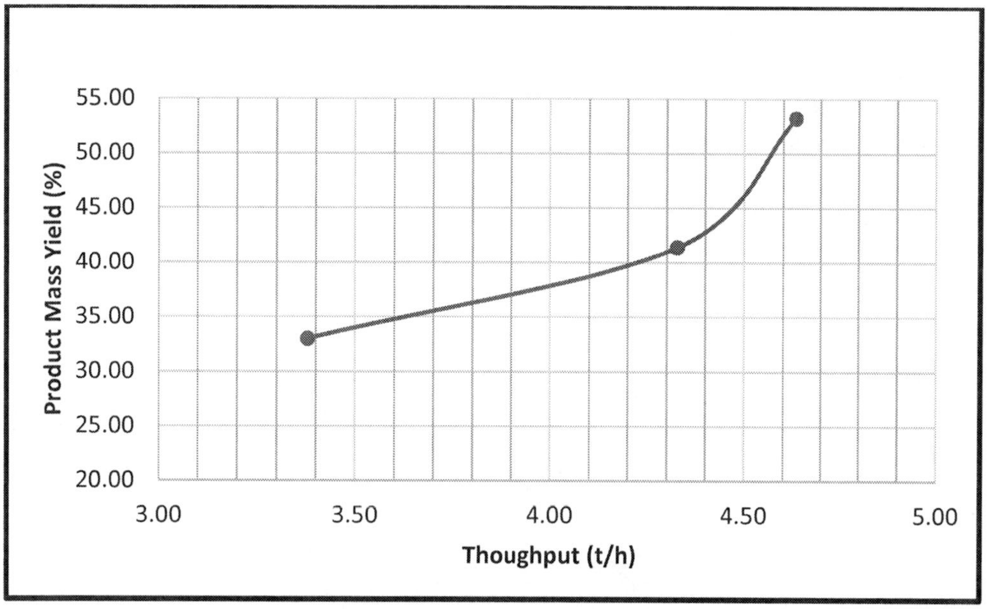

Figure 14: Relationship between Throughput and Mass Yield to Concentrate

The results show that when the throughput is increased on the discard sample, the mass yield to the product increases significantly. It can be observed that there is a constant gradient between a throughput of 3.3 tph and 4.3 tph while between a throughput of 4.3 tph and 4.66 the gradient becomes steeper indicating a massive increase in mass yield.

Figure 15 below shows the relationship between the solids concentration and the mass yield to concentrate.

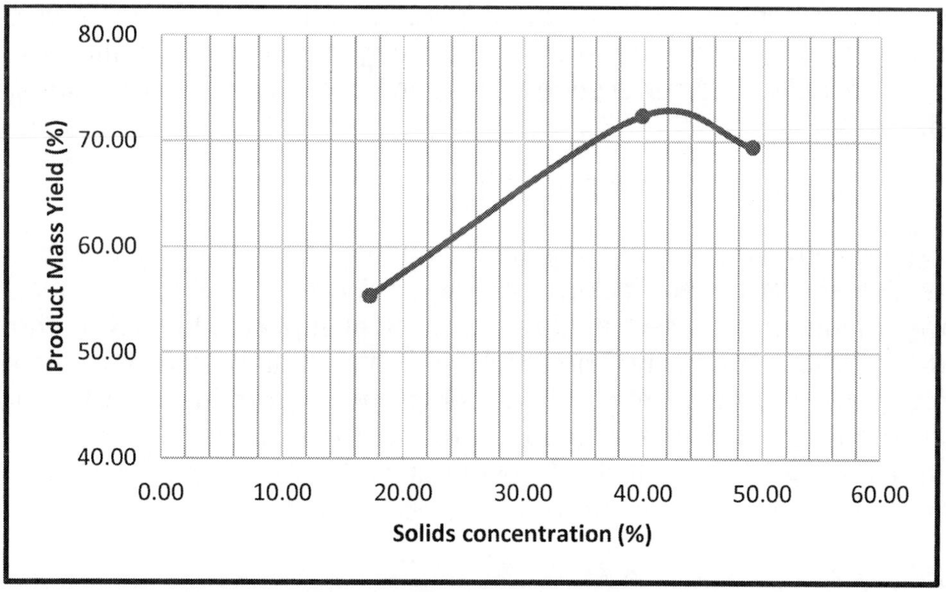

Figure 15: Solids concentration versus Mass Yield to Concentrate

Figure 15 shows that when the solids concentration is increased, the mass yield to the product increases significantly. When the solids concentration is increased further than 40 %, the mass yield decreases. An increase in solids concentration of higher than 40 % results in a decrease in mass yield at a solids concentration.

Figure 16 shows the relationship between throughput and product quality for a typical spiral feed treatment.

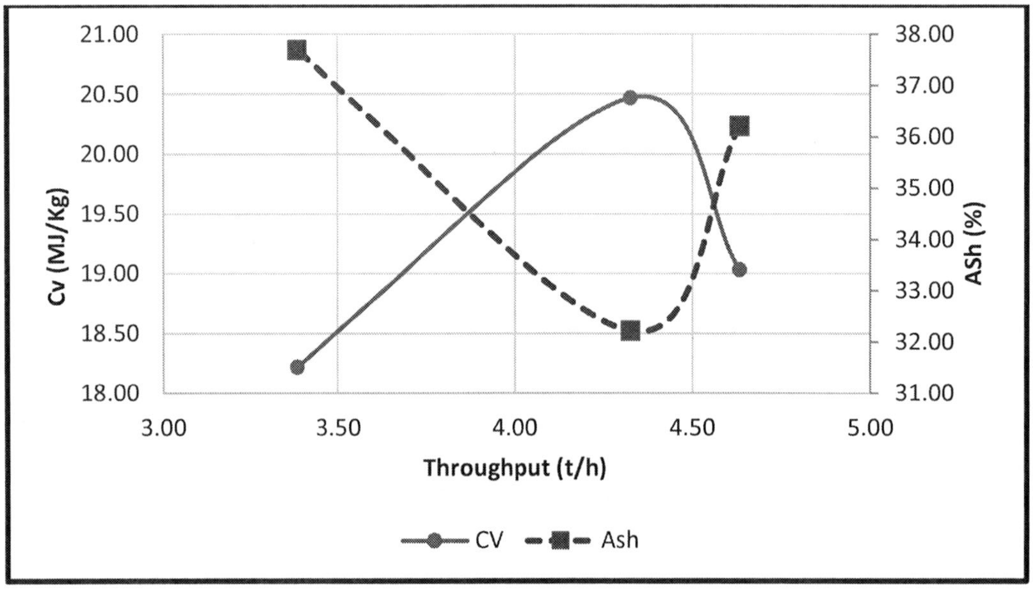

Figure 16: Relationship between Throughput and Product Quality

Figure 16 shows that when the throughput increases, the calorific value also increases up to a throughput of 4.3 tph. By further increasing the throughput, the product quality decreases significantly. The curve suggest that both the CV and ash are sensitive at high throughputs.

CONCLUSION AND RECOMMENDATION

The SX10 spiral shows a lower cut point as compared to the MX7 spiral in both samples. The results also shows better efficiency on region A coal than the MX7 spiral. The MX7 spiral shows a higher mass yield at higher cut densities, while the SX10 shows a lower mass yield at a lower cut density. The results indicates that the spiral can be operated at higher throughputs of about 5tph without compromising product quality when running fresh spiral feed material. It was also observed that the spiral can treat discard at higher throughputs of about 4.3 tph. If the throughput and solids concentration is increased further, the product quality is compromised, especially the ash content.

REFERENCES

Arnold, B., Erasmus, W., Bornman, F. & Bethell, P., 2018. *Development of a low specific gravity cutpoint spiral,* Pennsylvania: PrepTech.

Atasoy, Y. & Spottiswood, D., 1995. A study of particle separation in a spiral concentrator. *Minerals Engineering,* 8(10), pp. 1197-1208.

de Korte, G., 2015. *Evaluation of low-cut low- density spiral at Umlalazi Colliery,* Pretoria: Coaltech.

Enslin, F., 2019. *Audit of coal mines in Witbank,* Spartan: Multotec.

Erasmus, W., Bornman, F., Rust, J. & Diketane, C., 2018. *Low cut point spiral.* New Delhi, India Coal Processing Society.

Holland-Batt, A., 1995. The dynamics of sluice and spiral separators. *Minerals Engineering,* 8(1/2), pp. 3-21.

Holland-Batt, A., 2009. A method for the prediction of the primary flow on large diameter spiral troughs. *Minerals Engineering,* Volume 22, pp. 352-356.

Holland-Batt, A., Hunter, J. & Turner, J., 1984. The separation of coal fines using flowing-film gravity concentration. *Powder Technology,* 40(1/3), pp. 129-145.

Holtham, P., 1990. Flow visualisation of secondary currents on spiral separators. *Minerals Engineering,* 3(3/4), pp. 279-286.

Honaker, R., Jain, M., Parekh, B. & Saracoglu, M., 2007. Ultrafine coal cleaning using spiral concentrators. *Minerals Engineering,* Volume 20, pp. 1315-1319.

Klima, M. & Benusa, M., 2006. *Fine coal spirals: Data from a plant circuit treating 0.15 x 0.053 mm Coal,* Pennsylvania State University: PrepTech.

Matthews, B., Fletcher, C. & Partridge, A., 1997. *Computational simulation of fluid and dilute particulate flows on spiral concentrators,* Sydney: University of New South Wales.

Ramsaywok, P. & Mathumbu, P., 2017. *A straight up single pass comparison of a low cut spiral and conventional spiral in the South African Coal Industry.* Secunda, Southern African Coal Processing Society.

Subasinghe, G. & Kelly, E., 1992. Predicting the cut specific gravity of a coal washing spiral. *Minerals Engineering,* 5(2), pp. 193-203.

Discard Waste Amelioration Approaches at Greenside Colliery

David Power, **Christian Swanepoel**
Anglo American Coal South Africa
Johannesburg, South Africa
david.power@angloamerican.com; christian.swanepoel@angloamerican.com

Michael Phali **Maynard Lombard**
Greenside Colliery Khwezela Colliery
Emalahleni, South Africa
michael.phali@angloamerican.com; maynard.lombard@angloamerican.com

ABSTRACT

Greenside Colliery is a South African coal mining operation that has been in operation since the 1940's; it is located close to Emalahleni (formerly known as Witbank) and has at various stages in its lifetime mined and processed the Witbank Nos. 1,2 4 and 5 Seams. In the past, the plant has had a smorgasbord of unit operations in terms of baths, cyclones, spiral and fines dense medium processing to name but a few to its current configuration of cyclone, 3-product cyclone and flotation methodology to addressing the processing requirements of the coal. The refuse generated from these processes was in the form of a typical South African co-disposal method; i.e. a compacted discard encapsulating a slimes dam which requires careful management of the mass balance in terms of total refuse generation and the management of the discard /slimes ratio in the Mineral Residue Deposit (MRD).

Late in the first decade of this century, it was identified there was a looming constraint in terms of placement of the final refuse. A simple solution (if available) would be to expand the facility but as in most such conundrums, this was not necessarily an easy first choice; although it was a consideration in the planning process until 2018. The eventual solution(s) adopted did not start out as an integrated approach but as a series of abatement measures but when eventually the momentum gained, the approach became an integrated plan managed by the coal processing team at Greenside. The resultant solutions presented in the paper focus on some of the processing initiatives which include the implementation of a flotation and tailings filtration solution, conversion from a conventional dense medium cyclone to a 3-product cyclone operation and discard retreatment.

KEYWORDS: amelioration, discards, slimes, processing, solution, MRD

1 INTRODUCTION

Anglo American acquired Greenside Colliery through a sale agreement with Goldfields in 1998. The plant had washed a combination of Witbank No 1, 2, and 4 Seams in several dense medium module plants with spiral concentrators. At one stage a fine dense medium module was in operation during Goldfields tenure, but this was discontinued. In the final stages of their tenure, Goldfields installed a modular design plant (3 modules) to treat Witbank 4 Seam Coal – known as the 4 Seam Plant. On acquiring the mine, Anglo American quickly decided to rationalise operations and focussed on the mining and treatment of the Witbank No 4 Seam through the DRA designed modular plant. In 2000, a decision was made to open a 5 Seam reserve and to commission a purpose-built plant to supply a sized metallurgical coal product to Highveld Steel (then an Anglo American affiliate), this plant is known as the 5 Seam Plant. In 2004, a flotation module was added to the 4 Seam Plant based on the Anglo Multicell technology utilising solid bowl centrifuges for product recovery. The decision making process for this configuration was faulty as it was based on work carried out at another operation which did not support this configuration based on the centrifuge technology available to Anglo American at that time. This resulted in very poor reconciliation of flotation recovery, which was a problem that bedevilled such installations in the Witbank Coalfield over the past 30 years – there have only been three sustainably successful flotation operations there to date.

Towards the end of the 2000's, Greenside had to face yet another challenge regarding the airspace for the placement of discard and slimes. The discard and slimes are placed in a facility constructed with discard encapsulating an inner dam of slimes. In this case the facility is limited by the N12 highway which is the main artery between South Africa and Mozambique as shown in Figure 1 and Figure 2.

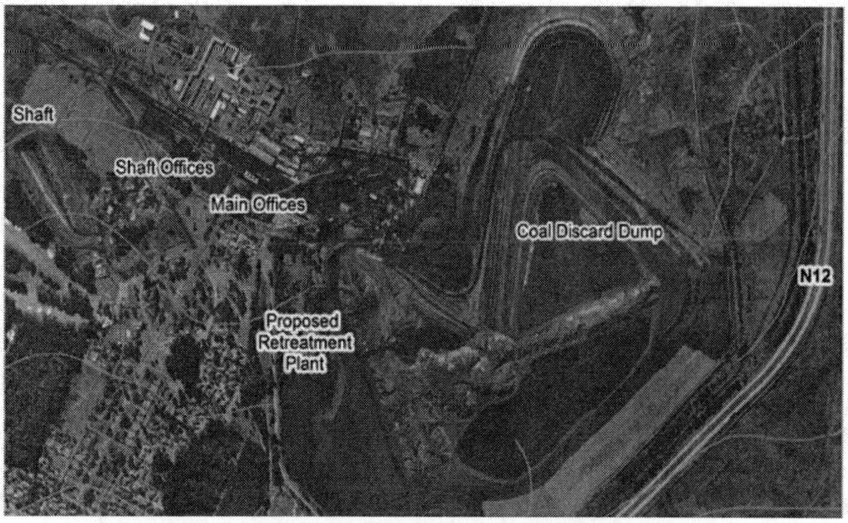

Figure 1: Aerial View of the Greenside Co-disposal facility

Figure 2: Plan view of existing facility with proposed location for new facility

The practicality of installing such a new facility on the other side of the highway was rather daunting but not impossible, there is a service bridge over the highway which could have served as a conduit for a conveyor, slimes line or a trucking option. The expense and likely licencing challenges led to other solutions being investigated.

2 AMELIORATION SOLUTIONS

The following illustrates some of the processing solutions devised by the Greenside team over the course of several years from 2012 onwards to address the issue:

- Reconfiguration of the flotation circuit with tailings filtration which eventually led to the elimination of a wet tailings dam
- Installation of 3-Product cyclones in the 4 Seam Plant
- Re-configuration of the 5 Seam Plant to enable the small-scale mining and retreatment of the discard dump
- Large scale mining of the discard dump

Mass Balancing techniques at the Anglo American mines is based on the methodology shown by (Power 2006) to determine the slimes discard ratio which enabled forecasts to be developed to understand the criticality of the airspace constraint. This guided the prioritisation of the project pipeline.

2.1 Reconfiguration of the Flotation Plant with Tailings filtration

As earlier described, the initial layout of the flotation plant had many flaws both with the technology (original Multicell) and the configuration of the ancillary equipment. A decision was taken with mine personnel to go back to the drawing board and address the shortfalls. The Multicell technology is an Anglo American concept described by (Esterhuizen 2008). An upgrade of the cell shown in Figure 3 designed by Enprotec which greatly enhanced the contact effectiveness of particle-bubble interface was chosen.

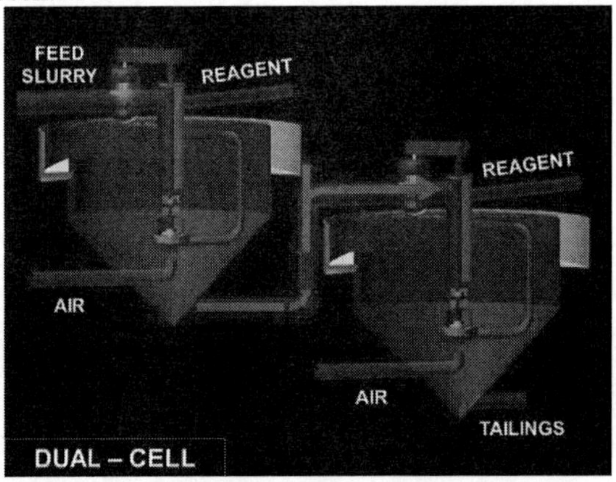

Figure 3: Layout of Dual Cell Flotation Cell

In partnership with Enprotec a cost-effective filtration solution was investigated in China and a decision was made to utilise the plate and frame technology manufactured by Jinjing which is capable of filtering high throughputs of material with a fine particle size distribution (ca. 60% - 44 micron). The plant was designed to recover coal with a calorific value of ca. 27- 27.2 MJ/kg with a recovery of ca.50- 60 % and this has been attained consistently since commissioning. The filtration plant enabled proper mass balancing to be carried out and it could be clearly demonstrated on a techno-commercial basis for capital closeout purposes and a clear reduction of refuse material being delivered to the dump.

After this configuration had proven successful, a decision was made to close the water circuitry of both 4 and 5 Seam plants with the installation of a tailings filtration plant using the Jinjing technology. A complete dry mass balance was enabled following this installation which provided more certainty to the life of mine forecasts for the operation.

2.2 Installation of 3-Product Cyclones

The 3-Product cyclone has been utilised extensively in the Chinese coal industry and its application in South Africa described by (de Korte 2014) in terms to the amenability for the technology to treat a typical Witbank No 4 Seam coal as shown in Table 1.

Table 1 Typical Washability of a 4 Seam reject washed at a specific gravity of 1.50

Float	Kg	%	%	%	%	%	%	%	%	%	MJ/kg	MJ/kg	%	%
	Mass	Yield	Cum		Cum		Cum		Cum			Cum		Cum
			Yield	IM	IM	VOL	VOL	ASH	ASH	FC	CV	CV	S	S
F@ - 1.300	0	0	0	*		*		*		*	*		*	
1.300 - 1.350	0	0	0	*		*		*		*	*		*	
1.350 - 1.400	0.141	0.35	0.35	2.8	2.8	27.4	27.4	9.5	9.5	60.4	29.6	29.6	0.57	0.57
1.400 - 1.450	0.16	0.4	0.75	2.9	2.8	23.2	25.2	12.4	11	61.5	28.28	28.9	0.48	0.52
1.450 - 1.500	0.382	0.95	1.7	2.9	2.9	21.5	23.1	17.7	14.7	58	26.16	27.36	0.67	0.6
1.500 - 1.550	1.155	2.88	4.58	3	2.9	21	21.8	19.7	17.9	56.3	24.77	25.73	0.44	0.5
1.550 - 1.600	2.989	7.45	12.03	2.8	2.8	21.2	21.4	24.2	21.8	51.9	23.1	24.1	0.62	0.58
1.600 - 1.650	4.517	11.26	23.28	2.6	2.7	21.4	21.4	28.1	24.9	47.8	21.18	22.69	0.98	0.77
1.650 - 1.700	4.01	9.99	33.28	2.3	2.6	20.3	21.1	33.4	27.4	44	19.35	21.69	0.79	0.78
1.700 - 1.750	3.152	7.85	41.13	2.3	2.6	19.8	20.8	35.4	29	42.5	18.77	21.13	1.76	0.97
1.750 - 1.800	2.571	6.41	47.54	2.1	2.5	19.4	20.6	40.3	30.5	38.2	17.24	20.6	2.63	1.19
1.800 - 1.900	5.355	13.34	60.88	2.2	2.4	19.6	20.4	43.2	33.3	35	15.1	19.4	3.65	1.73
1.900 - 2.000	2.329	5.8	66.68	2.1	2.4	20.6	20.4	44.8	34.3	32.6	13.63	18.9	5.65	2.07
2.000 - @S	13.371	33.32	100	1.3	2	18.6	19.8	60.4	43	19.7	9.17	15.65	9.97	4.7

For this washability, at a target CV of 27.2 MJ/kg there is little material left to be recovered but if one peruses the data at a cut of 1.70, a middling product of ca. 21.5MJ/kg with a sulphur less than 1.0% is recoverable. The 3-Product cyclone as shown in Figure 4 at Umlalazi had proven very successful in this application on 4 Seam Coals and based on this work there, a decision was taken to retrofit the units to replace the existing DSM cyclones in the 4 Seam Plant.

Figure 4: 3 – Product Cyclone installation at Umlalazi

The principle of the operation relies on the pressure differential arising within the barrel unit of the first stage being the driving force for the second stage cyclone. In the second stage, a separation takes place at approximately 20 points higher than the first stage separation with absolutely no control of this cutpoint and resultant quality of the middlings product (it must be noted that subsequently the control algorithm has been adjusted to allow for dilution water to be added at the second stage to enable some improvement of product quality by impacting the cutpoint on the second stage although at a cost of a higher magnetite consumption).The Umlalazi installation was also a retrofit and provided learning experience for the Greenside installation in terms of design (mass flow, medium flow, screen capacity and wear characteristic due to the abrasive nature of South African coals). In Figure 5, the layout for 2 of the modules is shown and the flows from the second stage to the re-configured sinks screen which is split to allow for recovery of middlings and final discards on the one unit.

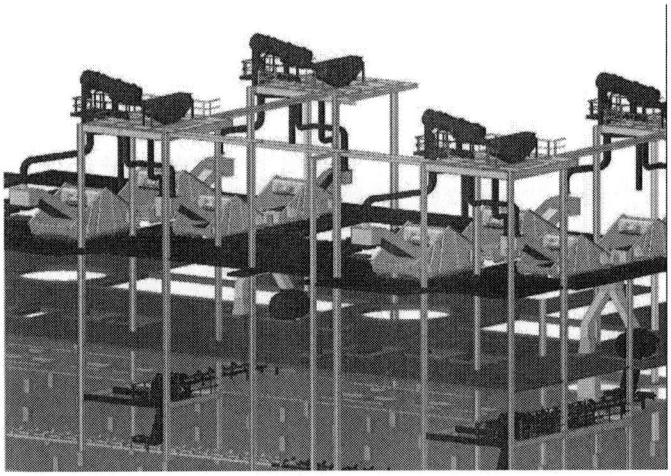

Figure 5: Retrofit of 3-Product Cyclone at 4 Seam Plant

A further complication for this intervention concerned the placement of the resultant middlings product within the environs of the plant as there was no real estate to install a separate stockpile. A novel approach was to fluidise this product and pump the mix across the plant complex to an area adjacent to the 5 Seam Plant where a new screen was installed. The effluent is classified (underflow to a dewatering screen and overflow to a thickener) to limit recirculating fines loading. This route is approximately 800m long (with a minimum velocity of 2 metres/second to prevent settlement in the line) and was designed by Paterson and Cooke, the pipeline layout is illustrated in Figure 6 (du Toit 2012).

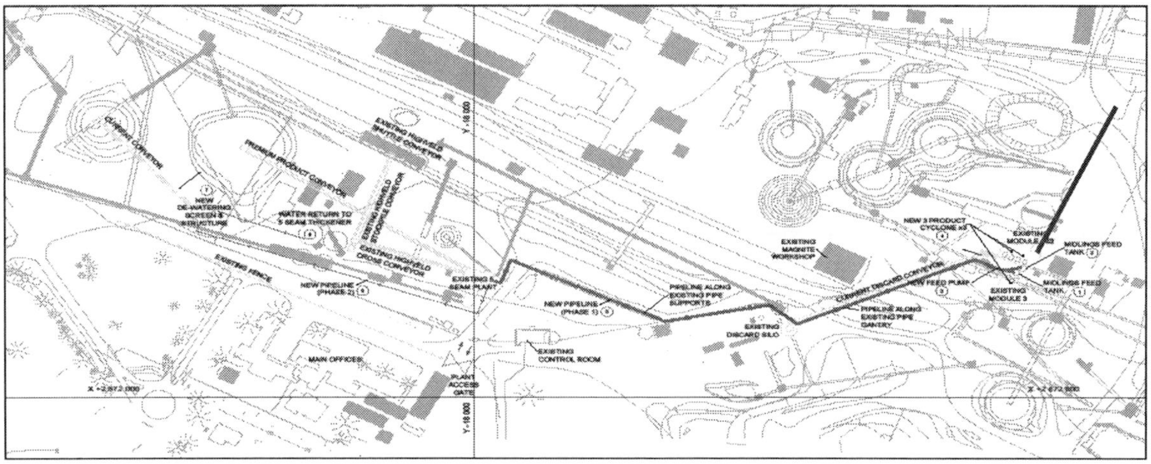

Figure 6: Route showing the middlings pipeline between 4 Seam and 5 Seam Plants

Again, this intervention led to a ca. 3.5- 4% recovery of the feed to plant which no longer reported to the discard facility thus relieving pressure on the available airspace.

2.3 Re-configuration of the 5 Seam Plant to enable small scale mining and retreatment of the discard dump

The 5 Seam resource was exhausted in 2012, which left the 5 Seam Plant potentially idle. As the airspace issue was ever present in the minds of the technical team at Greenside, an examination of the circuit, yield and size envelopes ensued to establish if the dump could be selectively mined to create usable airspace on the dump. To put the challenge of utilising this capacity into perspective, one has to understand that the coals being treated in the 5 Seam and 4 Seam Plants were washed at similar specific gravities but the 5 Seam Plant produced a sized washed product (-100mm) through a combination of dense medium bath/cyclones whereas the 4 Seam Plant utilises cyclones only (-50mm); the constraint for treating 4 Seam Discard in the 5 Seam Plant is highlighted in Figure 7, where it can be seen that the 4 Seam Discard is finer than the 5 Seam feed.

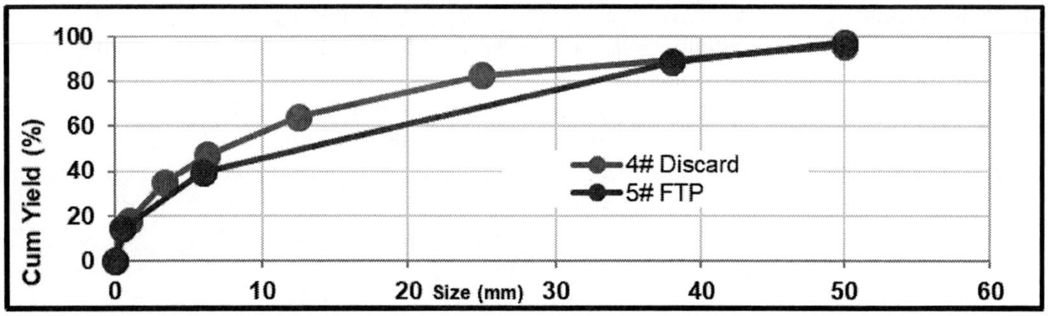

Figure 7: Particle Size Distribution of 4 Seam arising discard & 5 Seam Feed

In terms of a typical near density, the discard was not found to be any more difficult to treat than a typical South African coal as shown in Figure 8 but the yield characteristic was somewhat lower than the typical for a typical 5 Seam Coal. Because of the finer particle size distribution and lower yield characteristic and inherent constraints in the 5 Seam Plant (Drum, Cyclone and sinks screen capacity), the split feeding the drum had to be made quite fine to maintain the feed rate to the plant at 150 tph.

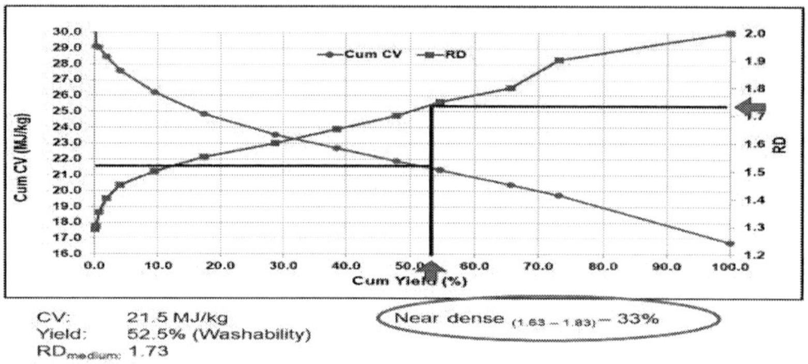

Figure 8: Washability characteristics of 4 Seam arising discard sample

During 2018, as the airspace projections for the dump became even more exacerbated, it was decided to abandon the dump extension due to both the life of mine economics and technical considerations. Further circuit changes were conceived for the 5 Seam Plant to be carried out in two distinct phases; firstly, to increase the head feed to plant from 150tph to 180tph by increasing the DSM Cyclone size from 610mm to 710mm with some associated ancillary equipment changes. This was implemented successfully at the end of 2018. The second phase involves the removal of the Drum and 710mm DSM Cyclone replacing them with a 900mm DSM Cyclone and associated ancillary equipment which will enable a further increase in the head feed to plant to 250tph. This will be a substantive change to the plant and requires alterations to filtration and thickener circuitry, conveyor and centrifuge capacity which will be commissioned in mid-2019.

2.4 Large Scale Mining of the Dump

Since 2003, several initiatives had been examined to commence the mining of the dump with a Black Economic Empowerment (BEE) entity for supply to the national power supply utility Eskom to no avail. However; the work concluded entailed the characterisation of the dump reserve to a measured status, the evaluation of transportation routes for mining and product removal and the establishment of a dense medium plant in the very limited real estate at the plant complex. Despite the lack of success to full implementation of these initiatives but because of the decision not to build a dump extension, a decision was made to avail of the opportunity of available toll washing capacity situated close by to mine the dump. This operation commenced in mid-2018 and is expected to peak at a rate of 1 million tons per annum from 2019 onwards. In Figures 9 and 10 the construction of the roadway from the MRD to the main road is shown.

Figure 9: View of the roadway off the MRD

Figure 10: View of the roadway to the main road

2.5 Summary

The accompanying Table 2 summarises all the initiatives taken including the processing initiatives taken above.

Table 2 Amelioration actions implemented to alleviate MRD air space constraint at Greenside

Amelioration Actions Implemented	Implementation Date	Impact on MRD Air Space
Relocation of the Greenside East Village	2011	Created 1300kt additional placement capacity
As-arising rejects from the 4 Seam Plant treated in the 5 Seam Plant	2012	Increased MRD capacity by nominal 1,5% of total head feed for the period May 2012 to Dec 2014 and generated additional revenue
Construction and Commissioning of Flotation & Filtration Plant	2012	Increased MRD capacity by nominal 3% of total head feed for Life of Mine and generates additional revenue

Construction and Commissioning of Tailings Filtration Plant	2014	Ensured "dry" disposal thereby allowing for MRD to be reclaimed as resource and reduced water consumption per feed tonne washed
Product change from 6000kcal/kg to 5850kcal/kg	2014	Increased MRD capacity by nominal 8% of total head feed for Life of Mine and generates additional revenue
Installation and commissioning of Three Product Cyclone in the 4 Seam Plant	2014	Increased MRD capacity by nominal 3,5% of total head feed for Life of Mine and creates additional revenue
Construction of a new pollution control dam	2014	Created 8300kt additional placement capacity
Discard retreatment in the 5 Seam Plant from the MRD facility	2015	Increased the MRD capacity by 1315kt over a 3-year period and generated additional revenue
Reclamation and raw discard sale of the MRD	2018	Will increase MRD capacity by nominal 1000kt/year and generate additional revenue
Upgrade of the 5 Seam Plant capacity from 150tph to 180tph head feed	2018	Increased the MRD capacity by 50kt in 6 months and generated additional revenue
Upgrade of the 5 Seam Plant capacity from 180tph to 250tph head feed	2019	Will increase MRD capacity by nominal 200kt/year and generated additional revenue

3 CONCLUSION

Hindsight is the perfect art, when one must deal with a number of legacy issues, logistic challenges and the daunting challenge of a physical constraint within an environment where commercial and legislative issues (both environmental and economic) dictate further conditions for your license to operate; it is difficult to arrive at a perfect solution i.e. one that is made in the most economic and timeous manner. In this case, the solutions embarked upon were correct and made sense at the time, but all were dictated by having a clear understanding of the mass balancing and a fluidity in interpretation of the mass balancing when undergoing the various scenario planning situations when performing life of mine studies accompanied by some well thought out solutions by the various processing teams at the operation over a period of eight years.

4 ACKNOWLEDGEMENTS

With any conundrum of this nature, it takes considerable patience and faith from a management team and with this uppermost in our minds we would like to thank the successive managements at Anglo American for working with us on this journey. As this work has proceeded over several years and with several iterations in approach, we would like to acknowledge the contributions of our technical colleagues at Anglo American and the various plant teams at Greenside Colliery. We would also like to acknowledge and thank the following engineering houses DRA, Enprotec, Fraser Alexander, Paterson and Cooke & Royal Haskoning DHV. In some cases, their respective work was implemented and sometimes not but in the latter the quality of input framed our thoughts as to how to proceed. A final acknowledgement is given to Coaltech through the support from Johan de Korte regarding the performance evaluation of the 3-Product Cyclones.

5. References

de Korte GJ., 2014, The Current Focus for R&D within the South African Environment, 15th Australian Coal Preparation Conference, Gold Coast, pp 2-14

du Toit F., 2012, Private Correspondence on Greenside 4 Seam Plant 3 – Product Cyclone Upgrade

Esterhuizen JJ., 2008, Coal Flotation – Goedehoop Colliery, 12th Australian Coal Preparation Conference, Sydney, pp 268 - 279

Power D., 2006, Metallurgical accounting for residues from coal preparation plants, XV International Coal Preparation Congress, Beijing, pp 707 - 717

DRY PROCESSING FOR COAL PREPARATION: A REVIEW

N Hughes, M Le Roux and Q.P Campbell
North-West University
Private Bag X6001, Potchefstroom 2520, South Africa
nikki11hug@gmail.com; marco.leroux@nwu.ac.za; quentin.campbell@nwu.ac.za

ABSTRACT

Conventionally coal beneficiation processes that employ water is favoured based on the sharp separation efficiency reachable; however, the substantial process water constraint proves wet beneficiation impractical and unsustainable in some parts of the world. As a result, research into the development of dry methods for effectively beneficiating coal has progressed; especially in the arctic, arid and also coal-rich regions. The intention of this review paper is to provide a comprehensive collection of pertinent information to dry coal processing globally, with respect to the trends and technologies available and also aid in understanding the applicability, performance, success or failure thereof.

Over the years, numerous dry techniques have been developed for processing a wide range of materials in the food, fuel, industrial minerals, mining and recycling sectors. The methods available for dry materials preparation, under each separation mechanism, with special regard to those applicable for coal beneficiation are detailed in terms of principle, applicability, limitations and performance. The dual energy X-ray sorter (DE-XRT), rare-earth-roll magnetic separator (RERMS) and tribo-electrostatic separator (TBS) were discussed. Density-based separation methods considered are pneumatic jigs (All Air jig), air oscillating tables (standard, FGX, and AKAflow), air fluidized bed separators (ADMFB), cyclonic suction separation (SEP-AIR), dry reflux classification and winnowing.

It was determined that dry coal beneficiation has potential to wholly or partially replace the current wet based methods with some necessary considerations to possible limitations set forth by the following: coal geology, application and product specification, particle size distribution (PSD), moisture, efficiency, capacity and cost. Dry coal processing may well be established in the near future through new plant installations and upgrades, in the arid and arctic regions; but may also be preferred for small and remote operations and the processing of fine material.

KEYWORDS: Dry coal processing, preparation, cleaning *or* beneficiation; sensor-based sorting; dry density separation; dry magnetic separation; tribo-electrostatic separation; microwave pre-treatment.

INTRODUCTION

Water plays a key role in effective coal preparation, however, many regions housing vast coal reserves and contributing majorly as coal producing and consuming countries, have experienced shortages of water available for processing (Rudolphy,

2017). In some of the arctic areas, problems associated with frozen process water arise, also hindering the efficiency in coal beneficiation plants (Chen & Wei, 2003). Coal continues to contribute as a key global source of energy and in the face of a water crisis, coal preparation may need to adapt so as to ensure sustainable utilization of both the valuable, water and coal resources (Chen & Yang, 2003; WEC, 2016).

By employing dry coal preparation processes, water usage can be significantly reduced or even eliminated. The opposing water based methods including dense-medium drums, cyclones, wet jigs, spirals and chemical floatation, currently dominate as coal beneficiation techniques. This is as a result of the near ideal separation (EP ~ 0.02) and optimal yield attainable at a wide density cut-point range (Yang *et al.*, 2013a). The dry coal beneficiation processes, however, have some obvious advantages over the wet options. Additional to the reduction in water usage, a dry product is obtained, removing the need for expensive dewatering post-processing. Moreover, also eliminating the treatment of potentially polluted water obtained from wet processing which proves costly (Lockhart, 1984). Most dry coal cleaning processes are modular and less expensive to implement and operate; proving favorable for pre and primary concentration, salvage operations, remote and under-developed locations and also for expansions or small, short term coal preparations plants (Arms, 1924; Weinstein & Snoby, 2007; Manoucherchi, 2003).

Dry coal processing is not a new concept and several methods have been researched and developed over the centuries (Arms, 1924; Houwelingen & De Jong, 2004; Dwari & Rao, 2007; Weinstein & Snoby, 2007; King, 2015). Despite that, and the many benefits offered, dry coal processing declined due to increased moisture levels in run-of-mine coal as a consequence of stricter dust suppression legislation (Weinstein & Snoby, 2007). Moreover, an increased demand for better coal qualities produced at high efficiency, large throughput and lower cut-point densities became necessary – piloting the processing of lower ore grades at finer grinds (Weinstein & Snoby, 2007). As a result, the more effective wet coal processing methods have prevailed and since commanded the coal preparation industry (Honaker *et al.*, 2008). Given the recent global water availability trends, a new drive for dry coal beneficiation has emerged, steering focus towards developing dry preparation practices that compete with the currently employed wet methods in terms of efficiencies, yields and product qualities. Many viable techniques have since been studied; improvements upon some of the previously abandoned units were made and some new methods emerged (Weinstein & Snoby, 2007; De Korte, 2013; Manoucherchi, 2003; Stepanenko, 2017; Morgan, 2017). Further investigation into dry beneficiation of fine coal material that is generally either utilized as un-treated or discarded, has also been undertaken (Dwari & Rao, 2007). Even water-based separation of fine fractions (i.e. flotation) proves difficult and costly in terms of dewatering, handling and storage. Dry beneficiation may provide the solution for processing these fines (Bada *et al.*, 2010).

OVERVIEW ON COAL PREPARATION

The inherent heterogeneous nature of coal often makes beneficiation necessary in order to remove the non-combustible and other impurities and obtain a product that meets specific, market-dependent requirements. These are usually in terms of calorific value, ash and sulphur contents along with some restrictions on the particle size range, amount of moisture and fine materials present (SACRM, 2011). The extent to which coal requires beneficiation, in order to be successful and profitable, is determined by the coal nature, the capabilities of the process and also the desired market (WCI, 2009; SACRM, 2011). The effectiveness of a beneficiation process is often determined by the extent of reduction in ash, sulphur and moisture contents of the coal; an improved calorific value, low Ecart Probable Mooyen (EPM) and wide cut-point density range.

Coal beneficiation is comprised of two fundamental operations: liberation and concentration (Wills & Napier-Munn, 2008). The former is accomplished by comminution, involving crushing and grinding to the coarsest possible particle size while releasing the valuable material from the host gangue. Complete liberation is rarely achievable in practice and most coal preparation plants merely reduce the ore to a workable particle size (Wills & Napier-Munn, 2008). The concentration processes are designed so as to produce a product made up of predominantly valuable mineral with an acceptable degree of locked gangue; a middling portion which may require further comminution and a tailings fraction that is mainly composed of the unwanted gangue (Wills & Napier-Munn, 2008). While every effort is taken into ensuring minimal fines generation during liberation, it is not entirely unavoidable. Generally, the particles finer than the minimum design particle size are removed before concentration and either discarded or beneficiated in a process that accommodates the fine nature of the particles. In some cases, a portion of the fine material can be added to the final product as un-beneficiated (De Korte, personal communication, 26 October 2018).

Most ores are concentrated, whether wet or dry, by methods that exploit differences in either the physical or chemical properties between the valuable component and its associated gangue. Dry beneficiation usually comprises techniques relating to the physical ore properties which include density, size, shape, lustrousness, frictional coefficient, magnetic susceptibility, and electric conductivity (Chen & Wei, 2003). The separation mechanisms utilizing these may be categorized, broadly, into the following (Wills & Napier-Munn, 2008):

i. Comminution and screening as a primitive form of concentration in which lump coal or gangue particles and harmful objects are separated in the oversize and either appropriately crushed or removed, respectively.
ii. Optical separation methods involve the appraisal of individual ore particles and the rejection of those particles that do not warrant further treatment.
iii. Elutriation or gravity (density) separation is based on the differences in density between the valuable mineral and gangue and the effect this has on the differential movement of particles in a fluid.
iv. Magnetic separation can be used to concentrate minerals from their gangues based on the susceptibility of either one or the other to a magnetic force.
v. Separation based on electrical conductivity is considered when attempting to separate conducting minerals from those that are non-conducting by introducing charged particles into an electric field

given that the material will either be repelled by or attracted to an electrode.

vi. Separation based on surface characteristics is implemented when a differential in surface characteristic (frictional coefficient) between valuable and gangue is present.

STATUS OF DRY COAL PREPARATION

Several ore concentrating techniques extending across the above listed categories have been developed on a wet and dry basis. As mentioned, the wet methods have attained much success and are readily implemented, globally, for a number of ore concentrating applications. Dry processing practices have also met success in certain food, recycling and mining applications with the sorting of gems, ferrous metals and fuels such as uranium, oil and shale making up the market in the latter sector (Von Ketelhodt, 2009).

However, due to the generally lower efficiency, higher cut-point densities and in some cases reduced throughput, much of the dry coal beneficiation applications implemented commercially to date (jigs, air-tables and sorters) are limited to pre-concentration and salvage operations (De Korte, personal communication, 26 October 2018). Dry separation of coal for achieving intermediate or final products are largely limited to operations requiring a minimal coal quality upgrade in order to meet the requirement (De Korte, personal communication, 26 October 2018). For this purpose, air-gravity separators and sorters may easily compete with the alternative wet methods on Northern hemisphere coal seams which are generally of good quality and easy to separate (Falcon & Ham, 1988). As a result, coal originating from countries such as the USA, China and Russia may be easily processed to intermediate or final product specification by dry means. Some of the notoriously difficult to separate Indian, South African and Australian coal seams (Falcon & Ham, 1988) may not be as successful. Important to note, however, is that the success of the process is determined by the market and product specifications. More specifically, dry processing for Southern hemisphere fields may not necessarily produce coal of high quality but could yield a product suitable for the domestic, thermal or synthetic fuels markets. In some cases, complete substitution by dry coal preparation plants, regardless of run-of-mine coal quality, are necessary for the extremely arid regions such as Australia, China, India, Mongolia and South Africa; where limited water is available and obtaining water by other means (i.e. transport) is uneconomical. Another extreme case validating complete dry coal processing plants is that the water utilized in wet coal preparation easily freezes in arctic areas (Canada, Mongolia and Russia) hindering process flow and production.

The applications of the currently commercially available dry coal processing units are generally limited to processing a good quality run-of-mine coal, in areas suffering from extreme drought or cold and also remote mining sites or small reserves, all at the possible cost of product quality, efficiency, yield and production. However, given that the extreme weather conditions continue to affect current coal processing techniques, numerous studies are still being conducted on dry coal beneficiation and some of the more recent developments may prove it an applicable replacement for, or addition to wet processing. Much of the research attempts to improve upon the product quality, efficiency, yield and throughput deficits experienced by the dry methods. Additionally, quite some attention has been placed on the dry processing of fine coal (Bada *et al.*, 2010) as most of the established processes are more successful in the larger (+6mm) circuit. This would allow for a significant reduction in the costly and difficult dewatering procedures associated with wet fine coal processing. Concentrating the fine coal may provide many benefits in that firstly, valuable material is not discarded and can be otherwise utilized; secondly, product quality is not reduced in the event of adding fines to the product and finally, less material for disposal and consequently reduced pollution.

DRY PROCESSING FOR COAL

In the remainder of this article, dry coal processing techniques applicable to each of the listed separation mechanisms is discussed.

Comminution and screening

Crushing and screening machinery that has been developed in an attempt to upgrade coal are the rotary breaker, scalping screens, and slate pickers (SACPS, 2015; Wills & Napier-Munn, 2008; Mahmoohdahbadi, 2015). However effective for crushing, screening, scalping and de-stoning, it is clear that these units cannot be singularly held responsible for beneficiation of coal to current product specifications. These methods rely quite heavily on a high degree of liberation which is nearly impossible to obtain at large particle sizes and specific particle shapes. Therefore, crushing, grinding and screening are used to reduce the ore particle size to an optimum for the most effective utilization in the downstream processes. This stage is also employed for the removal of harmful objects and fines from the system and so, can act as a slight form of pre-process concentration (SACPS, 2015).

A high degree of liberation may not always be necessary in certain processes as high recoveries are obtainable for minerals, especially coal, that are completely enclosed by gangue. This is as a result of the pronounced difference between middling and gangue in a selected physical property of the ore. The processes making up the optical and elutriation separation mechanisms do not require extensive liberation and are consequently ground to a predetermined optimum particle size for processing, regardless of the degree of liberation attained (Wills & Napier-Munn, 2008). The magnetic and electric processes, however, demand quite a fine particle size in order to achieve effective separation; this inadvertently coincides with a relatively high degree of liberation.

Sensor based sorting

Sensor-based sorting (SBS) is an umbrella term that covers any processes involving the appraisal of individual particles and the acceptation or rejection of those that, either, do or do not warrant further treatment, respectively (Wills & Napier-Munn, 2008). Sensor based sorters make up the optical separation mechanism and similar to handpicking, electronic sorters are designed to inspect particles in order to determine the value of a selected property and eject those particles that meet some criterion. This is accomplished in a sorting machine that identifies and analyses the selected particle property with suitable sensors and computational processors. Separation of the selected particles then occurs by means of an ejection device

(Manoucherchi, 2003). Most sorters make use of vibratory feeders or direct slide chutes that feed the material into a free-fall or onto a slow-moving belt where the sensors are placed optimally so as to allow effective detection and identification (Manoucherchi, 2003). A variety of ore properties can be used as the basis of automated sorting and with proper selection of detector type, sensor-based sorting can be applied to the concentration of numerous minerals. A summary of the sensors available along with the associated ore properties and minerals that can be concentrated thereby is contained in the work by Harbeck & Kroog (2008).

Principle of sensor based sorting for coal

Of the above, the developments in x-ray transmission (XRT) and electromagnetic (EM) sorting have demonstrated most effective for coal beneficiation. The EM sorting technology is based on the fact that some minerals show variable electrical conductivity and magnetic interaction. As a result, these minerals can be inspected and separated by a device that is able to detect their electrical or magnetic properties (De Jong *et al.*, 2004). Studies considering magnetic and electric separation techniques for coal have proven them effective to some extent and therefore viable properties for sensor-based sorting (De Jong *et al.*, 2004). However, EM sorters have only been investigated on small and laboratory scale and so, industrial scale tests are required for further evaluation (De Jong *et al.*, 2004).

The XRT technology is quite familiar and works in a similar fashion to that of the diagnostic tests performed in a doctor's office or airport baggage scanners. With these, the transmission and attenuation of hard x-rays are observed as they pass through solid matter (Robben *et al.*, 2014). X-ray transmission is applicable to ore sorting due to the differences in atomic absorption displayed by the different elements present in a particle. Resultantly, x-ray attenuation is dependent on atomic density and so it is made possible to discern between the density differing minerals. Denser particles are capable of absorbing larger amounts of energy, as such fewer x-rays transmit through the sample to the detector array (Strydom, 2010; Lessard *et al.*, 2014). The heavier elements can, therefore, be differentiated from the lighter elements based on their relative transmission of x-rays (Lessard et al., 2014). As with both XRT and EM sorters, once the classified data image or voltage is obtained, a command is sent to the control unit to open the appropriate valves for ejection of the desired particles. The sensitivity of this actuation may be varied and thereby the product can be changed to contain more or less of the unwanted gangue. This is accomplished by setting a threshold (e.g. atomic number or voltage). Changing the threshold would affect the product quality either positively if made extremely sensitive, or negatively if relaxed but would then either reduce or increase the machine throughput, respectively (Strydom, 2010).

Applicability of sensor based sorting

Sorting, including the EM and XRT, has been quite extensively studied and implemented in numerous countries around the globe for various applications. This technology may generally be used for (1) pre-concentration of the plant feed, (2) production of an intermediate product, (3) production of a finished product or, (4) salvage operations (Wotruba, 2006). An inherent advantage of XRT over other sensing technologies is the ability of this method to sample the entire mass of stone (Strydom, 2010, Lessard *et al.*, 2014). Many of the alternatives are surface techniques which are only capable of identifying minerals present in the surface of a sample. XRT, on the other hand, allows the sorter to see through the entirety of the stone and the dual energy sensor array permits the mitigation of geometry effects on the signal. Therefore, the manner in which a stone is presented to the sorter is inconsequential to the efficiency of the sorter (Strydom, 2010).

Specifically, in the case of coal, dual energy x-ray transmission (DE-XRT) sorters are useful for de-stoning coarse feed from the mining face and for processing coarse waste dumps containing a fair amount of in-seam and out-of-seam dilution (Tomra, 2014). The sensory system used within DE-XRT sorting can also be applied to evaluate particle size and shape along with material thickness and texture. Consequently, a DE-XRT sorter is a highly efficient system for classifying different materials or minerals. For this reason, DE-XRT technology may also be utilized for the removal of pyrite-bearing coals and separation of torbanite from coal in addition to the popular de-shaling application (Von Ketelhodt & Bergmann, 2010). In some cases, sorters can also be utilized in areas other than ROM preparation such as implementation after a milling circuit for removal of barren pebbles to prevent them from entering the mill through the recycle (Wotruba, 2006). Furthermore, some additions of sorters after other separation units is also a possibility to improve upon separation (De Jong *et al.*, 2004).

Limitations

DE-XRT sorting machines are capable of handling throughputs in the range of 70 – 240 tph with a general PSD of 10 – 120 mm (King, 2015). Treatment of smaller particle sizes results in a lower mass processed and therefore a noteworthy reduction in throughput. This is as a consequence of the monolayer feed presentation requirement and also the optical and time related capabilities of the detectors (Strauss, 2016). Therefore, DE-XRT sorters only prove competitive in the +30 mm size fraction with those smaller than 10 mm being uneconomical to process (Wotruba, 2006; King, 2015; Strauss, 2016). When approaching the upper size limit of 120mm, the sorters may suffer from dilution issues as a consequence of surface area and x-ray penetration depth (Strauss, 2016). With the larger rocks, inadequate flow or force during ejection of the rock may be problematic; however, some air jets are capable of ejecting particles up to 15 kg (Strauss, 2016). Furthermore, for effective concentration, the ratio of largest to smallest particle sizes present in the system must remain less than 3 to 1 in order to ensure proper particle detection and sufficient force exerted by the jets or mechanical plungers (Strauss, 2016). EM sorters for coal preparation have only been studied on +20 – 50 mm material at a throughput of 10 – 20 tph but are predicted to handle tonnages of 21 – 90 tph (De Jong *et al.*, 2004).

Clearly, preparation of the feed particle size and size ratio is essential for an effective EM and XRT sorter and material will, therefore, require adequate crushing, screening and blending. These PSD restrictions may limit the capability of sorting as an overall dry beneficiation process. Oversize particles can easily be crushed and reprocessed whereas the undersize would require slower processing and a more sensitive threshold setting which would ultimately affect production (De Korte, personal communication, 24 October 2018). Furthermore, inability to ensure the proper monolayer feed presentation and correct particle speed, may lead to non-effective beneficiation due to entrainment and thereby misplacement of material

(Wotruba, 2006). Given that all types of automated sorters can be operated successfully for coarse coal with, or without the use of water, the coarse feed ore needn't require extensive drying prior to processing as these sorters can be utilized, with no negative effect on the efficiency, in all weather conditions (dry, wet and freezing) (Tomra, 2014; Von Ketelhodt & Bergmann, 2010). In some moist cases where small and fine particles are present (either wholly or partially), adhesion of particle surfaces may result in entrainment, misdetection and misplacement which could affect the efficiency and product quality. This provides additional incentive for proper feed preparation in screening for the removal of undersize particles or particle washing (either with water or air).

Performance

The performance and success of a sorter is quite variable and may rely on, firstly the application and secondly, on the feed quality in addition to the PSD limitations as described. Von Ketelhodt and Bergmann (2010) noted that the cut-point of separation is controlled by the threshold sensitivity of the program. As the sensitivity is increased the efficiency of separation improves at the possible cost of throughput. Resultantly, the efficiency of this process is generally lower than that attained in a typical wet beneficiation plant (De Korte, 2013). The applicability and performance of DE-XRT sorting in de-shaling, separation of coal from torbanite and removing pyritic sulphides was studied by Von Ketelhodt and Bergmann (2010). This study concluded the following:

- Potential as a de-shaling application with and upgrade of ash value to 16.70% in the product from 22.38% in the feed for a +20-40mm South African coal from the Witbank coalfield.
- Further investigation into a +20-100mm range shows a significant reduction in ash content from 26.20% in the feed to 16.56% in the product with a product CV of 27.21MJ/kg and product yield of 79.6% at a throughput of 80 tph.
- Two tests conducted on pilot scale show good separation was achieved. A +60-120mm PSD was considered at 120 tph and achieved a reduction in ash value from 37.30% in the feed to 17.80% in the product at a product yield of 67%. Sulphur was reduced from 1.34% in the feed to 0.58% in the product and the product CV was improved to 24.96MJ/kg. Similar experiments on a +12.5-50mm PSD produced a product of 22.7% ash value from a feed of 31.90% at a yield of 66% and product CV of 23.83MJ/kg with product sulphur content of 0.83%.
- DE-XRT is best for applications in which coal needs to be differentiated from the density-similar torbanite mineral which is corroborated by studies conducted by Strydom (2010).
- The removal of pyritic sulphides and mercury is possible with DE-XRT as proven of a Texan lignite coal. These tests show a potential to yield an 84.73 percentage point reduction in pyrite with a 79.17 percentage point reduction in mercury along with improved product calorific and ash values.

Further studies conducted by De Korte (2013), Tomra (2014) and Robben *et al.* (2014) on selected coal types, feed qualities and particle sizes show similar results in that the DE-XRT has the potential to produce a coal with reduced ash and sulphur contents and consequently improved CV. These studies however also prove that the separation efficiency, cut-point densities and product yields obtained may vary considerably depending on the feed quality, particle properties and desired application (de-stoning or final concentration). Table 1 provides a summary of the pertinent data obtained in these studies.

Table 1: Summary of the separation efficiency data for DE-XRT

	De –Korte (2013)	Tomra (2014)	Robben *et al.* (2014)
PSD (mm)	+50	+50-120	+20-100
EPM	0.29	0.1	0.04-0.13
Cut-point density (SG)	2.1	-	1.65
Product yield (%)	47.58	75	36.7-84.5
Capacity (tph)	-	150	50-100
Feed ash value (%)	71	38	27.7-43.0
Product ash value (%)	59.5	18	19.5-25.7

Some efficiency tests were conducted on the EM sorter by De Jong *et al.* (2004) and showed that EPM values of 0.1 to 0.15 are achievable in cut-point densities of 1.5 to 2.3 SG, respectively. It was also found that the coal and gangue distinction was the same for both dry material and coal with added moisture. This competes with the data obtained for the DE-XRT sorter but since the studies were only conducted on small scale, further investigation is required.

Success and pitfalls

Even though a good separation is attainable in a DE-XRT sorter with the potential to deliver a product of competitive ash contents and acceptable product yields, some drawbacks have to be considered. The EPM values are quite low in some of the investigations detailed but still average in 0.14 with the feed quality significantly affecting the efficiency and consequently product yield. Although quite low in some studies, the EPM still doesn't quite compete with that obtained from wet processing. Clearly, the separation obtainable from a DE-XRT sorter is ore body specific and test work is therefore essential to understand the washability characteristics of the different coals (Strauss, 2016). Every coal deposit is unique in the style of mineralization, and so liberation varies too, making it so that not all ores are amenable to sorting (Strauss, 2016). This deposit variability and heterogeneity leads to a requirement of quite complex analyses and discrimination algorithms which, if incorrectly implemented, may lead to failure in not only the sensor based sorters but also other dry beneficiating techniques (Wotruba, 2006). Von Ketelhodt and Bergmann (2010) note that the amount of waste present in the feed is correlated to the throughput and in so a better quality feed will result in a higher tonnage.

Additionally, the throughput for coal with a relative density of 1.4 to 1.6 SG will be typically less than most other ores at the same volume. It is crucial to ensure whether the capabilities and limitations of the sorting unit are acceptable for the desired application and product outcome (De Korte, 2013).

The misplacement of particles is another important issue to consider. This is commonly noted in the lower density fractions and attributed to mechanical misplacement during the sorting process (Von Ketelhodt & Bergmann, 2010). Theoretically, if a larger discard particle is ejected, it is possible that small lighter coal particles could be dragged over the splitter by the turbulent ejection. For this reason, it is essential that the fine material is removed prior to sorting. It may also be useful to reduce the size range coefficient of the feed and to consider ejecting the product especially when the yield is in the region of 50% (Robben *et al.*, 2014). Beyond proper liberation and the consistent detection and ejection of the particles, most other pitfalls related to the use of sorters can be overcome.

Gravity separation

When a pronounced difference in the density of two minerals is observed, it is possible to separate them by gravity or density concentration also referred to as elutriation. This method of separation has its roots in antiquity and has long been used in iron ore and diamond processing, the concentration of metalliferous ores and coal beneficiation (Wills & Napier-Munn, 2008). Consequently, numerous process and unit designs have been introduced commercially for coarse and fine particle sizes. Both wet and dry based gravity separators are available. Density separation is well understood and accepted as an extremely successful concentration technique with the related theories well documented in the published literature (Kelly & Spottiswood, 1982; Wills & Napier-Munn, 2008: SACPS, 2015).

Principle of gravity separation for coal

Particles of unique density behave differently in a fluid (liquid or gaseous) medium, depending on the forces of gravity, friction and centrifuge acting upon them. These forces affect the differential movement between different density particles, related to the resistance of motion by a viscous fluid or medium (Wills & Napier-Munn, 2008). The relative particulate movement can be described as an effect of either floating or sinking in the medium according to Archimedes law (Luo and Chen, 2001). Depending on the chosen medium, density separation can be categorized as either wet (operating with water or suitable liquid) or dry (operating with air or suitable gas); both similar in concept (Wills & Napier-Munn, 2008). Noteworthy is that density is not the only parameter affecting the possibility and extent of separation achievable by elutriation. Certain properties of the particle and fluid medium also play role i.e. particle size, shape and medium type, density and viscosity (Wills & Napier-Munn, 2008; Sahu *et al.,* 2013; Zhao *et al.,* 2015).

Many minerals exhibit differences in density when compared to each other and their host rock. Coal is one such ore where the mineral gangue constituents report to generally higher density fractions than the associated maceral elements (Xie, 2015). As a result, density separation techniques are a suitable means of beneficiating coal to remove the undesired, high-density ash forming minerals (SACRM, 2011). Most currently operational dense medium separation techniques utilized for coal beneficiation are water based (Yang *et al.*, 2013). The coal can be upgraded by means of wet jigs, dense medium drums, cyclones and spirals depending on the particle size range applicable. Drum separators and wet jigs are commonly employed for coarse coal, whereas cyclones and spirals are applicable for the intermediate and fine size fractions, respectively (SACRM, 2011). Froth floatation is implemented for the ultra-fine coal fractions but is based on particle surface characteristics and not density.

Often a very dense, finely sized material such as magnetite or sand is used as a medium in addition to the respective fluid. Therein the dense medium is fluidized, forming a fine particle suspension in the fluid. The dense medium suspension behaves fluid-like and conforms to a unique, uniform and stable density and viscosity. Upon addition to the medium suspension, coal particles heavier than suspension (i.e. shale) sink, while those lighter (i.e. clean coal) would float. The addition of medium has been proven to aid in preventing particle back mixing (Prusti *et al.*, 2015) and also inhibits vigorous bubbling inside of the unit (Fan *et al.*, 2001).

Wet beneficiation processes obtain a very efficient separation but use significant amounts of water, yielding high production rates of good quality (low ash) coal that is high in moisture and requires extensive dewatering (Yang *et al.*, 2013). Common EPM and product yield values for a dense medium coal beneficiation plant are 0.023 and 71%, respectively (De Korte, 2013). Alternative to the currently employed density based wet methods, a number of dry elutriation technologies have been introduced to the coal preparation industry (Chen & Wei, 2003; Dwari & Rao, 2007, Zhao *et al.*, 2015). When considering the techniques available for dry processing, those proven applicable to density separation for the upgrade of coal are:

- Air jigs such as the All Air Jig
- Pneumatic oscillating tables such as the standard air table, FGX, KAT and AKAFlow
- Fluidized bed separators such as the ADMFB
- Vortex or cyclonic fluidization separators such as the SEP-AIR and AMSEP
- Dry reflux classifier
- Winnowing

Each of the above-listed technologies and variations has shown promise in the successful beneficiation of coal whether on research and developmental level or available commercially. Literature on the individual working principle on each of the above listed units is readily available (Luo & Chen, 2001; Chen & Wei, 2003; Dwari & Rao, 2007; Weinstein & Snoby, 2007; Wotruba *et al.*, 2010; Macpherson *et al.*, 2011; Zhang *et al.*, 2011; Ghosh, 2013; De-Korte, 2014; Chavaldi *et al*, 2016; Davaasuren *et al.*, 2016; Joanna *et al.*, 2016; Oshatani *et al.*, 2016; Stepanencko, 2016; Sinha *et al*, 2016; Sahkre *et al.*, 2018).

Applicability of gravity separators to coal preparation

Dry density separators can easily compete with the currently preferred wet methods in cases where coals are easy to separate or do not require a large degree of upgrading (De Korte, 2013). Additionally, the sample needs to contain as little near dense material as possible (Weinstein & Snoby, 2007); this may be countered by the units utilizing dense medium suspensions. These units can further be applied to de-stoning and pre-concentration applications, for salvage processes and remote mining sites or small plants (Weinstein & Snoby, 2007). Utilization of dry density separators is also useful when desulphurization by removal of pyritic sulphides is required (Sampaio *et al.*, 2007; Weinstein & Snoby, 2007). Some instances of using dry gravity methods for beneficiation of young lignite coals are found in the literature (Sampaio *et al.*, 2007; Weinstein & Snoby, 2007). The use of dry processes is crucial in this instance due to the degrading of the highly porous coal as a consequence of moisture (Sampaio *et al.*, 2007).

Limitations

The size and capacity ranges of the eleven dry density separators discussed in this chapter are summarized in Table 2.

Table 2: Summary of the applicable size ranges and capacities of the dry density separators

Dry separator type	Particle size range (mm)	Capacity (tph)
All Air Jig	0.5 - 50 [1]	50 - 120 [1]
FGX	6 - 80 [2,3]	10 - 350 [2,4]
Air table	0 - 6 [13]	5 [13]
AKAflow	0.05 - 3 mm [5,14]	5 - 25 [5]
KAT	1 - 10 mm [15]	3.7 - 31 [15]
ADMFB	6 - 300 [3]	40 - 150 [7,8]
ADMFB*	0 - 6 [3]	0.15 - 8 [7,9]
SEP-AIR	1 - 100 [10]	45 - 220 [10]
Reflux classifier*	0.25 - 8 [11]	50 - 200 tracer particles [11]
Winnower*	6 - 100 [12]	6 kg - 10 kg [12]

* Experimental scale

[1] Weinstein & Snoby (2007); [2] De Korte (2013); [3] Dwari & Rao (2007); [4] Zhang *et al.* (2011); [5] Wotruba *et al.* (2010); [6] Chavaldi *et al.* (2015); [7] Sahu *et al.* (2009); [8] Chen & Wei (2003); [9] Hughes *et*

al. (2017); [10] Gormash Export (2016); [11] Macpherson *et al.* (2011); [12] Sahkre *et al.* (2018); [13] Gupta (2016); [14] Wotruba *et al.* (n.d.); [15] Davaasuren *et al.* (2016)

From Table 2 it becomes clear that the *All Air jig*, *FGX*, ADMFB, *SEP-AIR* and winnower units are applicable to coarse coal processing. Some units such as the All Air Jig, FGX and SEP-AIR may be applicable for smaller size ranges (<6mm) but processing these may come at the cost of throughput. Research conducted into the density separation of the smaller coal has found that the *ADMFB*, reflux classifier and winnower may be utilized in the fine fraction processing, but the results are on an experimental scale and require upscaling and pilot tests. The *AKAflow* and KAT air table concentrators are already commercially available but only suitable for the small and fine coal fraction. With some units - maintaining a narrow PSD range is essential whilst others are more robust and can easily process a much wider size range. The *All Air Jig* is one such unit that is capable of processing feeds with a particle size distribution ratio of 10 to 1 (Weinstein & Snoby, 2007). The FGX, ADMFB and SEP-AIR concentrators can also be designed so as to accept a similarly large particle size range (Dwari & Rao, 2007; Chen & Wei, 2003; Gormash Export, 2016). The AKAflow, reflux classifier and winnower, however, require a narrower PSD ranging in 2 to 1 (Wotruba *et al.*, 2010; Macpherson *et al.*, 2009; Sahkre *et al.*, 2018). For the batch studies conducted on the ADMFB for fine coal processing, it is observed that narrow size ranges (3 to 1) may also be necessary (Chen & Wei, 2003).

The presence of a large amount of material that is beneath or above the specific PSD range for a given unit may severely affect the equipment performance (Weinstein & Snoby, 2007). This is particularly true for all of the processes listed in Table 2. Particles that are either significantly larger or smaller than allowed by the operating rage may be displaced on account of their relative mass. Fine discard particulates tend to be lifted or blown in with the clean coal product and coarse clean coal particles may report to the discard (Sahu *et al.*, 2009). Those units operating with an autogenous or dual density medium such as the *FGX* and *ADMFB* may benefit from the presence of a small number of coal fines in the feed. These fines aid in the development of the fluidized medium suspension by which the density separation is affected (Ghosh, 2013). The amount of these fines in the feed must be carefully controlled (below approximately $10\%_{wt}$) in order to prevent the altering of the stability and properties (i.e. density) of the fluidized bed suspension (He *et al.*, 2016c) or equipment plugging (in the presence of moisture) (De Korte, 2014). Removal of the fine material in the feed additionally reduces the amount of dust that has to be captured and treated in the dust-handling system (De Korte, 2013).

When considering the "near-dense" classification of many Southern hemisphere coals, beneficiation may prove difficult. The presence of a large amount of near dense material significantly affects the conduct of the dry separating units by providing an increased probability of the material being misplaced (De Korte, 2013). When considering dry processes, the EPM is generally higher and given that the amount of near-dense material is defined by the amount of material present in the density range of ±2 x EPM, more near dense material is present. Consequently, higher amounts of misplaced material results; lowering the product yield and making it exceedingly difficult to control the product quality (De Korte, 2013).

Dry density separators can typically process feeds where the extent of moisture present does not plug the existing material handling systems. In other words, the ore particles are considered dry free-flowing (Weinstein & Snoby, 2007). Units such as the *All Air Jig*, *FGX*, *SEP-AIR* and winnower could easily process coarse coal with a relatively high moisture content in the absence of fine material (De Korte, 2014). When considering the finer particles agglomeration usually occurs in the presence of moisture and causes particle misplacement. Agglomerates will report to either the clean coal or discard discharge (Weinstein & Snoby, 2007). This is also observed in the density separators utilizing dense medium such as magnetite. The magnetite particles adhere to one another, the coal and vessel walls. This affects the suspension stability and density; hence producing a poorer separation efficiency (Sahu *et al.*, 2009). Magnetite-coal agglomerates reporting to either discharge will require extensive cleaning in order to recover the magnetite and ensure that the products are not contaminated when used. Consequently, moisture contents lower than $3\%_{wt}$ may be required for the plants using a dense medium material (Sahu *et al.*, 2009; He *et al.*, 2015).

Feed preparation is required in order to ensure the correct feed moisture, particle size distribution and minimal fines present. As a result, drying, crushing, screening and surge storage equipment is essential (Weinstein & Snoby, 2007). Feed presentation is however not of great importance to most dry density separations as long as the correct feed flow is maintained; usually by vibratory feeders with star gates (Weinstein & Snoby, 2007). Given the belt driven feed of the SEP-AIR unit, feed presentation in a monolayer may be of some importance. The ADMFB, also, does require initial fluidization of the dense medium until stable before addition of the coal particles.

A number of successful studies have been conducted in an ADMFB with differing dense media such as silica (or sand), zircon, fine coal and magnetite. In some cases, mixtures of these have also been investigated (Lagner *et al.*, 2016; Firdaus *et al.*, 2012). Magnetite is preferred due to the fact that medium recovery and cleaning can be made easier by magnetic separation (Fan *et al.*, 2008). Dense medium size and moisture content must be carefully controlled additional to the pre-requisites of the coal feed. Studies by He and colleagues (2016a) proved that magnetite in a size range of +74 - 300 µm produced a stable fluidized bed. The moisture content of the magnetite should be on par with that required for the coal (<3%wt) (Hughes *et al.*, 2017). The moisture content of the feed air to an ADMFB should also be relatively low in order to prevent condensation in the bed which may lead to improper fluidization and problems with medium recovery (De Korte, 2016). The dense medium applicability and limitations discussed for the ADMFB above also relate to the dry reflux classifier if utilizing dense medium.

Performance

Numerous parameters determine the performance of dry gravity concentration depending on the type of separator implemented. Several studies have been conducted on the various methods of elutriation relating to the critical process variables along with the PSD, capacity, and moisture requirements discussed in the previous section. Given the above, typically available performance data for the EPM values, product yields, ash reduction and cut-point densities for the commercially available *All Air Jig*, *FGX*, air table, *AKAFlow*, *KAT*, *ADMFB*, and *SEP-AIR* units are summarized in Table 3.

Table 3: Typical performance data for different commercial air-gravity separators

Method	EPM	Product yield	Ash reduction	Cut-point density
All Air Jig (0.5 - 50mm) (50 – 120tph)	0.16 - 0.27 [1]	70% [2]	Feed: 15 - 25% [1] Product: 10 - 18% [1] Discard: 50 - 60% [1]	1.95 - 2.20 SG(D_{50}) [1]
FGX (6 - 80mm) (10 – 350tph)	0.12 - 0.23 [3]	93.1 - 95.54% [3]	Feed: 30% [3] Product: 10.8 - 19.4% [3] Discard: 23.4 - 61.3% [3]	1.90 - 2.03 SG(D_{50}) [3]
Air table (0 - 6mm) (5tph)	0.18 - 0.2 [6]	40 - 60% [6]	Feed: 48.7 - 54.9% [4] Product: 23.7 - 29.9% [4] Discard: 63 - 80% [4]	N/A
AKAFlow (0.05 - 3mm) (5 – 25tph)	N/A	73.5 - 80% [5,10]	Feed: 29.5 - 47.5% [5,10] Product: 9.5 - 18% [5,10] Discard: 61.4 - 82.5% [5,10]	1.30 - 1.96 SG(D_{50}) [10]
KAT	N/A	40.9 - 56.3 [11]	Feed: 35.6 - 48.4% [11]	2.0 SG(D_{50}) [11]

(1 - 10mm) (3.7 – 31tph)			Product: 9.4 - 12% [11] Discard: 66.1 - 75.5% [11]	
ADMFB (6 -300mm) (40 – 150tph)	0.05 - 0.07 [7]	41 - 55.5% [7]	Feed: 39 - 46% [7] Product: 16.4 - 18.2% [7] Discard: 63.8 - 67.5% [7]	1.3 - 2.2 SG(D_{50}) [9]
SEP-AIR (1 - 100mm) (45 – 220tph)	0.10 - 0.17 [8]	85.4 - 93.0% [8]	Feed: 13.1 - 19.0% [8] Product: 7.0 - 14.8% [8] Discard: 51.0 - 69.6% [8]	1.78 - 1.90 SG(D_{50}) [8]
[1] Weinstein & Snoby (2007); [2] Boylu *et al.* (2015); [3] Zhang *et al.* (2011); [4] Gupta, 2016; [5] Wotruba *et al.* (2010); [6] Chavaldi *et al.* (2016); [7] Luo & Chen (2001); [8] De Korte (2018); [9] Dwari & Rao (2007); [10] Wotruba et al. (n.d.); [11] Davaasuren *et al.* (2016)				

Quite large reductions in ash content are obtainable in the processes listed in Table 3 at relatively high process yields. However, the EPM values still do not compete with the average 0.023 achieved in a wet dense medium plant (De Korte, 2013). Additionally, high D_{50} cut-point densities are observed for some of the separators.

Some studies have investigated air-gravity separators for the desulphurization of coal by removal of pyritic sulphides. Sampaio *et al.* (2007) found that dry jigging was able to eliminate pyrite nodules and liberated clays from Brazilian coal of - 20mm. A reduction in ash and sulphur contents from 51% and 1.8% in the feed to 47% and 0.7% in the product were obtained, respectively. Moreover, the specific coal utilized for this study was of extremely low rank and had a very high porosity; making it impossible to beneficiate by wet means due to possible degradation of the coal by moisture (Sampaio *et al.*, 2007). The investigation into the *FGX* for sulphur reduction yielded product sulphur contents of 3.05% and 4.08% for two samples relative to 4.17% and 5.13% in the feed for both samples (Zhang *et al.*, 2011). Patil and Parekh (n.d.) found that a 43 percentage point reduction is possible using an air table - correlating to a 2.65% sulphur in the feed and 1.5% sulphur in the product. Sahu and colleagues (2009) conducted studies into the removal of pyritic sulphides in an ADMFB. They determined that a pyrite reduction to 0.51% in the product from

4.6% in the feed is possible with a synthetic mixture of coal and pyrite of 153 μm and 74 μm, respectively.

The performance data pertaining to the air-gravity separators that are still in the experimental phase viz. *ADMFB's* for fine coal, *dry reflux classifier* and *coal winnower* is summarized in Table 4.

Table 4: Performance data for experimental phase air-gravity separators

Method	EPM	Product yield	Ash reduction	Cut-point density
Dual density ADMFB (6 - 13 mm)	0.06 -0.13 [1]	N/A	N/A	1.50 - 1.90 SG(D_{50}) [1]
Vibrated ADMFB (1 - 6 mm)	0.175 - 0.225 [2]	Clean coal: 37 - 38% [2] Middling: 23% [2]	Feed: 40% [2] Product: 14 - 16% [2] Discard: 70 - 71% [2]	1.52 - 1.89 SG(D_{50}) [2]
Magnetically stabilized ADMFB (0.5 - 6mm)	0.065 [3]	61.3% [3]	Feed: 31% [3] Product: 12.62% [3] Discard: 61.5% [3]	1.516 SG(D_{50}) [3]
Dry reflux classifier (0.25 - 8 mm)	0.06 - 0.46 [4]	77% [5]	Feed: 24 - 30% [5] Product: 10 - 11% [5]	1.42 - 2.13 SG(D_{50}) [4]
Coal winnower (2 - 100 mm)	0.065 - 0.25 [6]	30 - 80% [6]	Feed: 28 - 46% [6] Product: 22 - 39% [6] Discard: 38 - 60% [6]	1.76 - 1.97 SG(D_{50}) [6]

[1] Wei & Zhao (2003) [2] Yang *et al.* (2013); [3] Luo *et al.* (2002); [4] Macpherson *et al.* (2011); [5] Macpherson & Galvin (2010); [6] Sahkre *et al.* (2018)

Clearly, these processes are able to beneficiate the coal with reasonable separation efficiency and product yields. The reduction in ash value indicates that an upgrade of the coal is possible within relatively good cut-point density ranges. Much of these studies are however related to the upgrading of small and fine coal on an experimental scale and may require further investigation and possible upscaling.

Success and pitfalls

Clearly, any of these are viable for beneficiation of an "easy to separate" coal with little near dense material or for a feed that does not require an extensive upgrade in quality to reach product specification. As a consequence, many of these units are in operation for specific applications in the especially arid or arctic regions around the globe including China, South Africa, India, Russia and the USA.

The ADMFB yielded the lowest EPM value and D_{50} cut-point density ranges with a significant reduction in ash from the feed to the product. This implies that the ADMFB performs better than the other processes and may be applied as a concentrator yielding a final saleable product; however, at the cost of product yield. Moreover, the ADMFB requires a dense medium which may prove expensive in some instances. Firstly, it is an additional raw material which contributes to the operating costs of the unit and the recovery and cleaning of the magnetite is often difficult and not completely effective (Wills & Napier-Munn, 2008). As a result, continuous addition of magnetite to the system is essential and so medium consumption can be relatively high (Sahu *et al.*, 2009). Contamination of coal product by magnetite and medium dilution by fine coal is possible; magnetite deterioration may also be experienced (Yang *et al.*, 2013). Additional process equipment is required for the recovery, cleaning, recycling and, possibly, drying of the medium; leading to increased capital and operating process costs (Yang *et al.*, 2013). Some studies have investigated the use of an autogenous medium wherein either fine coal or a mixture of fine coal and magnetite is used. This could be a solution for the availability, recovery and cleaning issues associated with the magnetite medium (He *et al.*, 2016c; Luo & Chen, 2001; Yang *et al.*, 2013b). The separation units operating without a dense medium such as the *SEP-AIR* may, however, be more attractive. This unit produced a coal product with a large degree of ash reduction at EPM values slightly above the ADMFB and at D_{50} cut-point densities lower than the *All Air Jig* and *FGX*. The wide particle size range capability and large capacity (Table 2) further make the *SEP-AIR* a viable dry coal beneficiation option (De Korte, personal communication 26 October 2018).

Magnetic separation

Magnetic separation (MS) is considered for the concentration of minerals by utilizing any differences in the magnetic susceptibility between the valuable minerals and gangue bearing components (Tripathy *et al.*, 2017). Separation by this means has received quite some research and developmental attention for a number of beneficiation applications, including that of coal. As such, much of the theory on magnetic separation is well documented in published literature along with reports of successful implementation in the iron, magnetite and wolframite industries, this will not be treated in extreme detail here (Morgan & Bronkola, 1991; Broek, 1995; Karmazin, 1997; Mostika et al., 1999; Wills & Napier-Munn, 2006; Iyer, 2011; Rao, 2011; Tripathy *et al.*, 2017).

Principle of magnetic separation for coal

Essentially, separation by magnetism is actuated by making use of the force exerted by a magnetic field upon magnetic materials in order to counteract the effects of gravity, viscosity or motion. Thus, forming different pathways for the magnetic and non-magnetic substances (Balasubramanian, 2015). All materials are affected (to differing extents) when entering a magnetic field; in most cases, however, the effects are insignificant and nearly undetectable. In terms of magnetism, materials can be divided into three universal groups based on the attraction or repulsion by a magnetic force (Wills & Napier-Munn, 2008):

- *Diamagnetic materials* are repelled to a point of lesser intensity in a magnetic field and are generally not concentrated by MS because the forces at play are minor.
- *Paramagnetic materials* are attracted to a point of greater intensity in a magnetic field and are able to be concentrated with high-intensity MS. Some examples of paramagnetic minerals which are commercially concentrated in magnetic separators are ilmenite, rutile, wolframite, monazite, siderite, pyrrhotite, chromite, hematite and manganese.
- *Ferromagnetic materials* are considered a special case of paramagnetism. These materials have significant susceptibility to a magnetic force and may retain some magnetism when removed from a magnetic field. It is possible to concentrate ferromagnetic substances with low-intensity MS. The main ferromagnetic mineral separated by this means is magnetite. Some applications for roasting hematite and siderite to form magnetite are known and the removal of tramp iron from ore is also considered a low-intensity magnetic separation.

Some elements are, in themselves, paramagnetic namely, nickel, cobalt, manganese, chromium, cerium, titanium and the platinum group metals. In most cases, however, the paramagnetic properties of a mineral can be attributed to the presence of iron in some or other ferromagnetic form (Wills & Napier-Munn, 2008). This is particularly true in the instance of coal ore where the pure or valuable coal is considered non-magnetic or diamagnetic. On the other hand, the mineral matter present within the heterogeneous ore may contain diamagnetic and/or paramagnetic material (Van Driel, *et al.*, 1984). This makes it possible to beneficiate the valuable coal from the associated paramagnetic gangue minerals in high-intensity MS.

Ideally, the magnetic separation for coal would proceed with the pulverizing of the sample to a suitable particle size with complete mineral liberation. The ground sample is supplied to the magnetic separator where the paramagnetic particles (mostly gangue) are removed by the strong magnetic forces as it exceeds the other forces acting on the particle (i.e. viscous, gravity or inertial). The non-magnetic particles remain unaffected by the magnetic field generating a product mixture of coal and diamagnetic mineral matter. Therefore, when considering the ideal case, coal can be produced containing less mineral matter where most of the paramagnetic minerals, such as pyrite (the sulfur-bearing mineral in coal), have been removed (Van Driel *et al.*, 1984). As previously noted, in practice complete liberation is unattainable and the pulverized coal sample would contain a middling fraction affecting the efficiency of separation.

Magnetic separators can be classified into low- and high-intensity machines which are capable of either dry or wet operation (Wills & Napier-Munn, 2008). This work takes a closer look at the high-intensity magnetic separators (HIMS) capable of inducing separation between the coal and gangue minerals. These processes possess higher magnetic field strengths imparted either by a permanent magnet or through induced magnetic fields. This allows for the separation of weakly or strongly paramagnetic material as a consequence of the magnetic susceptibility differential present (Gunther, 1909; Parker, 1977; Svoboda, 2004). There are a number of concentrator designs that account under the HIMS category. These have been comprehensively detailed in terms of principle, variable effects and application in the work by Tripathy *et al.* (2017). For implementation in dry coal processing, the permanent roll magnetic separator (PRMS) and rare earth roll magnetic separator (RERMS) prove most viable in terms of an up gradation of the coal due to ash content and pyrite reduction (Fujita *et al.*, 1981; Hucko & Maronde, 1982; Male, 1984; Order *et al.*, 2003; Tripathy *et al.*, 2017).

Applicability of magnetic separation in coal preparation

Most of the magnetic separators described in this work can be implemented with or without the use of water making it an attractive alternative for the current wet coal beneficiation practices. HIMS operated on a dry basis is well accepted in selected industries and used to concentrate paramagnetic material, for impurity separation and recovery of minerals. The removal of ash-bearing minerals and pyrite from coal ore in the RERMS is an example of a utilization relevant to the topic at hand. PRMS and RERMS operations are predominantly utilized in dry coal processing but some works have been conducted on a wet basis (Rylatt & Popplewell, 1999). These units are integral in primary and scavenging processing and are also applicable as a means of pre-concentration (Tripathy *et al.*, 2017). Furthermore, the availability of rare-earth permanent magnets and their affordability extends the applicability of magnetic separators to medium and weakly magnetic materials. The roll magnetic separators are efficient, selective and are easily adjusted to feed variations (Tripathy *et al.*, 2017).

Common diameters of the magnetic roll can extend from 72 – 300 mm with magnetic zone widths of 250, 500, 1000 and 1500 mm available (Tripathy *et al.*, 2017). The capacity of any roll magnetic separator depends on the roll diameter which is selected based on the particle size to be treated during the design phase (Svoboda & Fujita, 2003). RERMS machines are capable of handling throughputs in the range of 2 to 3 tons per hour per meter roll effective for particle sizes of 75μm. Some designs have proven a throughput of up to 10 tph per meter of roll but depend quite heavily on the application (Mineral technologies, 2018). Most dry RERMS are capable of processing coal with slightly higher moisture contents provided that the particles are free-flowing and do not stick to one another or the belt (Mineral technologies, 2018). Separation of particles up to 25 mm is possible with larger roll diameters and thereby increased capacities (Tripathy *et al.*, 2017).

Limitations

The production rate may however also be affected by the particle size, particle density, proportion of magnetic material in the feed, liberation and exposure of magnetic material and finally, product specifications. Most dry magnetic separators are inefficient while treating sizes finer than 75µm which can be successfully concentrated with wet magnetic methods. Furthermore, a study conducted by Ibrahim *et al.* (2017) concluded that a narrow PSD range may prove more efficient in a RERMS.

Numerous studies have shown the effects of particle size on separation in the RERMS separator and most findings come down to the trajectory of a particle in a magnetic field (Gehauf, 2004; Tripathy *et al.*, 2015; Ibrahim *et al.*, 2017). Diamagnetic particles profile according to the centrifugal force actuated by the roll and drop unhindered while classifying into particle size as expected. Larger particles would travel a further distance and are therefore typically processed with lower roll speeds. When considering the magnetic particles, strongly attracted magnetic particles will adhere to the roll until being released from the magnetic field but weakly attracted particles may only experience a measure of deflection by the magnetic force. These particles are then altered from their normal path and overlap of large weakly attached particles may occur with small non-magnetic particles (Gehauf, 2004; Ibrahim *et al.*, 2017). The importance of a narrow particle size range along with correct splitter setting and roller speed is hereby made very clear and contribute to some of the limitations of RERMS. Additionally, these operations require the removal of the minute particulates present on the surface of coarser particles, stringent dust control measures and monolayer feed presentation (Tripathy *et al.*, 2017). This limitation can easily be overcome by splitting the feed into closed size fractions before processing and utilizing multiple separation stages to enhance the recovery (Tripathy *et al.*, 2017).

Additional to the particle properties and roll diameter as discussed above, some design and process variables affect the performance of the separator. The design parameters considered are the magnetic field intensity, magnet design and magnet to steel ratio, type of belt, belt thickness and number of stages. The process variables of importance are the roll speed, splitter position, feed rate and finally, depth of the material on the belt (Tripathy *et al.*, 2017). Some studies indicate that by increasing the roll diameter, residence time and magnetic flux density, the separation efficiency increases. Additionally, increasing the roll speed and diameter while retaining an optimum residence time, a higher throughput may be achieved.

Performance

Numerous studies have been conducted on these separators in terms of coal beneficiation for the separation of ferruginous-rich maceral, pyritic sulphides and shale from the coal. These studies considered varying the main process parameters including particle size, roll speed, splitter position and feed rate (Fujita et al., 1981; Hucko & Maronde, 1982; Van Driel & de Kerk, 1983; Male, 1984). Many of the older studies include the use of HGMS on coal beneficiation which is presently not considered optimal. Separation using the preferred RERMS has been investigated in order to remove the shale contributing to the higher ash content in coal. These have concluded the following:

- Potential to reduce the ash content in a lignite coal sample from 35.6% in the feed to 12.3% in the product with an air jig and magnetic separator combination (Order *et al.*, 2003); corroborated in the work by Celik (2002) which reported an ash value reduction from 39.49% in the feed to 14.2% in the product.
- Potential to reduce the sulphur content from 2.49% in the feed to 0.39% in the product for a low rank semi-coked lignite coal sample (Yildinm *et al.*, 1996); corroborated by in the work by Celik (2002) with final sulphur content of 0.41% in the product from a feed of 2.09%.

In some cases, treating a coal sample with heat has been known to improve some of the properties such as magnetic susceptibility for one. Pyrolysis is considered such a pre-treatment and studies have been reported that by including a pre-treatment of lignite coal by means of pyrolysis can enhance the magnetism of the mineral matter. Separation thereafter yielded an ash value reduction from 36% in the feed to 22.7% in the product and sulphur reduction from 2.4% in the feed to 0.88% in the product (Koca *et al.*, 2000). Another study on magnetic separation on low-rank lignitic coal with pre-treatment by heating was conducted by Celik & Yildirim (2000). Utilizing microwave energy for the treatment of coal prior to magnetic separation has emerged as a favourable technique for the removal of pyritic gangue from coal fines (Uslu *et al.*, 2003; Uslu & Atalya, 2004; Zhang *et al.*, 2015). The studies concerning pre-treatment of the coal before magnetic separation are mostly confined to laboratory scale and require further development and enlargement. Also, upon improving the magnetic susceptibility of the material, the low-intensity magnetic separators may be more appropriate for the beneficiation (Tripathy *et al.*, 2017). These processes may, in future, provide a cost-effective unit for pyrite removal in high sulphur coals if further enhanced by optimizing and scaling.

Success and pitfalls

Although quite efficient separation is obtainable by magnetic methods, the process may suffer some pitfalls. Given the fine nature of the particles processed by magnetic means, some problems arise in terms of feed moisture and presentation. The coal need only be dry enough such that flow of the particles across the belt remain unhindered but, the presence of moisture severely affects the free flow of fine and ultra-fine particles. Flocculation or agglomeration of the particles may negatively affect the operation of the separator and consequently produce an inefficient separation (Wills & Napier-Munn, 2008). As a result, an almost bone dry feed is required for magnetic separation in especially the very fine material and extensive drying may be necessary prior to separation (De Korte, personal communication 24 October 2018). This may prove difficult and uneconomic. Larger particles (e.g. 25mm), however, may not require as much or any drying but may suffer from issue related to dilution or exposure of the magnetic portion in the coal.

The presentation of the feed may also prove difficult as a monolayer of particles across the belt and magnet is essential to avoid any entrainment. This significantly reduces the capacity of the machine (Wills & Napier-Munn, 2008). It is imperative that the feed is presented evenly across the belt width and at the correct velocity to also inhibit particle bouncing and scattering. Additionally, each particle should enter the magnetic zone at the same velocity so as to make certain that the trajectory of individual particles is determined by the magnetic force and not affected by the speed (Wills & Napier-Munn, 2008; Svoboda, 2004). Furthermore, the presence and amount of magnetic material in the feed along with the relative degree of magnetic susceptibility determines to a great extent whether magnetic separation is applicable. As such RERMS (and other MS designs) are mostly utilized for coals with high a pyritic sulphur content and for significant amounts of paramagnetic mineral matter present in the feed. For this reason, magnetic separation on RERMS would require some extensive petrographic analyses conducted on the feed before implementation.

Electrostatic separation

Electrical separation (also referred to as high tension separation) is a means of separating charged particles under the influence of an electric field and utilizes the difference in electrical conductivity between various minerals to do so (Wills & Napier-Munn, 2008; Inculet, 1984a; Ralston 1961). All minerals exhibit some difference in conductivity making electrostatic separation a universal means of ore concentration which is frequently combined with gravity and magnetic separating methods (Wills & Napier-Munn, 2008; Kelly & Spottiswood, 1989). Electrical separation has been utilized since 1800 and so, quite some case studies have been conducted thereon with the relevant theories well documented by Kelly & Spottiswood (1989), Wills & Napier-Munn (2008) and Dwari & Rao (2007).

Principle of electrostatic separation for coal

Separation of two substances by electrical means is based on a difference in the ability to develop and maintain charge along with the related particle movement when placed in an electric field (Dwari and Rao, 2007). The ability is determined by the electrical resistivity and conductivity as well as the difference in the electronic surface structure of the mineral and organic matter (Kelly & Spottiswood, 1989). Separation is achievable because the differences in the properties listed above enable particles to develop two different charged phases which may behave differently in the presence of an electric field (Bada *et al.*, 2010). Different types of electric separators are defined and detailed in Kelly and Spottiswood (1989).

Particles are electrostatically charged before entering a separation stage which can be most successfully accomplished by (1) ion or corona bombardment, (2) conductive induction and (3) contact or tribo-electrification (Dwari and Rao, 2007). The electro-physical properties of a particle determine the mechanism by which the particle surface may be endowed with charge (Knoll & Taylor, 1984; Kelly & Spottiswood, 1989). Other methods of charging include freezing, ion and electron-beam thermionics, photoelectric field emission, radioactive decay charging and also pyroelectric and photoelectric charging (Hendricks, 1973). Of these charging methods, tribo-electrification has received most attention for coal preparation practices in fine and ultra-fine sizes and some investigation into this direction has been made but has not yet achieved commercial status (Dwari & Rao, 2007). The triboelectric method is deemed most suitable for ultra-fine materials and for minerals with relatively similar and varying electro-physical properties (Manouchehri *et al.*, 2000; Kelly & Spottiswood, 1989).

In tribo-electrostatic separation, the particles are charged by contact or friction with other particles or another material (such as a container wall) and then separated in an electric field (Mohanta *et al.*, 2016). Electrons are transferred from one surface to the other when two dissimilar particles are rubbed against one another until the energy of each material has equalized at the interface. The result is that one particle is negatively charged and the other obtains a positive charge (Dwari & Rao, 2007). Early triboelectric separators induced charging of the particles by sliding or transport through chutes, pipes and nozzles (Dwari & Rao, 2007). Tribo-charging is now popularly achieved in a rotary drum or tube, cyclone, vibratory feeder, pneumatic conveyor or fluidized bed system due to better particle-particle and particle-wall interaction. El-Mouloud Zelmat *et al.* (2013) provides a discussion of the different tribo-charging systems and some references to combinations of these are also found in literature (Dwari & Rao, 2008; Dwari & Rao, 2009; Dwari & Rao, 2007; Soong *et al.*, 2001; Fricke, 1997; Haga, 1995; Cross, 1987; Inculet, 1984; Masuda *et al.*, 1983).

When focusing on the electro-physical properties of coal, many researchers agree that carbon particles charge positively whereas the associated gangue minerals such as kaolinite, illite and quartz obtain a negative charge by frictional contact (Dwari & Rao, 2008; Dwari & Rao, 2009; Carta *et al.*, 1976; Lockhart, 1984; Alfano *et al.*, 1988). Furthermore, Inculet *et al.* (1982) found that different maceral types acquired different charge polarities when beneficiated electrostatically. The predominant vitrinite maceral mostly obtained a positive charge whilst inertinite charged negatively. The electric conductivity of the material also proves of importance and once charged it is essential that the polarity is maintained and not dissipated through the particle or to other particles and surfaces. Coal is less conducting than mineral matter except in the case of brown coal due to the higher moisture and ion contents (Lockhart, 1984). Resultantly moisture, humidity and temperature may significantly affect the conductivity of the material and consequently the separation in an electrostatic separator (Wills & Napier-Munn, 2008). Of the compounds most commonly found in coal, pyrite is the most conducting and always takes a negative charge (Ciccu *et al.*, 1991). On the other hand, Ciccu *et al.* (1991) also noted that coal matter may acquire either positive or negative charge depending on the presence of either carbonate or silicate gangue minerals, respectively, but that good separation is achievable regardless.

Applicability of tribo-electrostatic separation for coal preparation

Aside from some studies conducted in a medium other than air (such as a dielectric liquid), electrostatic separation is only concerned with a feed that is perfectly dry (Wills & Napier-Munn, 2008; Kelly & Spottsiwood, 1989). This makes it a viable option for dry processing of materials with differing electro-physical properties such as the separation of coal from its associated gangue minerals. Separation by this means mostly acts upon the materials that are conductors and so are best for applications where small amounts of conducting material are to be separated from a large amount of non-conducting material. It is, however, possible with certain designs to concentrate the non-conducting materials with different electrical properties (Kelly & Spottiswood, 1989). Electrostatic separators are most applicable for fine material and are capable of processing particle sizes ranging in 50 - 1000 µm (Kelly & Spottiswood, 1989).

Although the tribo-electrostatic separation is still in the developmental stages, some potential industrial applications can be envisioned. The concentration of carbon from associated gangue minerals, the recovery of unburnt carbon from fly-ash, the elimination of sulphur bearing compounds from pulverized coal and recycling of plastics from waste are some (Carbini *et al.*, 1981; Gidaspow *et al.*, 1986; Finseth *et al.*, 1993; Inculet *et al.*, 1980; Gente *et al.*, 2003; Pascoe and Connell, 2003). Investigations by Finseth and coworkers (1993) indicate that sulphides and silicate impurities can be efficiently removed by tribo-electrostatic beneficiation for a variety of coals. One other application of tribo-electric separation, as suggested by Stencel and coworkers, includes implementing such a separator before the coal burners after pulverizing. At this point, substantial mineral liberation has occurred and it may prove economically and environmentally beneficial to remove the remaining mineral matter from the combustibles within the pipes leading to the burner. This application would decrease the need for water-based cleaning plants distances from the pulverizing unit; water pollution may be reduced and expensive and difficult fine coal dewatering would be eliminated. Work conducted by Stencel *et al.* (2010) shows that an 80% combustible recovery is possible with an associated 30% decrease in the mineral matter of the feed coal ash value. This study further states that pulverized coal particles in transport to the burners carry sufficient and differential charge to effect efficient electrostatic separation under the correct conditions.

Limitations

Narrower particle size distributions are common and are desired as the size influences the behaviour during charging and separation (Kelly & Spottiswood, 1989). The charge obtained on the surface of a coarser particle is lower (when considering its mass) as that of a fine particle. As a result, a coarse non-conducting particle may be displaced into the conducting material as a consequence of weaker attraction to the rotor, plate or in the electric field. Similarly, the finer material is more readily influenced by the surface charge and the conducting fraction may then be contaminated by the fine non-conducting material (Wills & Napier-Munn, 2008; Kelly & Spottiswood, 1989). Entrapment of particles is another concern; multiple particle layers on the rotor surface may cause particle misplacement due to the fine conductor particles being entrained among the non-conductors as a result of their lower mass. Resultantly, these separators are operated in multiple stages of cleaning and are thus manufactured in banks (Kelly & Spottiswood, 1989). In many cases, the separator is designed so as to produce a middling fraction in addition to the conductor and non-conductor products. This produces a fine conductor, coarse non-conductor and unliberated material products which can be treated again in order to obtain a clean overall separation (Kelly & Spottiswood, 1989).

Moisture and temperature are of importance during charging of particles. In studies by Mazumder *et al.* (1995) and Kwetus (1994), it was proven that the moisture content reduced the degree of charging and lowered the separation efficiency. It was however not clear whether the bone dry particles maintained the best charging properties. A subsequent study by Mazumder *et al.* (2015) shows that the resistivity of coal particulates depends largely on the ash content and moisture in the feed coal. Thus it is likely that different coal types would contain differing resistivity. Moreover, the charge of a particle or surface decreases with increasing temperature irrespective of the air pressure (Dwari & Rao, 2007). The mineral temperature also has a significant effect on the conductivity. Fraas (1962) found that the conductivity of some minerals (zircon and monazite) remain unchanged but that of rutile increases significantly. Herein is an indication that the separation of these minerals may be enhanced by operating at elevated temperatures (>90°C). Most electrostatic separators are fitted with heating coils in the feed hopper and heating lamps in the electric field zone (Kelly & Spottiswood, 1989). It has been made clear that surface moisture is undesired and the removal thereof is essential, however, some difficulty in maintaining low surface moisture is noticed when the relative humidity is high. Increased humidity reduces the selectivity of a given separation but is found to affect high tension separation less than electro-static concentration (Kelly & Spottiswood, 1989).

It is essential to ensure that the particle surfaces are free of moisture, organic materials and dust. De-sliming may also be necessary prior to the drying before electrostatic separation (Fraas, 1962). Some additional operating parameters are considered when optimizing the electrostatic separator performance. These are the feed rate, rotor speed, voltage, electrode position and product splitter. In determining the optimal settings, the strategy is to increase the feed rate until the minimum acceptable product quality is reached (Kelly & Spottiswood, 1989). Negative electrodes are preferred commercially and most separators are supplied with 40kV of direct current power achieving a current flow that varies from 5 to 15 A per meter of rotor length (Kelly & Spottiswood, 1989).

Performance, success and pitfalls

As mentioned, much work has been conducted on tribo-electric separation which has shown that it has potential in coal beneficiation practices especially in the finer size ranges. Much data proving successful separation has been reported (Bada *et al.*, 2010). Studies by Inculet and colleagues (1980) indicated a successful reduction in ash value while retaining a similar calorific value as relative to the feed. Furthermore, the recovered product quality was on par with that obtained through wet processing techniques. A summary of the performance is provided in the list below.

- A study by Masuda *et al.* (1984) shows an ash value reduction from 15.2% in the feed to 7.1% in the product with a clean coal recovery of 90%.
- The removal of pyritic sulphur from coal was illustrated in the work by Gidaspow and co-workers (1987). A decrease in sulphur content from 2.4% in the feed to 0.86% in the product was observed in a particle size range of -40 micron.
- Coal beneficiation in a copper tribo-charging pipe yielded a clean coal product of 2.1% ash which was obtained from a feed of 24.4% ash. A sulphur reduction from 1.45% in the feed to 0.6% in the product was observed with a total carbon recovery of 38.8% (Finseth *et al.*, 1993).
- Different charging devices were studied by Dwari and Rao (2006) and this work indicated that copper is the preferred material due to its high electrical conductivity. A decrease in ash content to 18% in the product from a coal feed of 43% was reported which supersedes

the ash value of 25% obtained in a parallel dense medium washability test.

- In the work of Trigwell and colleagues (2003b), tribo-electrostatic separation is proved for removal of gangue minerals before combustion. A reduction of both ash and sulphur contents were observed for two USA coal samples in the range of 0 – 75 micron. This study further investigates the effect of particle size and shows that a finer grind produces a better quality product but not necessarily an improved yield. Trigwell *et al.* (2003b) further shows that a two stage beneficiation may yield improved separation efficiency which is confirmed by the studies of Bada and co-workers (2010 and 2012).
- Dwari and Rao (2008) investigated the beneficiation of Indian coal and found that a reduction in ash content from 25% in the feed to 15.6% in the product is achievable with a 69% product yield. Some further investigation (Dwari & Rao, 2009) established that PSD, tribo-charging time, voltage, gas flow rate and fluidization time affect the separability in a cylindrical fluidized bed tribo-charger and tribo-separating unit. The results show a decrease in ash content from 43% in the feed coal to 18% in the clean coal product with a 30% product yield for -300μm particles. A higher ash value of 33% was obtained at a higher product yield of 67%. The conventional wet washability tests yielded a final coal product of 25% ash at a yield of 65% for the same coal sample.

From the above observations, some studies investigating the modification of the particle surface were undertaken. This was to determine if the separation efficiency would improve upon alterations to the energy structure of the particulate surface. Zhou and Brown (1988), found that a degree of improvement is obtainable after chemical treatment of the particles ranging in 75 - 150 μm. Vapors of methyl acetate, acetone and acetic acid were considered in a fluidized bed charger with a dry nitrogen atmosphere. Some further studies by Trigwell and coworkers (2003) on the charge properties of coal when treated with chemicals showed that acetone slightly enhances beneficiation; ammonia is considered detrimental to upgrading and sulphur dioxide proves ineffective. This branch of research is ongoing with investigations of fine coal being treated in gaseous forms of acidic and basic organic solvents (Dwari & Rao, 2009).

Tests conducted on South African coal by Bada *et al.* (2010) too show promising results for coal particles ranging in -177 μm. Two power station coal feed samples were investigated and a reduction in ash value from 36% and 31% to 14.9% and 12.2% were obtained for both coals respectively. Sulphur content was also reduced from 2.1% and 2.8% to 0.9% and 0.4%. The study further entailed research regarding a two-stage separation which produced a clean coal of 10.76% ash from feed ash of 30%. An improvement in the calorific value was noted as 20.149MJ/kg from 27.479MJ/kg. The single stage also produced a clean coal of 14.67% ash, 25.7MJ/kg CV and 25.22% combustible recovery. Bada *et al.* (2010) confirmed a similar separation performance conducted on another high ash power station coal sample and achieved clean coal of 17% ash and combustible recovery of 50%. Some examinations into the effect of co-flow were also made and it was found that the best product is obtained while operating at the lowest co-flow; yielding a product ash value of 8.9% from a coal feed ash value of 30.4% (Bada *et al.*, 2010). Similar results were obtained by Tao *et al.* (2009), claiming that an increase in co-flow is detrimental to the separation. Bada *et al.* (2010) further undertook some qualitative (XRD) studies to determine the differences between the feed and final product coals. The analysis confirms that the South African feed coals are mostly made up of crystalline minerals (quartz and kaolinite).

Further studies conducted by Bada *et al.* (2012) on No. 2 and No. 4 Seam South African coal samples in a two stage rotary tribo-electrostatic separator show good results for the -177 µm size fraction. Reductions in ash value from 30.4% and 36% in the feed to clean product of 8.9 - 13.1% and 10.8% ash for the No. 2 and No. 4 seam samples were observed, respectively. This further adheres to a respective 9.6% and 6.0% combustible recovery. Petrographic tests show a considerable improvement in vitrinite content along with a reduction in visible minerals. The No. 2 and No. 4 seam samples report 53% and 54% product vitrinite from a feed vitrinite of 21%, respectively (Bada *et al.*, 2012). Also confirmed is that separation may take place between suites of macerals and also between maceral and mineral. This suggests that tribo-electrostatic separation targets specific components within differing maceral suites, more specifically vitrintie. The decreased minerite (rock) and carbominertite (rock-vitrinite middling) contents provided by the petrographic analysis are proof of this (Bada *et al.*, 2012). Additionally, a significant reduction is sulphur content was obtained in the two-stage process from 1.6% in both No. 2 and No. 4 seam coal samples to 0.7% and 0.9% in the product, respectively (Bada *et al.*, 2012). This study also investigates the effects of splitter position in a single-stages separation, charger rotation speed in a double-stage separation, co-flow and feed rate. From this it is noted that a cleaner coal product can be obtained with the splitter positioned closer to the negative electrode and that the charger rotation speed may require variation when considering coal type. Moreover, a low co-flow provides the best separation efficiency coupled with increased feed rates (Bada *et al.*, 2012).

The data as summarized above indicates that both single and double stage tribo-electrostatic separation are effective in producing a clean coal product with reduced ash and sulphur contents. The two-stage process may lead to cleaner coal product but is, however, detrimental to the coal carbon recovery obtainable (Bada *et al.*, 2010). It must be noted that tribo-electric separation suffers some pitfalls. The monolayer feed presentation severely restricts the capacity and throughput for the especially fine coal and narrow size ranges allowable (Wills & Napier-Munn, 2008). The sensitivity to coal type, particle size, particle size range, surface moisture, temperature and relative humidity may prove maintaining an effective separation difficult. Additionally, the de-sliming and drying of the ultra-fine feeds prior to processing may be tricky and uneconomical.

Microwave treatment

Microwave heating technologies show potential in reducing the energy requirements and also environmental impacts of some coal beneficiation processes. These treatments can be utilized in the removal of minerals or water prior to combustion or to improve upon grindability. Some other applications include microwave processing for coking, liquefaction, enhancing fluid flow in coal beds and coal characterization

(Binner *et al.*, 2014). Numerous studies have been conducted in this regard - exploring the potential for microwaves to improve coal processing efficiency (Binner *et al.*, 2014). Resultantly many of these techniques have been proven to work on an experimental scale. However, few of these developments have managed to reach commercial scale and may require another look with special regard to upscaling. This may be attributed to the efficiency in comparison to conventional heating and also the intricacy of the multi-disciplinary approach required for successful microwave processing (Binner *et al.*, 2014).

Principle of microwave treatment

In conventional heating, the material is heated from the body surface toward the center by means of conduction, convection and radiation. Microwaves are capable of penetrating the core of most non-metallic materials and therefore provide the advantage of heating volumetrically (Binner *et al.*, 2014). The heating mechanism depends on the dielectric properties of the individual constituents originating in the ore (Meredith, 1998). In the case of a material that contains significantly differing components with respect to the dielectric properties, selective heating occurs. These volumetric and selective heating capabilities often reduce heating times to less than 1% of traditional heating techniques (Meredith 1998). Microwave heating can further result in phenomena such as superheating at atmospheric pressure and materials fracturing which is impossible with conventional heating (Kingman, 2006). The shorter treatment times allow for a significant reduction in equipment size and therefore microwaves can be used to great advantage in some applications. The theory of microwave treatment is well documented in studies by Chen et al., 1984; Meredith, 1998, Whittaker, 2005a; Scaife, 1989; Metaxas & Meredith, 1983; Salsman, 1991 and many more.

Microwave treatment for coal

The applications pertaining to microwave treatment of coal are summarized as follows:

i. Removal of mineral matter through microwave processing

In coal beneficiation, microwave processing has been specifically targeted for the removal of pyrite or desulphurization. Most coal associated minerals are transparent to microwave radiation due to the low dielectric properties (Marland *et al.*, 2001). Pyrite, however, reports higher conductive losses when compared to coal and the other associated gangue minerals (Huang & Rowson, 2001). The dielectric properties and heating rates of coal and pyrite have been extensively described in the works by Marland *et al.* (2001), Shuey (1975) and Fanslow *et al.* (1980); these show that the conductive losses of coal and pyrite are considerably different allowing pyrite to absorb more electromagnetic energy and heat faster than coal. As a consequence, the magnetic susceptibility of pyrite in coal can be significantly improved by heating and separation can be achieved through magnetic separation (Rowson & Rice, 1990a).

ii. Removal of moisture with microwave treatment

Much research has been conducted into the removal of water from coal. The significant calorific gains of said removal can result in immense cost savings. The majority of this work is in the utilization of waste heat and steam drying, mechanical and thermal compression and also some acoustic treatment. It is possible for microwave energy to be used in this respect since the water phase has the highest dielectric loss factor. Therefore, the water will be selectively heated while the bulk coal remains microwave transparent. The drying rates of brown coal using microwave heating have been determined to be in one or two orders faster than conventional thermal drying at 80°C (Standish *et al.*, 1988). Some studies, however show that the energy required to remove the bulk and loosely bound water in a coal sample of 60% moisture is comparable to thermal drying processes (Binner *et al.*, 2013).

Microwave processing is also applicable in other areas of coal processing, such as improvement of grindability, coke making, liquefaction pre-treatment, enhancing fluid flow in coal beds and characterization of coal. Numerous studies have been conducted in this regard but will not be dealt with in this article body. One specific concern with microwave treatment is the possible alteration of the coal structure (Binner et al., 2014). Low-rank coals are more susceptible to thermal modification than high-rank coal but may depend on power density, residence time and cavity design. Analyses of microwave dried samples of sub-bituminous coal (600W) showed no major structural changes except for small changes in the aromaticity (Miknis *et al.*, 1996). Wilson (2009) showed that higher powered treatments resulted in the heating of the entire coal sample generating additional transformations.

Some attempts at scaling the microwave drying process are found in the literature. In these it was found that coal fires are a safety concern and continuous monitoring or possible removal of smoking coal lumps may be required (Wilson, 2009). Issues relating to the entrainment of fine particles out of the bed by the rapid evolution of water have been reported (Allardice *et al.*, 2004). A commercial unit known as the *Drycol* located in Kentucky USA was capable of continuously processing 15 tph of coal. Loss of volatiles and coal fires were avoided by maintaining the coal at a temperature below 90°C (Binner *et al.*, 2014). In producing a commercial size unit, the main challenge is in the design of an appropriate microwave cavity that minimized residence time through the use of high power densities while avoiding bulk heating. The *Drycol* shows no changes in the proximate characteristics apart from the desired 50% reduction in moisture as a result of microwave processing. Some coal microwave drying systems have also been developed to create upgraded fuels for pulverized coal combustion and gasifiers. The processes claim to increase the calorific value by 50% through moisture removal operating at 120 000 ton per year (Wilson, 2009).

CONCLUSION

Dry coal processing holds many advantages over the currently employed wet beneficiation techniques and may prove indispensable in the near future. These methods provide low investment, operating and maintenance cost with the benefits of being environmentally friendly, safe and easy to operate. The units are also compact, containerized and modular generating a smaller mining footprint and allowing for simplified plant erection, disassembly and relocation. The plant circuit is significantly reduced to crushers, screens, conveyors and storage with no extensive and expensive product and fines dewatering systems required. The above advantages create opportunities for beneficiating small or remote coal reserves and also discard facilities (stockpiles and tailing dams) that prove uneconomical to process by other means. This increases the life of mine and revenue obtainable from processing. Additionally, beneficiation is made possible near or in the mine site which reduces haulage costs and allows for the use of discard as aggregate and backfills in the rehabilitation of the site upon closing.

The advantages of dry coal processing must, however, be measured against the limitations thereof. Although successful separation is obtained in numerous dry coal processors, the efficiency does not compare to that of a wet coal beneficiation plant. As a result, coal beneficiation plants operating exclusively on dry techniques is a goal that is yet to be achieved and most commercial dry methods are currently limited to pre-concentration, separation of an "easy to clean" coal with little near dense material, concentration of a coal that does not require a large upgrade to meet product specification and scavenging operations.

Coal preparation practices will soon have to adapt to the current resource and environmental trends as water becomes an increasingly scarce commodity and needs to be preserved. Water demands may become a major problem for present and future developments in mining, beneficiation and new plant construction. It is predicted that stricter laws regarding water usage and wastage will be introduced and that it may increase in expense in the future. It is vital to ensure maximum throughput at reduced operating costs and in the face of declining ore grades and environmental concerns. It is clear that a significant commercial and environmental drive exists for the development of an alternative approach to water-based beneficiation of coal. Consequently, the demand for effective dry beneficiation technologies is increasing, irrespective of the technical and economic feasibility of the wet methods. The research and development of dry coal beneficiation are therefore considered extremely important most specifically to countries with abundant coal resources, relying on the production and consumption of coal and also facing extended periods of drought or arctic conditions.

Several dry coal processing techniques are available; each capable of successfully beneficiating coal to varying degrees for specific particle size, capacity and moisture limitations. These are as listed and discussed in the article body. Generally, technologies belonging to the optical and elutriation categories are utilized for coarse coal (6 mm - 100 mm) processing and have larger capacities and low processing cost; but may yield poorer separation efficiencies due to the certain limitations pertaining to the separation mechanism. In contrast, techniques forming part of the magnetic and electrostatic categories have a relatively high separation efficiency which is suitable for producing high-quality coal in the fine fraction (45 μm - 6 mm) but at the expense of operational cost. Additionally, the magnetic and electrostatic processes have strict requirements in terms of coal properties and liberation; thus capacities may be limited and so these methods are only used for special purposes. In certain cases, it may be useful to combine magnetic and tribo-electric separation with a microwave pre-treatment of the coal prior to beneficiation. Microwave treatment is known to enhance either the magnetic susceptibility or dielectric properties of pyrite and in so providing a more effective magnetic or tribo-electrostatic separation, respectively. Microwave treatment of coals has also proven successful on an experimental scale for dewatering and improvement of grindability but may suffer from extremely large energy requirements.

As mentioned, the processes utilized for beneficiating coarse coal typically belong to sensor-based sorting and gravity separation or elutriation. Some of the technologies in these categories have been adapted commercially and on the experimental scale for the fine coal fraction (1 mm - 6 mm) but suffer from throughput related issues and upscaling requirements, respectively. The Dual Energy X-ray Transmission (DE-XRT) sorter may not always be able to upgrade coal to the required quality and control of the product quality may, at times, prove difficult. The selectivity of the sorter can easily be made more sensitive to improve upon separation efficiency and processing of smaller particle sizes but at the cost of throughput. Sorters further have strict requirements in terms of feed preparation and presentation which can be intricate and affect unit efficiency and productivity. As a result, sorters may not prove a competitive coal beneficiation technique and consequently, sorters for coal processing find use as a pre-concentration or scavenging in addition to a wet coal treatment plant or for the scarce cases where the coal is easy to process with little near dense material. The DE-XRT does also show promise in the separation of torbanite and pyritic sulphides from coal.

In gravity separation the *All Air Jig*, *FGX*, *ADMFB*, *SEP-AIR* and coal *winnower* are utilized for +6 mm coal. The *All Air Jig* and *FGX* units yield the poorest separation efficiencies of those listed as determined by the higher EPM values of 0.16 - 0.27 and 0.12 - 0.23, respectively. The ADMFB produced the best separation efficiency with EPM values ranging in 0.05 and 0.07. This process, however, requires the use of a dense medium such as magnetite and intricate and possibly expensive recovery and cleaning systems therefor. Additionally, narrow particle size distributions may be required with minimal fines material and moisture present. The coal *winnower* and *SEP-AIR* units obtained larger EPM values in comparison to the ADMFB but lower than the *All Air Jig* and *FGX*. These processes may be attractive since no dense medium and related handling systems are required and feed moisture effects may not be so significant. The coal *winnower* is still in the developmental stages of research but shows promise in dry coal beneficiation for +6 mm coal along with some research into coal as small as 2mm. The *SEP-AIR* unit may only yield an EPM value of 0.10 - 0.17, but has the advantage of processing coal down to 1mm in size. This provides the opportunity to beneficiate the previously removed +1mm - 6mm material, producing an upgraded fine fraction that can be added to the product without negatively affecting the quality. Gravity separation adapted for -6mm particle treatment includes the *air table*, *AKAFlow*, KAT, variations of *ADMFB* and *dry reflux classifier*. These may provide efficient separation but show reduced throughputs due to the small particle size and have some feed preparation requirements with respect to narrow size range and feed moisture content. Moreover, the dry *reflux classifier*, *coal winnower* for <6mm particles, dual density *ADMFB*, vibrated *ADMFB* and magnetically stabilized *ADMFB* are still in the research and developmental stages with further investigation into upscaling required.

The novel and upcoming *SEP-AIR* and *coal winnowing* units seem most attractive as dry coal beneficiation techniques for coal particles above 1 mm and 6 mm, respectively. The *coal winnower* requires further investigation and upscaling whereas the SEP-AIR is already commercialized in Russia and has undergone several efficiency and productivity tests on various minerals. In conclusion, given the applicability and advantages of the *SEP-AIR* pneumatic complex, it appears to be the most suitable process for dry coal processing in coal sizes of 1 mm - 100 mm, irrespective of the lower efficiency achieved when compared to the alternative dry (i.e. *ADMFB*) and wet coal beneficiation processes. For the <1 mm coal fraction, tribo-electrostatic separation proves most applicable as dry coal beneficiation technique especially in combination with microwave pre-treatment for de-watering of the fine coal (if necessary) and improving the dielectric properties of the coal (with special regard to pyrite) but operating cost and ease of processing need to be considered.

ACKNOWLEDGEMENTS

The authors would like to acknowledge the financial and technical contribution made by the CoalTech Research Association NPC Ltd., and the expertise and guidance provided by study supervisors and mentors in the Centre of Excellence in Carbon-Based fuels, North-West University, Potchefstroom, South Africa. This study is supported in part by the National Research Foundation (NRF) of South Africa (Coal Research Chair Grant numbers: 86880, UID115228, UID85632). The opinions, findings and conclusions or recommendations expressed is that of the author(s) and the NRF accepts no liability whatsoever in this regard.

REFERENCES

Alfano, G., Carbini, P., Ciccu, R., Ghiani, M., Peretti, R., Zucca, A. 1988. Progress in triboelectric separation of minerals. In: Forssberg, E. (Ed.), XVI International Mineral Processing Congress Amsterdam, Netherlands, Elsevier B.V., Amsterdam, 833–844.

Allardice, D. J., Chaffee, A. L., Jackson, W. R., Marshall, M. 2004. Chapter 3. Water in brown coal and its removal. In Advances in the Science of Victorian Brown Coal, Li, C. Z., Ed. Elsevier: Oxford.

Arms, R. W. 1924. Dry coal cleaning. American Institute of Mining and Metallurgical Engineers. New York, USA.

Bada, S. O., Falcon, R. M. S., Falcon, L. M. 2010. The potential of electrostatic separation in the upgrading of South African fine coal prior to utilization-a review. *The Journal of the Southern African Institute of Mining and Metallurgy, 10:691-702.*

Bada, S. O., Falcon, L. M., Falcom, R. M. S., Du Cann, V. M. 2012. Qualitative analysis of fine coals obtained from triboelectric separation. *The Journal of the Southern African Institute of Mining and Metallurgy,* 112:55-62.

Balasubramanian, A. 2015. Overview of mineral processing methods. Centre for advanced studies in earth science. University of Mysore, Mysore.

Binner, E., Lester, E., Chaffee, A. L., Kingman, S. K. 2013. Microwave treatment of Morwell brown coal to determine the viability of microwave coal drying as an industrial concern. 14th International Conference on Microwave and High Frequency Heating, Dimitrakis, G., Ed. AMPERE: Nottingham, UK, Sept. 16-19:43-46.

Binner, E., Lester, E., Kingman, S., Dodds, C., Robinson, J., Wu, T., Wardle, P., Matthews, J.P. 2014. A review of microwave coal processing. *Journal of Microwave Power and Electromagnetic Energy*, 48(1):35-60.

Boylu, F., Cetinel, T., Guven, O., Karakas, F., Cinku, K., Karaagaclioglu, I. E., Celik, M. S. 2015. Optimum separation route for semi-bituminous coal using semi-pilot pneumatic stratification jig. *Physiochemical Problems of Mineral Processing,* 51(2):559 - 573.

Broek, M. 1995. The distribution of magnetic susceptibility in crushed ores. *Magnetic and Electrical Separation*, 6:213–228.

Carbini, P., Carta, M., Ciccu, R., Del FA, C., Ghiani, M., Rossi, G. 1981. Beneficiation methods for coal desulphurization, in *Proc. 64th CIC Coal Symp.*, Halifax, NS, Canada, May 31–Jun. 3, 1981, 164–169.

Carta, M., Del Fa, C., Ciccu, R., Curreli, L., Agus, M. 1976. Technical and economic problems connected with the dry cleaning of raw coal and in particular with pyrite removal by means of electric separation. In: Partridge, A. (Ed.), Proceedings of 7th International Coal Preparation Congress. Adept Printing Pvt. Ltd., Australia, 1–35. K2.

Celik, M. S. 2002. Dry cleaning of low rank coals by low temperature carbonization and magnetic separation. *Coal Preparation*, 22:311–322.

Celik, M. S. & Yildirim, I., 2000. A new physical process for desulfurization of low-rank coals. *Fuel*, 79:1665–1669.

Chavaldi, G., Singh, R. K., Sharma, M., Singh, R., Das, A. 2016a. Development of a generalized strategy for dry beneficiation of fine coal over a vibrating inclined deck. *International Journal of Coal Preparation and Utilization*, 36:10-27.

Chavaldi, G., Singh, R. K., Das, A. 2016b. Processing of coal fines using air fluidization in an air table. *International Journal of Mineral Processing*, 149:9-17.

Chavaldi, G., Singh, R.K., Sharma, M., Das, A. 2015. Recovery of combustibles from fine coal through controlled air fluidization over inclined deck. *Separation Science and Technology*, 50:2032-2040.

Chen, T. T., Dutrizac, J. E., Haque, K. E., Wyslouzil, W., Kashyap S. 1984. Relative transparency of minerals to microwave radiation. *Canadian Metallurgical Quarterly*, 23 (3):349-351.

Chen, Q. & Wei, L. 2003. Coal dry beneficiation technology in China: the state-of-the-art. *China Particuology*, 1(2):52-56.

Chen, Q. & Yang, Y. 2003. Development of Dry Beneficiation of Coal in China. *Coal Preparation*, 23:3–12.

Ciccu, R., Ghiani, M., Peretti, R., Serci, A., Zucca, A. 1991. Tribocharging studies of ground coal matter. In: Dugan, D.R., Attia, Y.A. (Eds.), Processing and Utilization of high Sulfur Coals IV, P.R. Elsevier Science Publishers B.V., Amsterdam, 223.

Cross, J.A. 1987. Electrostatics: Principles, Problems and Applications, Adam Hilger, Bristol, England.

Davaasuren, J., Kim, B. G. Lee, J. H., Davaatseren, G., Bazarragcha, M. 2016. Dry coal preparation of fine particle by KAT process, in (Litvinenko, V.), XVIII International Coal preparation congress Saint-Petersburg, Russia 28 June – 1 July 2016, Springer, Switzerland, 1171-1176.

De Jong, T. P. R., Houwelingen, J. A., Kuilman, W. 2004. Automatic sorting and control in solid fuel processing: opportunities in European perspective. *Geologica Belgica*, 7(3-4):325-333.

De Korte, G.J. 2013. Dry processing versus dense medium processing for preparing thermal coal. CSIR (Council for Scientific and Industrial Research). Pretoria, South Africa.

De Korte, G.J. 2014. Dry processing of coal - status update. Coaltech Research Association NPC Ltd. Secunda, South Africa.

De Korte, G.J. 2016. Dry dense medium processing of raw coal – progress report [correspondence]. 14 June, Potchefstroom.

De Korte, G.J. 2018. SEP-AIR dry coal separator: efficiency determination carried out in Russia. CoalTech Research Association NPC Ltd. Secunda, South Africa.

Dwari, R. K. & Rao, K. H. 2007. Dry beneficiation of coal - A review. *Mineral Processing & Extractive Metallurgy*, 28:177–234.

Dwari, R. K. & Rao, H. K. 2008. Characterising Electron Transfer Mechanism in Tribo Electrification of pyrite through Contact Angle Measurements. *The Open Mineral Processing Journal*, 1:6–17.

Dwari, R.K. & Rao, K.H. 2009. Fine coal preparation using novel tribo-electric separator. *Minerals Engineering*, 22:119-127.

Dwari, R.K., Mohanta, S. K., Rout, B., Soni, R. K., Reddy, P.S.R., Mishra, B.K. 2015. Studies on the effect of electrode plate position and feed temperature on the tribo-electrostatic separation of high ash Indian coking coal. Advanced Powder Technology, 26:31-41.

El-Mouloud Zelmat, M., Rizouga, M., Tilmatine, A., Medles, K., Miloudi, M., Dascalescu, L. 2013. Experimental Comparative Study of Different Tribocharging Devices for Triboelectric Separation of Insulating Particles. IEEE Trabsactions on industry applications, 49(3):1113-1118.

Falcon, R. & Ham, A. J. 1988. The characteristics of Southern African coals. *Journal of the South African Institute of mining and metallurgy*, 88(5):145-161.

Fan, M., Chen, Q., Zhao, Y., Luo, Z. 2001. Fine coal (6-1mm) separation in magnetically stabilized fluidized beds. *International Journal of Minerals Processing*, 63:225-232.

Fan, M., Chen, Q., Zhao, Y., Guan, Z.L.Y., Li, B. 2002. Magnetically stabilized fluidized beds for fine coal cleaning. *Powder Technology*, 123:208-211.

Fanslow, G. E., Bluhm, D. D., Nelson, S. O. 1980. Dielectric heating of mixtures containing coal and pyrite. *Journal of Microwave Power*, 15 (3):187-191.

Finseth, D. H., Newby, T., Elstrodt, R. 1993. Dry electrostatic separation of fine coal, in *Proc. 5th Int. Conf. Util. High Sulfur Coals*, Lexington, KY, USA, 1993, 91–98.

Firdaus, M., O'Shea, J.P., Oshitani, J., Franks, G.V. 2012. Beneficiation of coarse coal ore in an air-fluidized bed dry dense-medium separator. *International Journal of Coal Preparation and Utilization*, 32:276-289.

Fraas, F. 1962. Electrostatic Separation of Granular Materials. U.S. Bureau of Mines Bulletin 603.

Fricke, G. 1977. The use of electrostatic separation processes in the beneficiation of crude potassium salts. *Phosphorous Potassium*, 90: 42–45.

Fujita, T., Hashimoto, H., Masuda, S., Oda, T. 1981. HGMS of Coal. In Proceeding of International Symposium on Powder Technology, 797.

Gehauf, R. 2004. A practical guide on selecting and optimizing rare earth magnetic separators. In SME Annual Meeting, Denver, CO, USA.

Gente, V., Marca, F. L., Lucci, F., Massacci, P. 2003. Electrical separation of plastics coming from special waste. *Waste Management*, 23(10): 951–958.

Ghosh T. 2013. Modelling of an air-based density separator. Theses and Dissertations-Mining Engineering. Paper 7. Obtained from http://uknowledge.uky.edu/mng_etds/7. Date accessed: 18 August 2017.

Gidaspow, D., Wasan, D. T., Saxena, S., Shih, Y. T., Gupta, R., Mukherjee, A. 1986. Electrostatic desulfurization of coal in fluidized beds and conveyors, in *Proc. AIChE*, Nov. 1986, vol. 83, 74–85, (Series no.255).

Gidaspow, D., Gupta, A., Mukherjee A., D. T. Wasan D. T. 1987. *Processing and Utilization of High Sulfur Coals II,* Y. P. Chugh and R. D. Caudle, Eds., Elsevier, New York, 271-281.

Gormash Export. 2016. SEP-AIR pneumatic concentration complex. Gormash Export Novosibirsk, Russian Federation. Accessed from http://www.gmexp.ru. Date of access: 4 June 2018.

Gunther, C. G. 1909. Electromagnetic Ore Separation. New York: McGraw-Hill Publishing Co.

Gupta, N. 2016. Evaluation of pneumatic inclined deck separator for high-ash Indian coals. *International Journal of Coal Science Technology*, 3(2):198-205.

Haga, K. 1995. Applications of the Electrostatic Separation Technique, in: J.S. Chang, A.J. Kelly, J.M. Crowly (Eds.), Handbook of Electrostatic Processes, Marcel Dekker Editors, New York, 365–386.

Harbeck, H. & Kroog, H. 2008. New developments in sensor based sorting. Montan University Loeben, Australia.

He, J., Zhao, Y., Zhao. J., Luo, Z., Duan, C., He, Y. 2015. Separation performance of fine low-rank coal by vibrated gas-solid fluidized bed for dry coal beneficiation. *Particuology*, 23:100-108.

He, J., Tan, M., Zhao, Y., Zhu, R. & Duan, C. 2016a. Density-based segregation/separation performances of dense medium gas–solid fluidized bed separator (DMFBS) for coal cleaning and upgrading. *Journal of the Taiwan Institute of Chemical Engineers,* 59: 252–261.

He, J., Tan, M., Zhu, R., Zhao, Y., Hong, P. & Jiang, Y. 2016c. Density-based separation performance of a secondary air-distribution fluidized bed separator (SADFBS) for producing ultra-low-ash clean coal. *Fuel*, 172: 178–186.

Hendricks, C. D. 1973. Charging macroscopic particles. In Electrostatics and Its Applications, (A. D. Moore, Ed.), New York: Wiley Interscience, 57–85.

Honaker, R. Q., Saracoglu, M., Thompson, E., Bratton, R., Luttrell, G. H. & Richardson, V. 2008. Upgrading coal using a pneumatic density-based separator. *International Journal of Coal Preparation and Utilization*, 28 (1): 51-67.

Houwelingen, J. A. & De Jong, T. P. R. 2004. Dry cleaning of coal: Review, fundamentals and opportunities. *Geological Belgica*, 7: 335–343.

Huang, J. H. & Rowson, N. R. 2001. Heating characteristics and decomposition of pyrite and marcasite in a microwave field. *Minerals engineering*, 14 (9):1113-1117.

Hucko, R. E. & Maronde, C. P. 1982. The Apphco-Lion of HGMS in Coal Preparation." In Proceedings on International Coal Preparation Congress, New Delhi, India, F2.

Order, R. R, Weinstein, R. S., and Snoby, R. J. 2003. Combined air jig and dry magnetic separator for cleaning coal. In Twentieth Annul International Pittsburgh Coal Conference, Pittsburgh, PA, Sept, 15–19.

Hughes, N., Le Roux, M., Campbell, Q.P. 2017. The dry beneficiation of South African small coal in a dense medium fluidized bed. North-West University Potchefstroom Campus Potchefstroom, South Africa.

Ibrahim, S. S., Farahat, M. M., Boulos, T. R. 2017. Optimizing the performance of the RER magnetic separator for upgrading silica sands. *Particulate Science and Technology*, 35(1):21–28.

Inculet, I. I., Quigley, R. M., Bergougnou, M. A., Brown, J. D., Faurschou, D. K. 1980. Electrostatic beneficiation of hat creek coal in the fluidized state, *Can. Inst. Mining Metall. Bull.*, 73 (822):51–61.

Inculet, I. I., Bergougnou, M. A., Brown, J.D. 1982. Electrostatic beneficiation of coal. In: Liu, Y.A. (Ed.), Physical Cleaning of Coal – Present and Developing Methods. Marcel Dekker, New York, 87–131.

Inculet, I. I. 1984. Electrostatic Mineral Separation. Wiley, New York. International Energy Outlook, 2005. <http://www.eia.doe.gov/oiaf/ieo/coal.html>, Retrieved on September 9.

Iyer, P. V. 2011. Magnetic and electrostatic separation. In SME Mining Engineers Hand Book, 3rd Edition, (P. Darling, Ed.). Society of Mining, Metallurgy and Exploration, 1533–1545, Colorado, USA.

Joanna, C. M., Iwaszencko, S., Bajerski, A., Janoszek, T. 2016. Novel dry sorter for coal processing and coal recovery from mine originating wastes, in (Litvinenko, V.), XVIII International Coal preparation congress Saint-Petersburg, Russia 28 June – 1 July 2016, Springer, Switzerland, 1149-1154.

Karmazin, V. V. 1997. Theoretical Assessment of Technological Potential of Magnetic and Electrical Separation. *Magnetic and Electrical Separation*, 8:139–159.

Kelly, D. J. & Spottiswood, E. G. 1982. *Introduction to Mineral Processing.* John Wiley & Sons, New York.

Kelly, E.G. & Spottiswood, D.J. 1989. The theory of electrostatic separations: A review Part I. Fundamentals. *Minerals Engineering*, 2(1):33-46.

King, I. 2015 (March 3). The importance of ore sorting technologies in the future. Powerpoint presentation at Mintek, South Africa.

Kingman, S. W. 2006. Recent developments in microwave processing of minerals. *International Materials Reviews*, 51 (1):1-12.

Knoll, F. S. & Taylor, J. B. 1984. Advances in Electrostatic Separation. SME Preprint, 71-84.

Koca, H., Koca, S., Kockar, O. M. 2000. Upgrading of Kutahya lignites by mild pyrolysis and high intensity dry magnetic separation. *Minerals Engineering*, 13(6):657–661.

Kwetus, B.A. 1994. Contact electrification of coal and minerals. *Journal of Electrostatics*, 32:271–276.

Langer, D. J., 2016. *The dry beneficiation of duff coal in a dense medium fluidised bed,* North-West University Potchefstroom Campus Potchefstroom, South Africa.

Lessard, J., De Bakker, J., McHugh, L. 2014. Developments of ore sorting and its impact on mineral processing economics. *Minerals Engineering,* 65:88-97.

Lockhart, N. C. 1984. Dry Beneficiation of Coal- Review Paper. *Powder Technology*, 40: 17–42.

Luo, Z. & Chen, Q. 2001. Dry beneficiation technology of coal with an air dense-medium fluidized bed. *International journal of mineral processing*, 63: 167–175.

Luo, Z., Zhao, Y., Chen, Z., Fan, M., Tao, X. 2002. Separation characteristics for fine coal of the magnetically fluidized bed. *Fuel Processing Technology*, 79:63-69.

Macpherson, S. A. & Galvin, K. P. 2010. The effect of vibration on dry coal beneficiation in the reflux classifier. *International Journal of Coal Preparation and Utilization*, 30:283-294.

Macpherson, S. A., Iveson, S. M., Galvin, K. P. 2010. Density based separations in the reflux classifier with air-sand dense medium and vibration. *Minerals Engineering*, 23:74-82.

Macpherson, S. A., Iveson, S. M., Galvin, K. P. 2011. Density based separation in a vibrated reflux classifier with air-sand dense medium: tracer studies with simultaneous underflow and overflow removal. *Minerals Engineering*, 24:1046-1052.

Mahmoodabadi, M. 2015. Korean anthracite coal cleaning by means of dry and wet based separation technologies. Dissertation in the fulfilment of Master of Engineering. University of Kentucky, USA. Theses and Dissertations--Mining Engineering.18. Accessed from: https://uknowledge.uky.edu/mng_etds/18. Date of Access: 15 June 2018.

Male, S. E. 1984. Magnetic susceptibility and separation of inorganic material from UK coals. *Journal of Physics D: Applied Physics*, 17:55.

Manouchehri, H. R., Hanumantha Rao, K., Forssberg, K. S. E. 2000. Review of electrical separation methods Part 1: Fundamental aspects. *Miner Metall. Process*, 17:1–14.

Manoucherhi, H. R. 2003. Sorting: Possibilities, limitations and future. Northland Oretech ConsultingCo. Docentvagen 26, SE-977 52, Sweden.

Marland, S., Merchant, A., & Rowson, N. 2001. Dielectric properties of coal. *Fuel*, 80 (13):1839-1849.

Masuda, S., Taraguchi, M., Takashashi, T., Haga, K. 1983. Electrostatic beneficiation of coal using cyclone-tribocharger. IEEE Trans Ind. Appl. (IA), 19(5):789–93.

Masuda, S., Taraguchi, M., Takashashi, T., Haga, K. 1984. Electrostatic Beneficiation of Coal Using a Cyclone-Tribocharger. *IEEE Transactions of Industry Applications*, 1A:19.

Mazumder, M.K., Tennal, K.B., Lindquist, D.A. 1995. Triboelectric separation of coal from mineral impurities. *Proceedings of the Annual Meeting of the Electrostatic Society of America*, Morgan Hill, CA, Laplacian, 59–70.

Meredith, R. 1998. Engineers' Handbook of Industrial Microwave Heating. The Institution of Electrical Engineers, London.

Metaxas, A. C. & Meredith, R. J. 1983. Industrial Microwave Heating. The Institution of Electrical Engineers, London.

Miknis, F. P., Netzel, D. A., Turner, T. F., Wallace, J. C., Butcher, C. H. 1996. Effect of different drying methods on coal structure and reactivity towards liquefaction. *Energy and Fuels*, 10:631-640

Mineral technologies. 2018. Rare earth roll magnetic separator, data sheet - RER. Mineral Technologies a Downer company, Carrara Gold Coast, Queensland.

Mohanta, S. K., Rout, B., Dwari, R. K., Reddy, P. S. R., Mishra, B. K. 2016. Tribo-electrostatic separation of high ash coking coal washery rejects: Effect of moisture on separation efficiency. Powder Technology, 294:292-300.

Morgan, D. G. & Bronkala, W. J. 1991. The selection and application of magnetic separation equipment: Part I. *Magnetic and Electrical Separation*, 3:5–16.

Morgan, L. 2017. *Winnowing as dry separation method for fine coal.* North-West University Potchefstroom Campus Potchefstroom, South Africa.

Mostika, Y. S., Karmazin, V. I., Shute, V. Yu., Grebenyuk, L. Z. 1999. About the equations of motion of a magnetic particle in a magnetic separator. *Magnetic and Electrical Separation*, 10:35–44.

Oshatani, J., Kazuhiro, T., Yoshida, M., Kubo, Y., Nakatsukasa, S., Franks, G. V. 2016. Dry beneficiation of fine coal using density-segregation in a gas-solid fluidized bed. *Advanced Powder Technology*, 27:1689-1693.

Parker, M. R. 1977. The physics of magnetic separation. *Contemporary Physics*, 18(3):279–306.

Pascoe, R. D. & Connell, B. O. 2003. Development of a method for separation of PVC and PET using flame treatment and flotation, *Minerals Engineering*, 6(11):1205–1212.

Patil, D.P. & Parekh, B.K. n.d. Beneficiation of fine coal using the air table. Centre for applied energy research, University of Kentucky Lexington, USA.

Prusti, P., Sahu, A.K., Biswal, S. K. 2015. Prediction of the position of coal particles in an air dense medium fluidized bed system. *Journal of Mining Science and Technology*, 25:421-427.

Rao, D. V. S. 2011. Magnetic separation. In Mineral Beneficiation – A Concise Basic Course, Boca Raton, FL: CRC Press, 109–114.

Ralston, O. C. 1961. Electrostatic separation of mixed granular solids. Amsterdam: Elsevier.

Robben, C., De Korte, J. G., Wotruba, H., Robben, M. 2014. Experiences in dry coarse coal separation using X-ray -transmission-based sorting. *International Journal of Coal Preparation and Utilization, 34:210-219.*

Rowson, N. & Rice, N. M. 1990a. Magnetic enhancement of pyrite by caustic microwave treatment *Minerals Engineering*, 3 (3/4):355-361.

Rudolphy, K. 2017. An overview of water resources and the uses of water. *Thought.Co*. Accessed from: https://www.thoughtco.com/what-are-water-resources-1435834. Date of access: 15 May 2018.

Rylatt, M. G. & Popplewell, G. M. 1999. Diamond processing at Ekati in Canada. Minerals Engineering, February, 19.

SACPS (South African Coal Processing Society). 2015. *Coal Preparation in South Africa*. Pietermaritzburg: Natal witness commercial printers.

SACRM (South African Coal Roadmap). 2011. Overview of the South African coal value chain. Johannesburg. Accessed from: http://www.sanedi.org.za/coal-roadmap/.Date of access: 25 April 2018.

Sakhre, D.K., Singh, S.P., Acharya, R.K., Katley, R.L., Prasad, A.L.V., Chayande, P. 2018. Coal winnowing: an innovative technique for dry coal beneficiation. *CPSI Journal*, 10(28):3-13.

Sahu, A.K., Bismal, S.K. & Parida, A. 2009. Development of air dense medium fluidized bed technology for dry beneficiation of coal – A review. *International Journal of Coal Preparation and Utilization*, 29: 216–241.

Salsman, J. B. 1991. Technique for measuring the dielectric properties of minerals as a function of temperature and density at microwave heating frequencies. *Materials Research Society Symposium Proceedings*, 189:509-515.

Scaife, B. K. P. 1989. Principles of dielectrics. Clarendon: Oxford.

Shuey, R.T. 1975. Semiconductor Ore Minerals. Elsevier Vol. 4.

Sinha, V.K., Dey, B.K., Baranwal, P.K. 2016. Coal Preparation in India: New business opportunities & need for dry separation technology, in (Litvinenko, V.), XVIII International Coal preparation congress Saint-Petersburg, Russia 28 June – 1 July 2016, Springer, Switzerland, 1089-1093.

Soong, Y., Schoffstall, M.R., Link, T.A. 2001. Triboelectrostatic beneficiation of fly ash. *Fuel*, 80:879–884.

Standish, N., Worner, H., Kaul, H. 1988. Microwave drying of brown coal agglomerates. *Journal of microwave power and electromagnetic energy*, 23(3):171-175.

Stencel, J. M., Schaefer, J. L., Ban, H., Li, T. X., Neathery, J. K. 1998. Triboelectrostatic cleaning of coal in-line between pulverizer and burners at utilities. International Journal of Coal Preparation and Utilization 19(1):115–129.

Stepanenko, A. I. 2016. SEPAIR pneumatic separation complex for dry coal beneficiation, in (Litvinenko, V.), XVIII International Coal preparation congress Saint-Petersburg, Russia 28 June – 1 July 2016, Springer, Switzerland, 1089-1093.

Strauss, T. 2016 (April 20). Ore sorting in mining. Powerpoint presentation at Merlyn Consulting. Dublin, Ireland.

Strydom, H. 2010. The application of dual energy X-ray transmission sorting to the separation of coal from torbanite. University of the Witwatersrand Johannesburg, South Africa.

Svoboda, J. 2004. Magnetic Techniques for the Treatment of Materials -Reviews of Magnetic Separators. Kluwer Academic Publishers, London, UK.

Svoboda, J. & Fujita, T. 2003. Recent developments in magnetic methods of material separation. *Minerals Engineering*, 16:785–792.

Tomra Sorting Solutions. 2014. Dry de-stoning of coal. Tomra Sorting Pty (Ltd). Ferndale, South Africa.

Trigwell, S., Tennal, K.B., Mazumder, M.K., Lindquist, D.A. 2003. Precombustion cleaning of coal by triboelectric separation of minerals. *Particulate Science and Technology*, 21:353–364.

Tripathy, S.K., Banerjee, P.K., Suresh, N., Murthy, Y.R., Singh, V. 2017. Dry high-intesnity magnetic separation in mineral industry - a review of present status and future prospects. *Mineral Processing and Extractive Metallurgy Review*, 38(6): 339-365.

Uslu, T. & Atalay, U. 2003. Microwave heating of coal for enhanced magnetic removal of pyrite. *Fuel Processing Technology*, 85:21–29.

Uslu, T., Atalay, U., Arol, A. I. 2003. Effect of microwave heating on magnetic separation of pyrite. *Colloids and Surfaces A: Physicochemical and Engineering Aspects*, 225:161–167.

Van Driel, C. P. & de Kerk, C. B. W. J. S. 1983. Coal upgrading by HGMS. In Proceedings on 3rd European Coal Utilization Conference, Amsterdam, 3:63.

Van Driel, C., Sikkenga, J., de Kerk, C. 1984. Coal cleaning by high gradient magnetic separation. *Journal de Physique Colloques*, 45 (C1):C1-775–C1-778.

Von Ketelhodt, L. 2009 (June 4). Sensor-based sorting new developments. Powerpoint presentation at Mintel, South Africa.

Von Ketelhodt, L. & Bergmann, C. 2010. Dual energy X-ray transmission sorting of coal. *The Journal of the Southern African Institute of Mining and Metallurgy,* 110:371-378.

WCI (World Coal Institute). 2009. The Coal Resource: A Comprehensive Overview of Coal. *World Coal Institute.* Accessed from: https://www.worldcoal.org/sites/default/files/coal_resource_overview_of_coal_report (03_06_2009).pdf. Date of access: 27 April. 2018.

WEC (World Energy Council). 2016. World Energy Resources, coal. Accessed from: https://www.worldenergy.org/data/resources/resource/coal/. Date of access: 11 May 2018.

Wei, L., Chen, Q. & Zhao, Y. 2003. Formation of Double-Density Fluidized Bed and Application in Dry Coal Beneficiation. *Coal preparation*, 23: 21–32.

Weinstein, R. & Snoby, R. J. 2007. Advances in dry jigging improves coal quality. *Mining Engineering*, 59(1):29-34.

Whittaker, G. 2005a. A Basic Introduction to Microwave Chemistry. Accessed from http://homepages.ed.ac.uk/ah05/ microwave.html. Date of access: 9 May 2018.

Wills, B. A. & Napier-Munn, T. J. 2006. Wills' mineral processing technology. Burlington: Elsevier Ltd.

Wilson, A. 2009. Super Coal. Vision Systems Design, 14 (7).

Wotruba, H. 2006. Sensor sorting technology - is the minerals industry missing a chance. XXIII International Mineral Processing Congress, Istanbul, Turkey.

Wotruba, H., Weitksemper, L., Steinberg, M. 2010. Development of a new dry density separator for fine-grained materials. XXV International Mineral Processing Congress (IMPC) Brisbane, Australia 6 - 10 September 2010.

Wotruba, H., Sampaio, C. H. Effective dry density beneficiation of fine coal using a new developed fluidized bed separator.

Xie, K. C. 2015. Coal gasification. In: Structure and reactivity of coal. Springer, Berlin, Heidelberg.

Yang, X., Zhao, Y., Luo, Z., Song, S., Duan, C., Dong, L. 2013a. Fine coal dry cleaning using a vibrated gas-fluidized bed. *Fuel Processing Technology*, 106: 338–343.

Yang, X., Zhao, Y., Luo, Z., Sing, S., Chen, Z. 2013b. Fine coal dry beneficiation using autogenous medium in a vibrated fluidized bed. International Journal of *Mineral Processing*, 125:86-91.

Yildinm, I., Yiice, H., Onal, G., Celik, M. S. 1996. Upgrading of low-rank semi-coked lignite by high intensity dry magnetic separator. In Proceeding of Changing Scopes in Mineral Processing, (Mevlut Kemal, Vedat Arslan, Aliakar, Mehmet Canbazoglu, Eds.), Balkema, Rotterdam, 137–142.

Zhang, B., Akbari, H., Yang, F., Mohanty, M. K., Hirschi, J. 2011. Performance optimization of the FGX dry separator for cleaning high-sulfur coal. *International Journal of Coal Preparation and Utilization*, 31(3-4):161-186.

Zhang, B., Zhao, Y., Zhou, C., 2015. Fine Coal Desulfurization by Magnetic Separation and the behavior of sulfur Component Response in Microwave Energy Pre-treatment. *Energy and Fuels*, 29(2):1243–1248.

Zhou, G. & Brown, D. 1988. Coal surface conditioning for electrostatic separation. *Canadian Journal of Chemical Engineering*, 66:858–863.

HORIZONTAL AIR FLOW FINE COAL WINNOWER

Quentin Campbell, Marco le Roux and Lee-Roy Morgan
North-West University
Private Bag X6001, Potchefstroom 2520, South Africa
quentin.campbell@nwu.ac.za; marco.leroux@nwu.ac.za; leemorgan777@gmail.com

ABSTRACT

Dry coal beneficiation has been investigated on several occasions and to date most of the viable options indicate positive results on coarser coal samples. Dry fine coal beneficiation is still a very new technology and most of the methods are still in the development phase.
Horizontal air winnowing has been used in the agricultural and pharmaceutical sector for decades and the method has been proven to be effective for small particle density-based separation. To prove that this method can be used to beneficiate fine coal, a study was done that included the modelling and simulation of the method and the construction of an experimental unit, consisting of a closed rectangular box utilising the horizontal flow of air.
During the development phase the simulations were used to establish the necessary parameters, such as the air flow velocity and box dimensions needed to achieve separation. The modelling was done using kinematic equations for projectile motion taking into account the drag force created by air resistance. During the experimental phase both tracers and coal particles were used to test the separation capability of the winnowing unit. The tracers and coal particles were also used to confirm the simulation results and size dependency as indicated by the simulation.
Although the partition data showed promising results the data also indicated that very narrow size ranges are needed in order to achieve good separation. A comparison of the results obtained by the simulation, tracers and coal particles indicate that the method is in fact viable. During the next phase the winnowing unit will be tested on different types of coal at more controlled and non-fluctuating airflow.

KEYWORDS: dry processing, fine coal, winnowing, simulation

1 INTRODUCTION AND BACKGROUND

For the past several years there has been a steady increase in the importance of dry fine coal beneficiation and with the decline in suitable high-quality coal (de Korte, 2015) and the ongoing decrease in usable process water, dry fine coal beneficiation has never been more important.
Fourie, et al., (1980) states that it has been customary to either discard the -6 mm coal or blend it into a steam coal product without the fine coal being washed, however, from both a financial and environmental point of view this practice is no longer feasible. In South Africa, fine coal beneficiation is currently dominated by wet separation processes such as dense medium cyclones (van Houwelingen & de Jong, 2004). The drawbacks of these methods are that they require large amounts of process water and in some cases the discarded slurry can contribute to water pollution (de Korte, 2015).
The increasing water scarcity is not only effecting South Africa but it is a worldwide concern (Kanjere, et al., 2014). In South Africa water scarcity is impacting the coal industry, the reason for this impact is that some of the larger coal deposits are

situated in arid and semi-arid regions which makes coal beneficiation difficult (Jeffrey, 2005). The alternative and possible solution to this problem is dry beneficiation and thus the research into these methods especially for fine coal is becoming crucial. Several dry beneficiation method such as the FGX and X-ray sorters have been proposed and tested. These methods use air instead of water to beneficiate coal, however most of these methods are still being developed or tested and in some cases where these methods have been implemented they are not able to compete with the water based methods in terms of either separation efficiency or throughput (de Korte, 2015).

A different approach to dry beneficiation is needed and by drawing inspiration from other sectors a new method can be developed. The agricultural sector has been using a water free separation technique for decades, this technique uses only wind and gravity to separate products from waste. This method known as winnowing separates the particles based on the difference in their density, where lighter particles will be blown away by the wind and the heavy particles will be left behind. Adapting this technique it is possible to develop a winnowing method that can be implemented for dry fine coal beneficiation. By introducing coal particles into a horizontal air stream and allowing the force to displace the particles along the horizontal axis, separation between light and heavy particles will occur creating a density separation. Researchers have already started to investigate the method in 2008 and the investigation still continues to find the best way to implement coal winnowing (Sakhre, et al., 2018). The main question that still remains, is whether winnowing is a feasible and efficient separation method for coal particles smaller than 6 mm.

2 EXPERIMENTAL EQUIPMENT AND PROCEDURE

2.1 Design of the winnowing box

A laboratory scale winnowing unit was designed from first principles. Using kinematic equations for projectile motion and factoring in the drag force created by air resistance the necessary parameters such as the dimensions of the unit, air velocity needed for separation, particle displacement and the velocity of the particles were predicted. The equations derived during the design phase were entered into both Excel and Python to help predict the parameters and simulate the particle behaviour. Using the concept as illustrated in Figure 1 several different scenarios were simulated by changing parameters such as air velocity, particle size and particle density for each case. The closed winnowing unit was designed to have a single air inlet and a single air outlet directly across one another to prevent a pressure build up within the unit. The particle inlet was placed as close to the air inlet as possible to allow the maximum force to be applied to the particle and to provide enough space in the unit for the required horizontal particle displacement. The air outlet provided the possibility for dust extraction if needed.

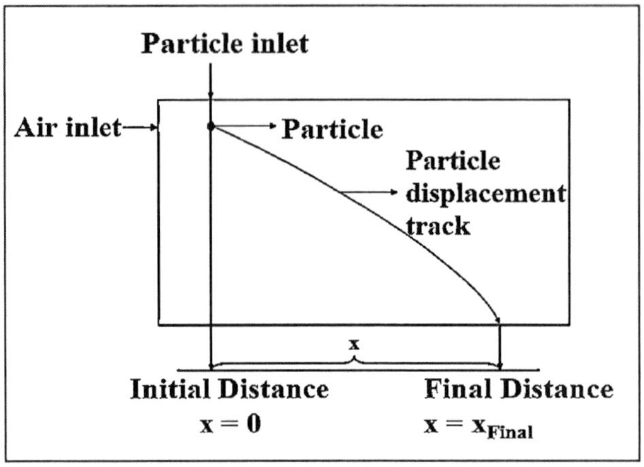

Figure 1: Design concept of the winnowing unit.

For the initial simulation particles were assumed to be either spherical or cubic with three size classes (2.0 mm, 4.0 mm and 6.0 mm). The density of the particles ranged from 1.30 – 2.20 gcm $^{-3}$ and the air velocity ranged from 12 to 26 ms $^{-1}$.
The data obtained from the simulations were entered into several prediction matrices similar to the one shown in Table 1, which was used to estimate the required velocity to achieve either size separation or density separation given certain particle properties. The initial model indicated that separation by both size and density was possible in theory. For example, a 6 mm particle with a density of 1.5 gcm $^{-3}$ can be expected to displace 21 cm when using a velocity of 20 ms $^{-1}$ while a similar 4 mm particle will be displaced 35 cm indicating that size separation is possible. For density separation, a 6 mm particle with a density of 1.5 gcm $^{-3}$ can be expected to displace 21 cm while a 6 mm particle with a density of 1.8 gcm $^{-3}$ will be displaced 16 cm.

Table 1: Horizontal displacement distances of particles for selected air flow velocities.

Particle size	Air flow velocity (ms^{-1})	Particle density (gcm^{-3})					
		1.3	1.4	1.5	1.6	1.8	2.0
6 mm	18	21	19	18	16	14	12
	20	**24**	**22**	**21**	**19**	**16**	**14**
	22	**28**	**26**	**24**	**23**	**19**	**16**
	24	34	32	30	28	24	20
4 mm	18	33	29	27	26	23	20
	20	**42**	**37**	**35**	**33**	**29**	**26**
	22	**52**	**45**	**42**	**40**	**36**	**33**
	24	62	54	51	48	43	39
2 mm	18	67	55	54	53	50	45
	20	**84**	**70**	**68**	**66**	**62**	**56**
	22	**101**	**84**	**82**	**79**	**75**	**68**
	24	121	101	98	95	89	81

Using the matrices and applying the design from Figure 1, the laboratory scale winnowing unit was constructed. The air flow was generated by a 1.6 kW domestic vacuum cleaner that was set to the blower function. The velocity could be adjusted between 19 ms $^{-1}$ and 33 ms $^{-1}$ which provided a wide range of velocities.
From the matrices it was found that the best separation (for both size and density) occurs at an airflow velocity of 21 ms $^{-1}$. This setting was used to do all further

experiments including the tracer and coal tests. The temperature and air velocity were both measured using a FLUS Professional ET-961 hot film anemometer, and these readings were taken at several different points in the winnowing unit to determine the temperature and air flow patterns within the box.

2.2 Tracer tests

To verify the computer simulated results and to mimic the separation of coal particles the laboratory scale winnowing unit was first tested using plastic cubical tracers. Three sizes were selected (2 mm, 4 mm and 6 mm), each with a density range from 1.30 to 2.00 gcm^{-3} with 0.05 increments. The tracer tests were conducted in three stages. During stage 1 the tracers were used to confirm the displacement of the particles as predicted by the models and computer based simulations. The tests confirmed that the predictions were accurate with small deviations which were possibly caused by the fluctuating airflow.

Stage 2 of the tracer tests tested the actual separation capability of the winnowing unit. This was done by installing several separation plates (as in Figure 2) at different distances along the floor of the unit. The divisions were at x = 28 cm and x = 50 cm allowing the particles to be separated into three bins. For each of the three sizes and 15 densities, nine tracers were used, resulting in a total of 405 tracers per test. Each test was repeated four times and the tracers in each of the bins were hand-sorted by size and density.

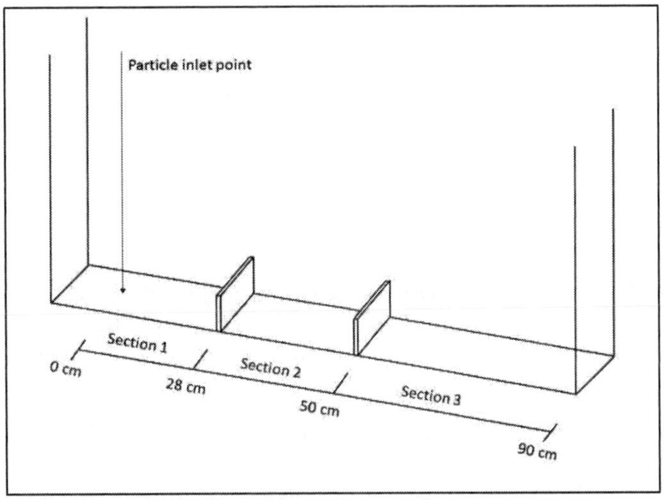

Figure 2: Sections for size separation.

Stage 3 was to test the winnower's capability to separate particles of a single size according to density. For this stage only the 6 mm tracers were used and the separation plates were set at 2 different positions (x = 20 cm and at x = 23 cm) that were predicted to give two density cut points of about 1.7 gcm^{-3} and 1.5 gcm^{-3} respectively. The data was used to construct partition curves for the two cuts, and the separation capability of the unit was examined by comparing the calculated cut points to the predicted cut points for the set positions of the separation plates.

2.3 Coal tests

Coal particles were tested in two stages. For stage 1 the initial preparation of the coal sample was done using the RhoVol BDD127SM - a separation unit owned by the De Beers Group that characterises and separates a population of particles based on various properties, like density, shape and size. To prepare the coal sample, it was first physically separated in the Rhovol by shape. It uses at sphericity definition called the compactness of a particle. For this experiment a shape class with an average compactness of 0.65 was used, meaning that the extremely flat and the mostly round particles were excluded from the experiments.

The selected shape class was then physically separated using the Rhovol by size from +3.0 mm -8.0 mm with a 0.5 mm interval. Then, the final preparation step was to separate each of the size classes obtained above by density in the range of 1.30 to 2.20 gcm $^{-3}$ with 0.1 increments. This meant that the RhoVol was used to effectively produce "coal tracer particles" of defined sizes and densities. These coal particles were introduced into the winnower at 21 ms^{-1} one-by-one and their behaviour was observed. Two size classes (+2.0 mm -4.0 mm and +4.0 mm -6.0 mm) over all densities were used and the bin into which the particle reported was recorded. The data was analysed, and the particle sizes contained in the contents of each of the two bins were determined.

Stage 2 was to confirm that the winnowing unit was capable of separating coal particles by density, at least for narrow sizes. For this test only the +6.0 mm -6.5 mm particles were used. The estimated required cut point densities were 1.40 and 1.50 gcm^{-3} with the separation plates set at x = 55 cm and x = 60 cm respectively. Again, particles were introduced one by one, and their behaviour recorded. By counting which particle at which density reported where, partition data could be determined.

3 RESULTS AND DICUSSION

3.1 Tracer size and density responses

For each size class the same number of tracers (36) was used for each density. This resulted in a linear feed densimetric curve making the separation difficulty "formidable" according to the Bird's scale. The feed, although unrealistic, was selected in this manner to assess the separation capability of the winnowing unit even though it did not resemble the washability characteristic that are normally seen for actual coal.

During the stage 2 test (tracers of all sizes and all densities at an air velocity of 21 ms^{-1}), size separation could be largely achieved in the three bins, almost independent of density (Figure 3). Bin 1 was closest to the feed side, and bin 3 the furthest. The majority of 6.0 mm tracers reported to the first bin with some overlap to the second, whilst most of the 4.0 mm and 2.0 mm tracers were collected in the 2nd and 3rd bins respectively. There was a slight overlap because of density.

This could be quantified further by examining the density distribution of particles in each bin.

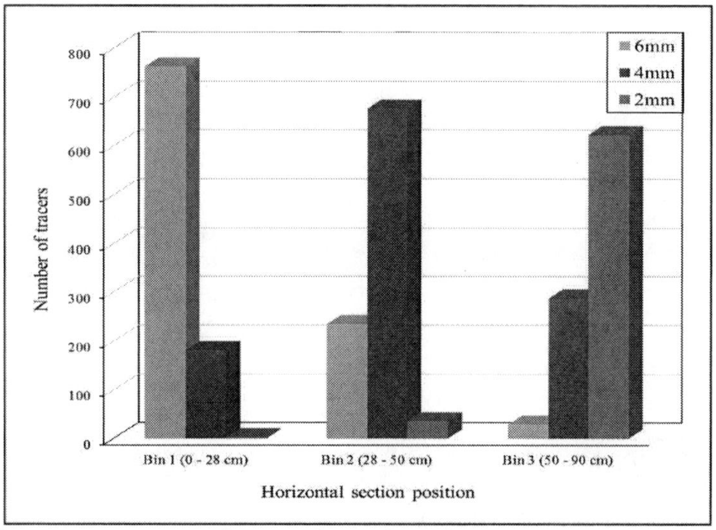

Figure 3: Size responses of complete density range tracers at 21 ms $^{-1}$ air velocity.

Figure 4 and 5 shows the density distribution in each of the bins for the 6.0 mm and 2.0 mm tracers respectively. From Figure 4 it can be seen that the 6.0 mm tracers with high densities mostly reported to bin 1, with the lower density tracers displaced towards bin 2. Figure 5 shows a similar but opposite result for the 2.0 mm tracers, where the lower density tracers were collected in bin 3, with an overlap into bin 2. The test showed good size separation between the 6.0 mm and 2.0 mm tracers, and the misplaced tracers that reported to Bin 2 could be contributed to the feed characteristics and fluctuating air velocity. This will of course result in a poor density separation sharpness. From this particular test the size sensitivity of the winnowing unit was confirmed and it is noted that a density separation, using a wide size distribution, is problematic.

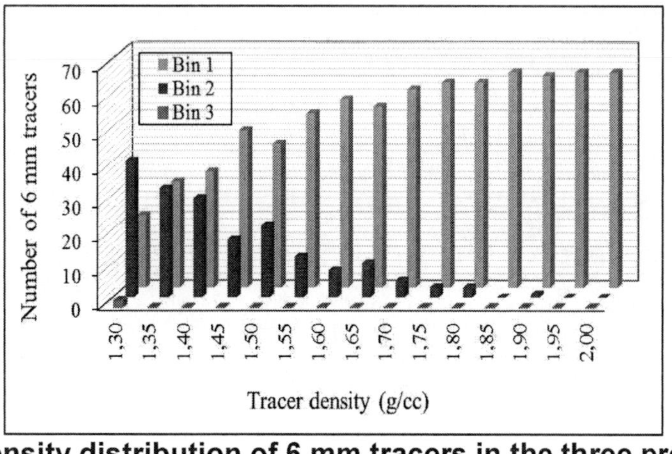

Figure 4: Density distribution of 6 mm tracers in the three product bins.

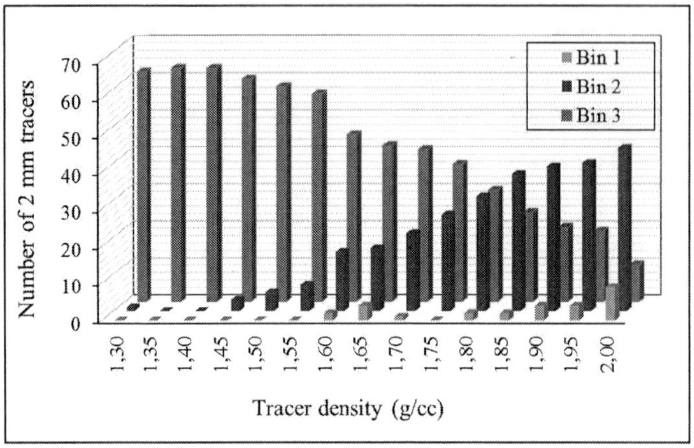

Figure 5: Density distribution of 2 mm tracers in the three product bins.

3.2 Partition data for 6 mm tracers

The third phase of tracer testing (done on only 6.0 mm tracers) aimed at determining the performance of the winnowing unit if certain cut point targets were set by the positions of the division plates. Table 2 and Figure 9 show the partition data obtained for these target cut points.

Table 2: Performance data for the separation of 6 mm tracer particles at 21 ms^{-1}.

Target cut point	1.70 gcm^{-3}	1.50 gcm^{-3}
Actual cut point	1.72 gcm^{-3}	1.58 gcm^{-3}
Plate position	20 cm	23 cm
Yield (by number of tracers)	58.7%	42.2%
E.p.m.	0.17	0.16

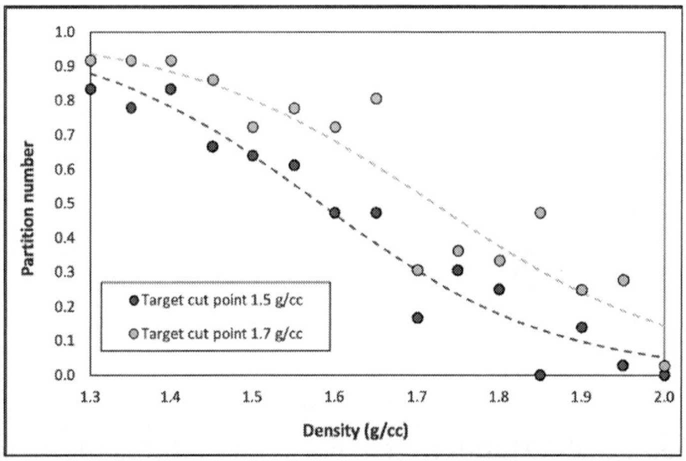

Figure 6: Partition curves for 6 mm tracers with the separating plates set at 23 cm (for 1.7 gcm^{-3}) and 20 cm (for 1.5 gcm^{-3}).

The calculated cut point (actual) and the target cut point values were similar, and the difference between the values can be contributed to the fluctuation in air flow (possibly turbulence). This can be seen by the scattered data in Figure 9 as well.

Nonetheless the phase 3 tests indicated that it was possible to predict a cut point with some degree of accuracy and with a non-fluctuating airflow (which was not tested in this phase) it is believed that the accuracy of the prediction will improve.

3.3 Feed data coal

The coal used for the second part of the experiments was a low density Moatize sample. The proximate analysis (air dried basis) showed that the coal had 1.60 % moisture with a feed ash and volatile content of 17.88 % and 32.20 % respectively. The densimetric curve for this sample as determined by a Rhovol analysis is shown in Figure 7. For this phase only one coal sample was tested and compared to the tracer results however different qualities of coal will be investigated at a later stage. The coal particles were prepared as described in section 2.3 and the separation capability of the winnowing unit was tested on a single particle response basis.

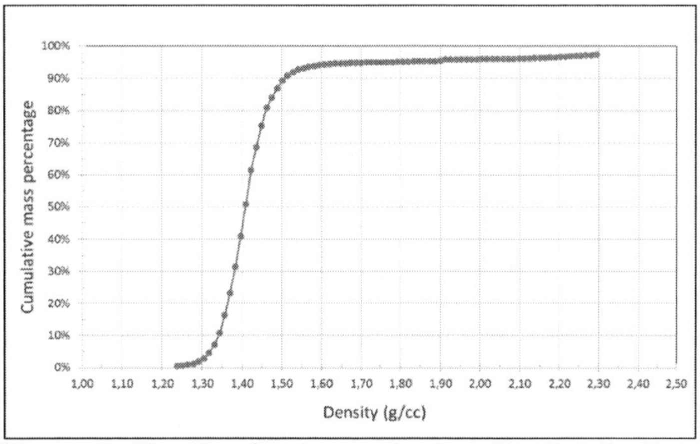

Figure 7: Moatize feed sample density distribution.

3.4 Size response

Stage 1 (for coal particles) tested the ability to separate coal particles by size irrespective of the density. For this test two size classes were chosen (+6.0 mm -6.5 mm and +4.0 mm -4.5 mm) and introduced one by one into the air stream. Visual observation was used to characterise the behaviour of each coal particle. It indicated that the separation plate distances, as predicted by the tracer tests, were inaccurate and adjustments had to be made to the separation plate positions. A single separation plate, resulting in only two receiving bins, was placed at 65 cm from the particle inlet and the test was restarted. Figure 8 shows the results of the size separation, the larger particles (+6.0 mm) were collected in the bin closest (bin 1) and the smaller (+4.0 mm) particles reported to bin 2. Some overlap occurred between the sections which was expected due to fluctuating air flow, and the wide density distribution.

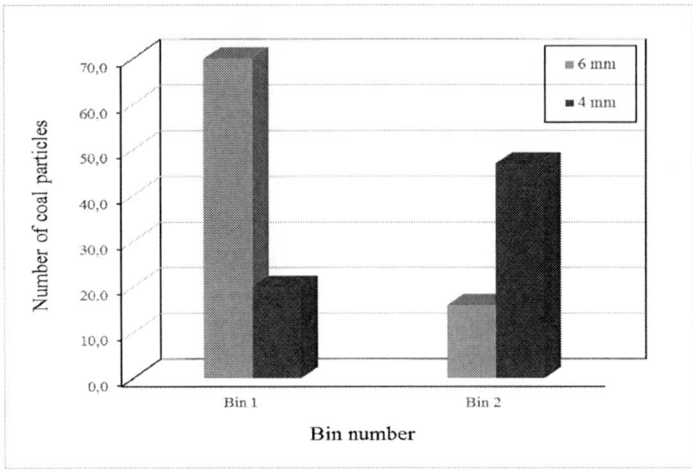

Figure 8: Size responses of complete density range coal particles at 21 ms^{-1} air velocity.

3.5 Partition data coal

Stage 2 of the coal tests was to determine the density separation capability of the winnowing unit and to test the prediction control with regards to setting certain cut points. A narrow size interval (+6.0 mm -6.5 mm) was used. Figure 9 shows the partition curves generated for both the 1.50 gcm^{-3} and 1.40 gcm^{-3} target cut points. The yield, actual cut point and sharpness of separation were calculated and are shown in Table 3. The data shows very accurate predictions for the cut points at the pre-set plate positions. Again, it is thought that the accuracy can be improved by somehow stabilizing the air flow in the winnowing unit.

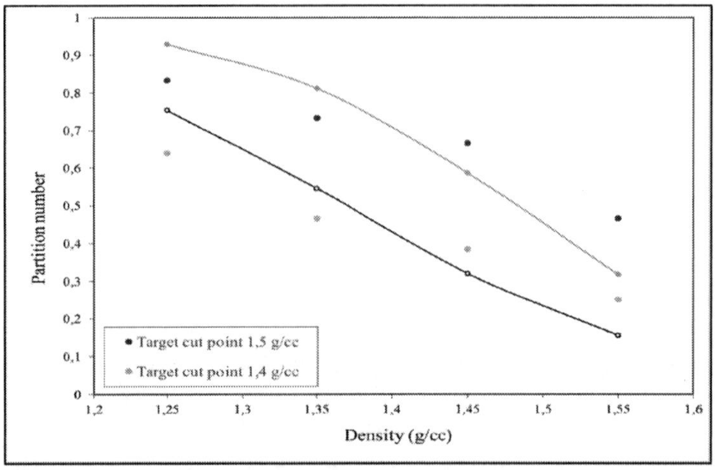

Figure 9: Partition curves for +6.0 mm -6.5 mm coal particles with the separating plates set at 55 cm (for 1.5 gcm^{-3}) and 55 cm (for 1.5 gcm^{-3})

Table 3: Performance data for the separation of +6.0 mm -6.5 mm coal particles at 21 ms^{-1}.

Target cut point	1.40 gcm^{-3}	1.50 gcm^{-3}
Actual cut point	1.37 gcm^{-3}	1.48 gcm^{-3}
Plate position	60 cm	55 cm
Yield (by number of particles)	55.56%	71.74%
E.p.m.	0.117	0.099

4 CONCLUSIONS

The results from both the tracer and coal tests prove that dry fine coal separation is feasible when using the method of winnowing. The results indicate that a narrow size distribution is indeed needed to achieve a sufficiently sharp separation, as is seen by the size sensitivity of both the tracers and the coal particle results. With this narrow distribution good separation can be achieved with coal e.p.m. values in the order of 0.117 or 0.099, however to improve on this separation better control of the air flow must be implemented. The tests also confirmed that it was indeed possible to predict the cut points with a fair degree of accuracy using the developed simple models and that a given sample can be separated by using a separation plate to create "clean coal" and a "discard" bins.
Future testing of the winnowing unit will include testing and comparing the separation efficiency of the winnowing unit on different qualities of coal. The influence of non-fluctuating airflow and particle shape on the separation efficiency of the winnowing unit will also be further investigated.
The results show that a two-stage separation is imperative in order to compensate for the size-density independency. The first stage will be a size separation from where the respective size classes can be fed to a number of parallel secondary units where the density separation will occur.

5 ACKNOWLEDGEMENTS

The authors wish to acknowledge the financial and technical contribution made by The Coaltech Research Association NPC Ltd., and the Centre of Excellence in Carbon Based Fuels, North-West University, South Africa.
This work is based on the research supported in part by the National Research Foundation of South Africa (Coal Research Chair Grant Numbers: 86880; and UID85643, UID85632). Any opinions, findings and conclusions or recommendations expressed in any publication generated by the NRF supported research is that of the author(s) alone, and that the NRF accepts no liability whatsoever in this regard.

6 REFERENCES

De Korte, G., 2015. Processing low-grade coal to produce. The Journal of The Southern African Institute of Mining and Metallurgy, Volume 115, pp. 569-572.

Fourie, P. J. F., Van der Walt, P. J. & Falcon, L. M., 1980. The beneficiation of fine coal by dense-medium cyclone. The Journal of The South African Institute of Mining and Metallurgy, pp. 357-361.

Jeffrey, L., 2005. Characterization of the coal resources of South Africa. The Journal of The South African Institute of Mining and Metallurgy, pp. 95-102.

Kanjere, M., Thaba, K. & Lekoana, M., 2014. Water Shortage Management at Letaba Water Catchment Area in Limpop Province, of South Africa. Mediterranean Journal of social Sciences, 5(27), pp. 1356-1360.

Sakhre, D. et al., 2018. Coal Winnowing: An Innovative Technique for Dry Coal Beneficiation. CPSI, 10(28), pp. 3-13.

Van Houwelingen, J. & de Jong, T., 2004. Dry cleaning of coal: review, fundamentals. Geologica Belgica, 7(3-4), pp. 335-343.

WEIR
Minerals
INNOVATIVE SLURRY PUMPING
SOLUTIONS
OVERCOME CHALLENGES
IN
COAL PROCESSES
SA COAL PROCESSING SOCIETY
GRACELANDS SECUNDA
WEDNESDAY 21 AUG 2019 SESSION 7
MURRAY JOOSTE

Abstract: Challenges encountered in DM Cyclone and Correct Medium Feed.

Two coal application case studies were evaluated.

Firstly a wet gland, expeller-less DM cyclone feed application, where throughput performance was significantly curtailed by the parts passing incapability of the impeller.

Secondly a dry gland expeller correct medium feed application, where gland sealing performance was curtailed by inadequate expeller hold-back characteristics.

Unlined metal/metal horizontal centrifugal slurry pumps with particle passing capability > 150 mm, utilising exceptional wear resistant materials were installed.

Closed three radial vane metal impellers, with a specific speed < 30 were utilised, incorporating turbulence reducing reverse layover vane-let designs.

Large impeller diameters maximised impeller aspect ratios.

Large expeller diameters maximised expeller / impeller diameter ratios provided the maximum expeller hold back capabilities.

Replacement modern wet ends retrofitted directly onto the existing mechanical ends.

Abbreviations used:

1, DMC = Dense Medium Cyclone
2. DM = Dense Medium
3. BEP = Best Efficiency Point
4. Cm = Concentration / Mass (-)
5. RD = Relative Density (-)
6. TDH = Total Developed Head (m)
7. NPSH = Net Positive Suction Head (-)

Application by Cm (-), RD slurry (-), h total developed head (m).

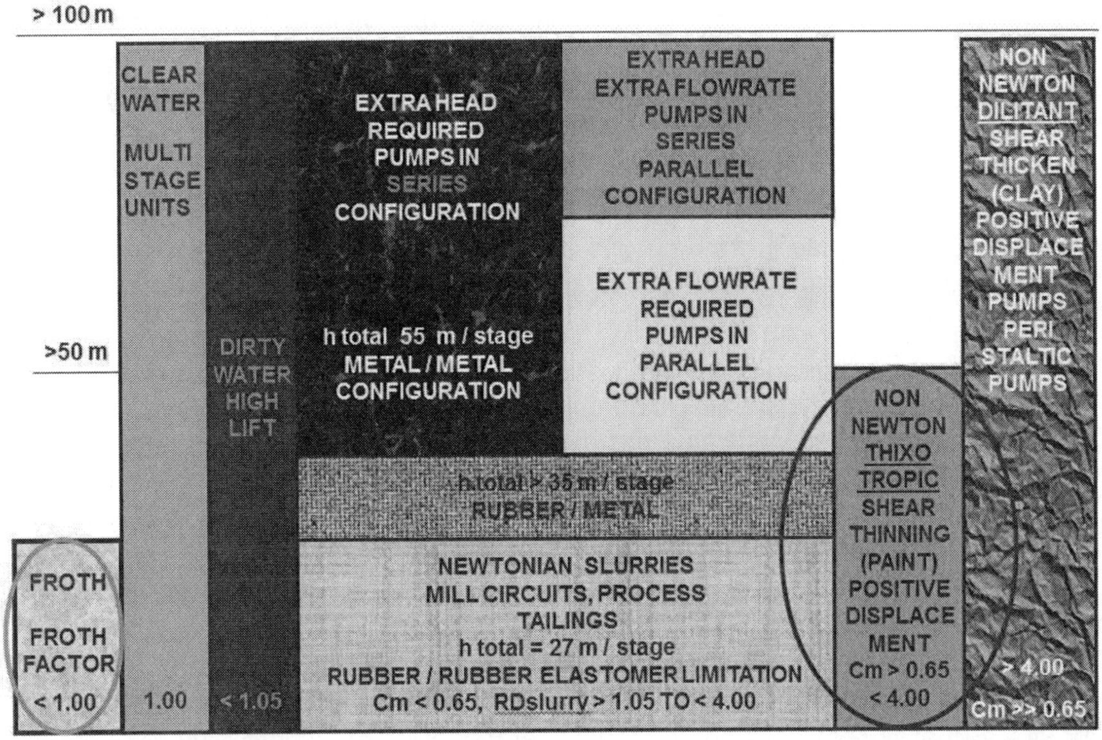

Customer Site System Curves interaction with OEM Pump Curves.

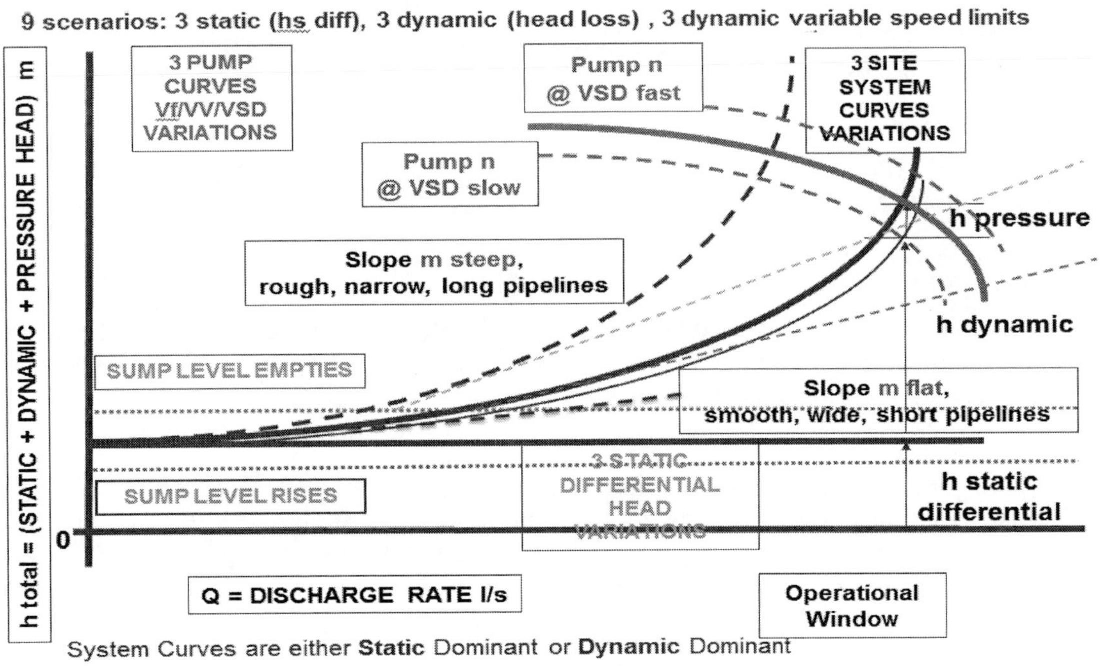

Slurry Pump Selection Case Study Criteria

Evaluation of the process metallurgical mass balance (3 phase configuration) is imperative. The RD dry solids and the concentration by mass Cm, determines the RD slurry. This provides the effective volumetric slurry yield of the pumping circuit in volume rate per mass rate (m^3/h) / (tonnes/h).Utilising the specified dry solids mass feed rate the resultant volumetric slurry flow rate can be established.

The heads up display of the RD slurry, Cm and the available total generated heads of the range of centrifugal slurry pumps are complimented with the addition of special purpose pumps applied in froth circuits and shear thinning (thixotropic) applications. Shear thickening (dilatant) applications are included for comprehensive information

The unique interaction between the customer site system curves and the OEM pump curves requires specific scrutiny. The customer system curve accumulates the static differential head component plus the dynamic components (friction head, pressure head, and velocity head). The OEM pump curve represents the total developed head of the pump at the specified discharge rate. This unique node is represented by the rotational speed of the pump, providing additional vital information regarding efficiency of the pump, power absorbed, NPSH required, impeller tip speed and operational point relative to the BEP.

Specified factors of safety fs are used in order to select the most appropriate electric motors at the highest possible full load operational points incorporation the optimal motor efficiencies and power factors.

The DM cyclone feed parts passing challenge as well as the correct medium expeller holdback challenge fall into the severe service class 4 category, with aspect ratios 3.3 > [D2 / D1] > 2.7

The specific speed Ns <30 with pump discharge diameters 457 mm > did > 304 mm. Radial vane types considered varied from closed 5 vane, closed 4 vane large particles passing, to open 3 vane configurations.

DM Cyclone Parts Passing Challenge Evaluation Matrices.

Two matrices are presented comparing all the vital parameters. Wet gland expeller-less pump circuits were monitored at two relative slurry densities of RD slurry = 1.79 and RD slurry = 1.88

Wet gland expeller-less pump circuits compared the behaviour between existing closed 4 vane, large particle passing, 965 mm impellers rotating at 425 r/min and the latest closed 3 vane, 1041 mm impellers rotating at 401 r/min. Phase amps of both the electric motors were measured as comparison, which verified that the closed 4 vane large particle predicted e-Curve pump efficiency values were overstated and the pump kW absorbed values were understated, primarily due to the severe particle blockages experienced within the impeller passages.

The predicted operational duty point position relative to the best efficiency point (BEP) is also over optimistic. Site experience concluded that the closed 3 vane, 1041 mm impeller option offered the correct solution.

The specified factor of safety used in the calculations was fs = 1.45, with Er = 0.90 and Hr = 0.90 The existing 450 kW motor was retained utilising all the existing drives, pulleys, cabling and VSD equipment, resulting in a considerable saving in component costs, rectification costs and commissioning costs.

DM Cyclone Feed Pump Performance Challenged by Parts Passing.

Parameters Of Pumps @ 2 DUTIES	Parts Passing max (mm)	# Vanes (-)	Impeller Diam (mm)	kW abs	Factor of Safety fs (-)	Aspect Ratio (D2/D1) (-)	Pump Speed (r/min)	Pump Efficiency (%)	Position wrt BEP (%)
14/12 AH LP – 1.79 WET GLAND	150.00	4	965.00	***258.82*** *e-Curve Predicted*	1.45	2.71	412.90	***68.86*** *e-Curve Predicted*	***94.94*** *e-Curve Predicted*
14/12 AH LP – 1.88 WET GLAND	150.00	4	965.00	***276.06*** *e-Curve Predicted* *300.10*	1.45	2.71	425.59	***68.16*** *e-Curve Predicted* *62.70*	***83.03*** *e-Curve Predicted*
300 MDC 1.79 WET GLAND	203.00	3	1041.00	270.16	1.45	2.92	387.08	65.97	79.30
300 MDC 1.88 WET GLAND	203.00	3	1041.00	291.45	1.45	2.92	401.65	64.56	69.05

Warman 300 MDC wet gland, expeller-less pump installed as replacement

Comparison of DM Cyclone Feed Pump Characteristics.

Para meters of Pumps	Inlet Diameter (mm)	Discharge Diameter (mm)	Casing Pressure (kPa)	RD slurry (-)	Discharge Flowrate (l/s)	Total Head (m)	Impeller Tipspeed (m/s)	NPSH Reqd (m)	Comparative Capability Comments
14/12 AH LP - 1.79 WET GLAND	356.00	305.00	2100.00	1.79	439.40	23.10	20.87	3.90	Inadequate Parts Passing Capability
14/12 AH LP - 1.88 WET GLAND	356.00	305.00	2100.00	1.88	397.00	25.70	21.51	3.76	Inadequate Parts Passing Capability
300 MDC 1.79 WET GLAND	356.00	305.00	900.00	1.79	439.40	23.10	21.11	3.38	**Large Parts Passing Capability**
300 MDC 1.88 WET GLAND	356.00	305.00	900.00	1.88	397.00	25.70	**21.90**	3.22	**Large Parts Passing Capability**

WARMAN 300 MDC CLOSED 3 VANE IMPELLER, WET GLAND EXPELLER-LESS DM CYCLONE FEED APPLICATION

Weir Minerals Africa (Pty) Ltd

PO Box 70, Isando 1600
31 Isando Road, Isando
Johannesburg, South Africa

Tel: +27 11 929 2800
Fax: +27 11 929 2950
www.weirminerals.com

Customer Details 07 June 2019
Project:
No:
Customer:
Application: HYDRO CYCLONE FEED SLURRY PUMP
Pump Item Number: 300 MDC (Pump no. 0)

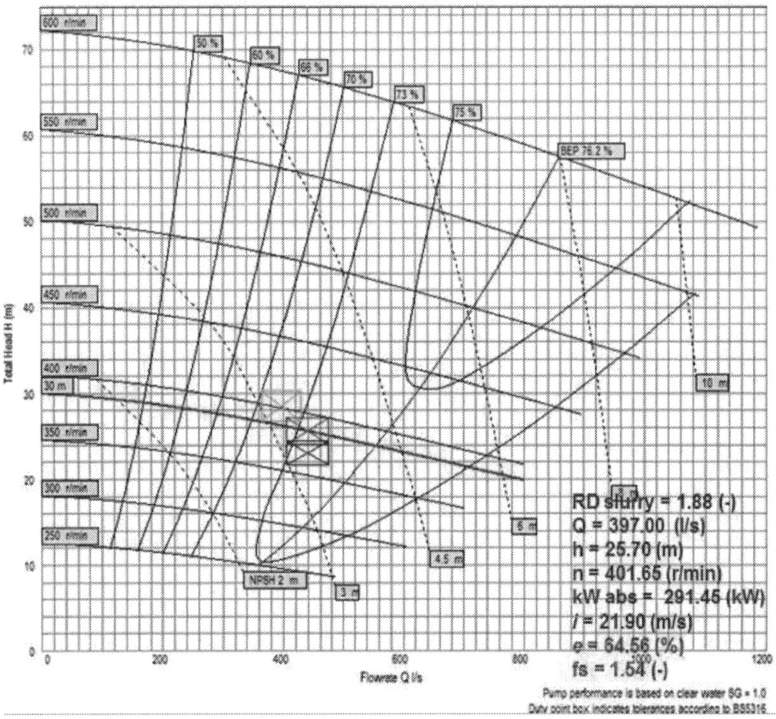

Warman MDC

Notes

Dense Media Feed pump
Efficiency Predicted
Add 5% to power for expeller pump

Duty Point	Primary	Secondary	
Specific Gravity SG:	1.79	1.88	
Flowrate Q:	439.40	397.00	l/s
Total Flowrate Q:	439.40	397.00	l/s
Head H:	23.10	25.70	m
Total Head (1 stage):	23.10	25.70	m
Pumps in Series:	1	1	
Pumps in Parallel:	1	1	
P Absorbed:	270.16	291.45	kW
P Required:	391.73	422.60	kW
P Installed:	400.00	450.00	kW
Specified Safety Factor:	1.45	1.45	
Actual Safety Factor:	1.48	1.54	
Pump Speed n:	387.08	401.65	r/min
Efficiency:	65.97	64.56	%
Impeller Tip Speed:	21.11	21.90	m/s
Position From BEP:	79.30	69.05	%
Minimum NPSH:	3.38	3.22	m
Gland Service:	0.00	0.00	kPa

Derating and Speed

Head Ratio:	0.90
Efficiency Ratio:	0.90
Froth Factor:	1.00

Pump Information

Suction Diameter:	356	mm
Discharge Diameter:	305	mm
Parts Passing:	203	mm
Casing Pressure Rating:	900	kPa
Impeller Diameter:	1041	mm
Number of Vanes:	3	

Curve Source

NT0043 rev 0

Pump

300 MDC

WARMAN 300 MDC 3 vane wet end installed onto the existing mechanical end.

A: DM Cyclone Feed Pump Challenged by Parts Passing Capability.

The DM cyclone feed became erratic, roping occurred, with a significant deterioration in the DM cyclone performance, operational pressure and wear life. Pump wet end component lifespan of 6 months was achieved.

Remedial action was undertaken by replacing the entire existing wet end components with the latest generation pump wet end components, with a large 1041 mm impeller, closed 3 vanes, parts passing capability < = 230 mm.

All replacement components were retro fitted directly onto the existing mechanical end and bearing barrel assembly without changes to the plinth, the overhead self-adjustable e-base, the wedge belt pulleys and drives, utilising the existing 450 kW electric motor, existing variable speed drive, existing cabling, switchgear and motor control centre.

The new pump, identical in size, operated at 69% wrt. BEP, less than the predicted 83% for the replaced pump. The additional shaft radial load deflection of the new horizontal cantilever centrifugal pump, compared with the replaced pump was not experienced in the field, confirming that the efficiency curve predictions were overstated in the case of the original replaced pump due to the persistent impeller choking effect caused by the oversized particle feed > 150 mm.

New pump efficiency of 65% (versus *calculated* 62%), kW abs of 291 kW (versus *calculated* 300 kW) were calculated from the measured phase currents in both applications, confirming improved efficiency and reduced electricity kW.h expenses.

The DM cyclone slurry feed rate and pressure stabilised, resulting in significant improvement in the DM cyclone cut and overall performance, eliminating persistent roping. Lifespan of the pump components doubled from 6 months to 12 months. Process system consistency and pumping system stability were achieved.

A significant reduction in maintenance, downtime, replacement component costs were experienced considering that the mechanical end and bearing barrel assembly replacement intervals were substantially extended by a 2:1 ratio.

Warman 300 MDC wet gland expeller-less pump was successfully installed as replacement

Correct Medium Feed Pump Performance Challenged by Expeller Action.

Parameters Of Pumps	Parts Passing (mm)	# Vanes (-)	Impeller Diam (mm)	Expeller Diam (mm)	Impeller Expeller Ratio (-)	Aspect Ratio D2/D1 (-)	Pump Speed (r/min)	Pump Efficiency (%)	Position wrt BEP (%)
14/12 AH EXPELLER DRY GLAND	90.00	5	965.00	333.00	2.90	2.71	382.00	60.00	48.00
14/12 AH UNDERCUT IMPELLER EXPELLER DRY GLAND	90.00	5	772.00 UNDERCUT IMPELLER	333.00	2.32	2.16	477.50	67.91	100.00
250 MDC EXPELLER DRY GLAND	176.00	3	890.00	307.00	2.90	2.91	301.65	72.27	91.14

Warman 250 MDC dry gland, expeller pump was successfully installed

Correct Medium Feed Pump Characteristics Comparison

Parameters Of Pumps	Inlet Diameter (mm)	Discharge Diameter (mm)	Casing Pressure (kPa)	RD slurry (-)	Discharge Flowrate (l/s)	Total Head (m)	Impeller Tip Speed (m/s)	NPSH Reqd (m)	Comparative Capability Comments
14/12 AH EXPELLER DRY GLAND	356.00	305.00	2100.00	1.60	**273.05**	9.74	19.30	**1.41**	Pump Oversized Expeller Ineffective
14/12 AH UNDERCUT IMPELLER EXPELLER DRY GLAND	356.00	305.00	2100.00	1.60	**273.05**	9.74	19.30		Undercut Impeller Expeller Effective
250 MDC EXPELLER DRY GLAND	**305.00**	**254.00**	900.00	1.60	**273.05**	9.74	**14.06**	**3.25**	Expeller Effective Correct Pump Size

Warman 250 MDC dry gland expeller pump was successfully installed

WARMAN 250 MDC CLOSED 3 VANE IMPELLER, DRY GLAND EXPELLER PUMP IN CORRECT MEDIUM APPLICATION

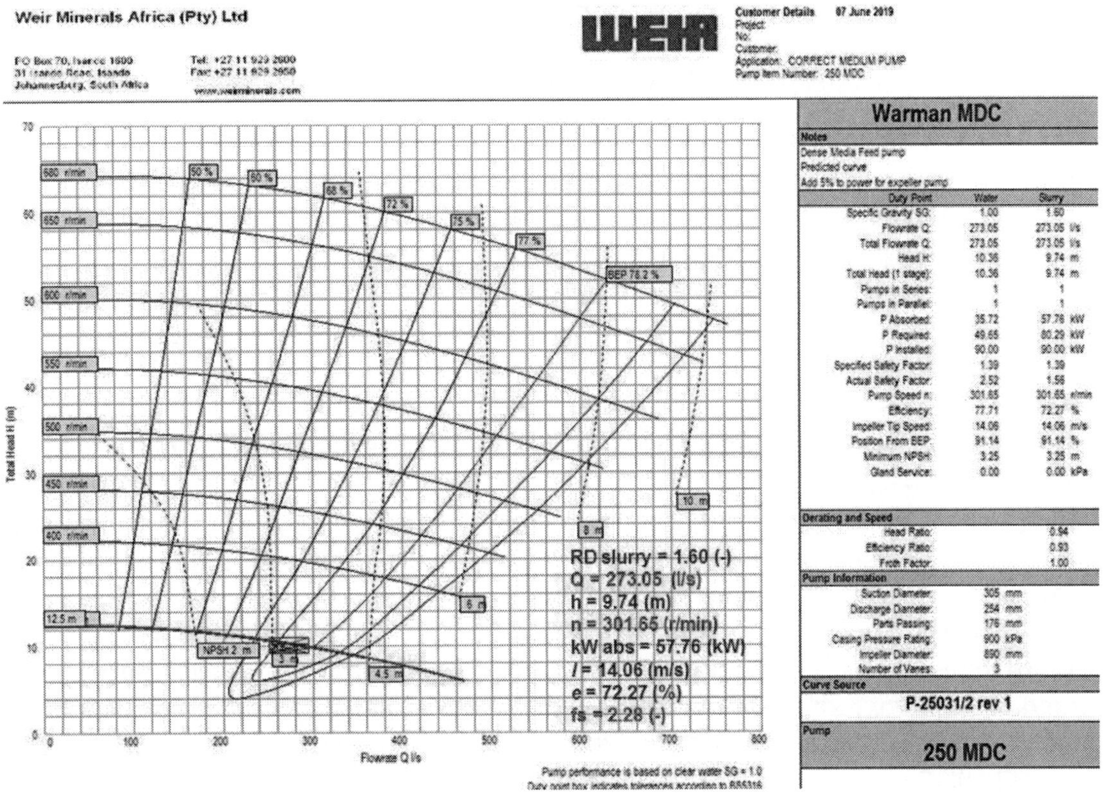

B: Correct Medium Feed Pump Challenged by Expeller Sealing Capability.

Erratic static flooded suction head conditions prevalent at the correct medium suction tank resulted in extensive spillage and ineffective pumping conditions. This was attributed to significant variations in the existing site process parameters, exacerbated by the inability of the dry gland expeller to generate adequate holdback characteristics against the continuous altering static suction head. This was further influenced by blockages in the launder between the wing tank overflow as well as erratic feed from the overhead raw coal screens via the return tank.

Expeller holdback characteristics depend on the required discharge rate of the pump and the rotational speed of the pump expeller. Remedial action was undertaken by reducing the impeller/expeller ratio from 2.90 by undercutting the impeller diameter by 20%, thus retaining 80% of the full size impeller. The undercut impeller/expeller ratio was reduced to 2.32 (versus 2.90) for the full size. The pump speed was increased to 477.50 r/min (versus 382.00 r/min) to achieve the desired duty point. The erratic flooded suction head conditions as well as the unacceptable excessive spillage conditions were curtailed at both the correct medium sump and the distribution box feeding the wing tank.

Next replacing only the wet end components with the latest generation pump wet end, where the replacement components were retro fitted directly onto the existing mechanical end and bearing barrel assembly with changes to the plinth, the overhead motor base, changes to the wedge belt drives and pulleys, utilising the existing 132kW electric motor, existing direct on line switchgear and existing soft starting facilities.

The replacement pump is one size smaller, thus operating at 91% of BEP (versus 48% of BEP).This represents a significant reduction in the radial load deflection experienced by a horizontal centrifugal pump (comprising of a cantilever impeller and shaft arrangement) subjected to variations in duty points, thus significantly reducing mechanical end bearing failures.

The efficiency of the pump increased to 72% (versus 60%), with the associate reduction in electrical energy consumption in (kW.h) as well as total electrical costs. A significant reduction in maintenance costs, downtime costs and replacement component costs were experienced considering that mechanical end and bearing barrel replacement intervals were extended from 4 months to 18 months (Ratio 4.50)

Feed and pressure stabilised as a result of the controlled slurry levels achieved in the distribution box. Significant elimination of spillage was achieved in both areas.

Aspect ratios improvements resulted (new pump installed versus existing pump with the undercut impeller), A1 = (2.91 / 2.16) = 1.347. Impeller tip speed improvement (new pump installed versus existing pump with the undercut impeller), T2 = (14.06 / 19.30) = 0.728, relating directly to improved impeller lifetime.

Referring to the diagram of the shaft radial deflection created by operating the pump at a duty point far removed from the best efficiency point indicates the significant increase in the pump shaft radial deflection that approached 7 per unit at the closed valve position.

Shaft Radial Deflection Created by Q duty point relative to Q BEP. (per unit)

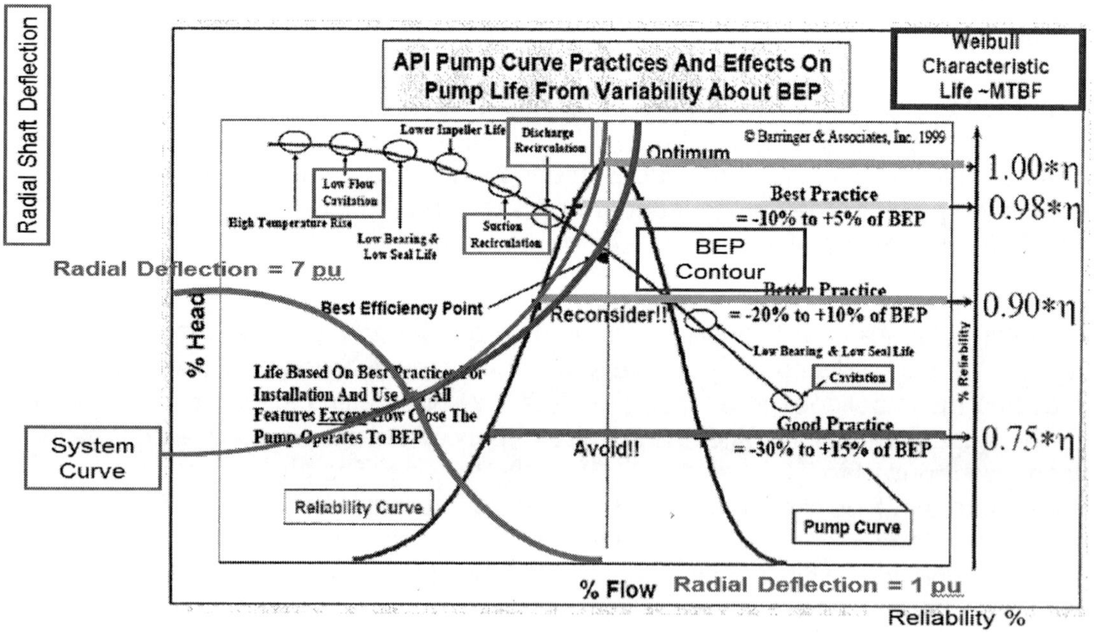

Source Material References

1. Slurry Pump Manual, 2nd ed. Salt Lake City, Envirotech, 1999.
2. Slurry Pumping Manual, Weir Minerals Australia Ltd, 2009
3. Slurry Pumping Manual, Weir Minerals Australia Ltd, 2007

Acknowledgements

4. Weir Minerals acknowledge all participant colleagues for their technical input, assistance, critique and advice in the preparation of this paper
5. Special recognition is given to our Middelburg Field Service Manager, Martin Smith, who was personally involved throughout commissioning, enabling practical site input and experience to be incorporated

Investigating the relationship of South African coal physical properties by using analytical data from the Rhovol particle analyser.

N. Tshiongo-Makgwe

Corresponding author. Tel.: +27(0) 11 559 6168 cell +27(0) 84 714 2162 E-mail address: ntshiongo@uj.ac.za

University of Johannesburg, Faculty of Engineering and the Built Environment, School of Mining, Metallurgy and Chemical Engineering, Department of Metallurgy, Minerals Processing and Technology Research Centre, South Africa, POBOX 526, Wits 2050, South Africa.

ABSTRACT

The South African coal industry has shown great and innovative investigative and analytical techniques over the past years, and the latter keeps rising. Among the new improvements is the use of RhoVol particle size analysis which provides informative data of coal sample per particle. The data generated by the RhoVol analyser is useful in understanding the various properties of coal that may not be obtained from any other single analyses. Such information includes density, volume, mass, shape and orientation of each particle. These parameters are only obtainable from multiple analyses and thus make the use of this equipment an important and innovative process that may be used to model and predict the behaviour of coal during the various handling and processing stages. The current work investigated the effectiveness of the equipment against the traditional float and sink analyses and also the correlation and relationship of the different parameters. The efficiency of the RhoVol particle analyser was found to be equivalent to float and sink analyses, though the float and sink analyses was found lacking as it can only separate materials according to one physical parameter (density). The data from RhoVol was modelled to find that the different parameters have an impact on the density.

Keywords: Coal, Physical properties, correlation assessment, RhoVol particle analyser

2. INTRODUCTION

Coal is and will continue to be the cheapest and most abundant sorce of effective power generation in the world. The South African power grid is no different as it depends highly of coal for the generation of electricity. About 60-80% of coal power plant operational cost is in coal mining and preparation (Papanicolaou et al. 2004), which includes the crushing and sizing of coal particles meaning that the particles parameters should be studied. Such inexpensive energy from coal leads to sustainable economic growth and development although it would present technological and environmental challenges (Papanicolaou et al. 2004). Coal is primarily produced from water based beneficiation methods that involves the float and sink analyses where the coal particles are separated according to density. There are however other properties that affects the efficiency of separation as these properties will affect the usage process of coal, where coal properties such as calorific value, ash content and composition, moisture content, volatile matter, fixed carbon other properties that affects the abrasion of mechanical parts of the boilers (Papanicolaou et al. 2004). There is a relationship between the density, abrasion resistance and durability of coal properties (Mishra et al. 2010)

Nevertheless, the research never focused on the physical properties of coal that would result in high abrasive properties like particle size and shape. Mineralogical compositions and structural properties are closely related to the density and strength of rocks. Density, strength and slake durability resistance decreases with increased clay content of rocks, and the effects of overall shape and surface roughness of the testing material on the outcome of slake durability indices (SDI) of rocks have been investigated (Kolay and Kayabali 2006) using the different shape and angular rocks to determing the analytical link between surface roughness and the slake durability indices. The effect of particle size on other properties of coal were investigated, where they found that the fixed carbon content decreased with the reduction in particle size. The relationship with particle size, mass and yields were found to be directly proportional (Yu et al. 2005). Slake durability depends on many factors like rock type, grain size, mineralogical composition and structural properties (Kolay and Kayabali 2006). Aggregate shape and surface roughness depends on two things being; the attrition amoung particles and the inner surface of the drum, which increased the abrasive stresses and results in disintergration of sample particles, and the higher surface area created by the rough surface aggregate.

Coal like any rock has many related properties that are important in defining it. Of those properties density and strength together with their relationship with surface roughness and shape are fundamental properties that can be used to describe the particles. Coal beneficiation is the primary technology of cleaning with wet beneficiation as the predominant method applied. Due to the worldwide shortage in water, there is a need to explore dry beneficiation technologies that can replace the wet methods. Dry coal beneficiation is not as efficient as wet beneficiation techniques as it has lower separation efficiency due to the motion of particles during separation. Research have been conducted on the motion of particles of different shapes (Zhao et al. 2015), and this includes studies of particle size, shape, surface and pore structure.

The currect work focused on the investigation of using a Rho-Vol particles analyser as an effective dry method separation of coal. This is done by investigating the different separation parameters that affects the separation effeciency and how they are related and affects each other.

3. EXPERIMENTAL

For Rho-Vol, a coal sample was sized and screened to remove the finer particle sized coal. The particles where loaded into a feeder from where they are fed (one particle per time). The particles are then divided automatically into 10 different bins that accept the particles of 10 pre-determined parameters where in this case density ranges was used to be compared with the traditional float and sink analyses performed in the coal industry. The equipment is fitted with various cameras that capture and record various parameters of each coal particle and the results were automatically recorded on the computer connected to the equipment. The same coal was subjected to float and sink analyses using the same densities as per Rho-Vol, and the mass percentage of the samples floating and sinking were recorded.

4. RESULTS

The current part of the paper shows the results obtained from the analyses done on the coal sample. The summary of the overall statistics of the bin density range, mass, volume, compactness, elongation and flatness of the individual particles is presented in the table below. Out of all the coal particles variables that were measures and analysed, there were no data missing on each variable per density range.

As for the float and sink analyses performed, the mass percentage and particle size analyses were the only parameters analysed for. When comparing the two methods, this was the major difference as the different parameters may affect the coal properties during operations. And understanding of such parameters would provide useful insights on the operation and also various usage processes.

4.1 The effect of sample density on the mass and volume of the particles.

When wet coal beneficiation is performed, coal particles are separated according to density. The density of any material is however affected by the mass and the volume of that material. The Rho-Vol material analyser is able to measure the mass and volume of each particle as they are transferred through the chutes to the colllection bins.

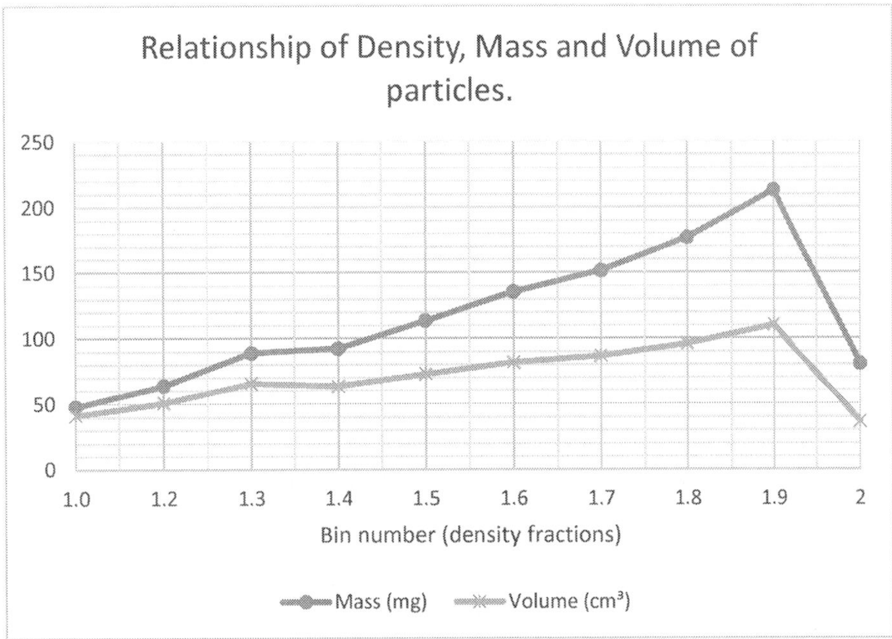

Figure 1: Results of particle Density, Mass and Volume as analyzed in the different bins.

The figure above shows the results from Rho-Vol classification of different bin samples that shows that there is an increase in the mass and the volume as the density ranges increases. The mass and volume of particles have the same trend of increase and a slight decrease noted at the end of the density range. This relationship indicated that when the volume of the materials is increasing, there is an increase in the mass and that is directed to the density, as this is a proportion of the particle mass and volume.

4.2 Relationship of density with particle dimension.

Apart from the mass, volume and density relationship, other particle parameters can be investigated to explain the efficiency of separation as they may be attributed to the performance of coal in operation. When conducting float and sink analyses, only particle size distribution is investigated and analysed for and this only considers the minimum square diameter as per the sieve design and oriented, but such leaves out other parameters like minimum size diameter that affects the ability of a particle to flow and move within the processing.

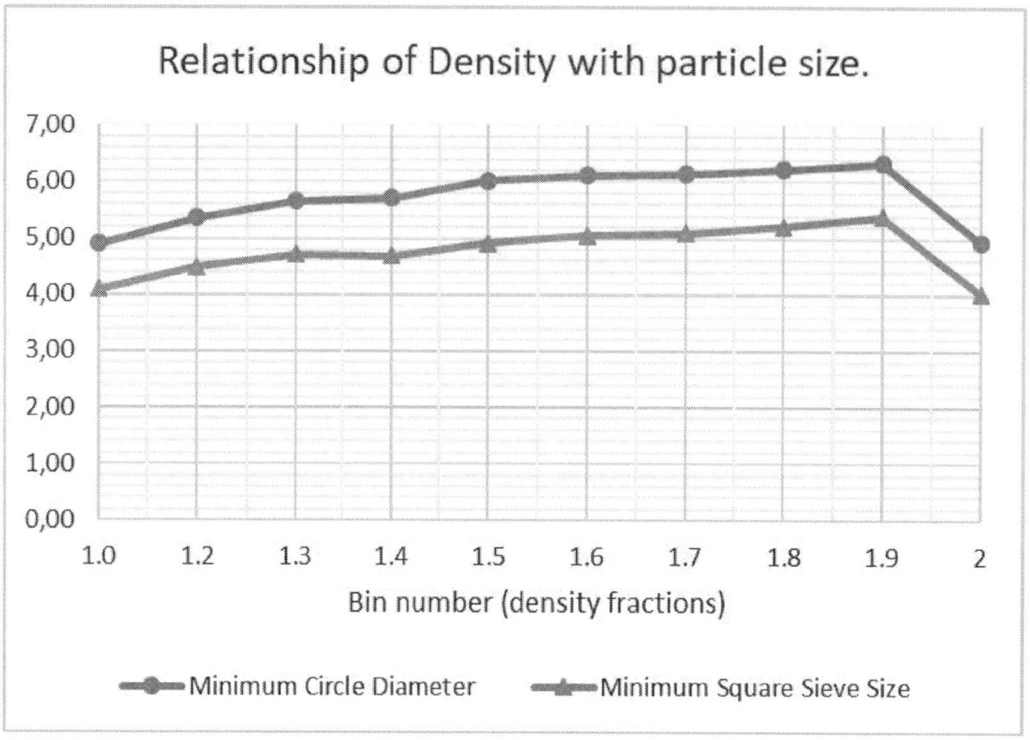

Figure 2: Results of particle shape orientation.

The minimum circle diameter and square sieve size were analysed for each bin (increase density), at it was found to be increasing with the density increase. This behaviour is similar to that noted for the mass and volume relationship. The importance of this is that it considers the square sieve size and the circle diameter where analyses of particle size using the screening process can only focus on one analyses per time and the results can not be intrepreted on the other. With this it can be concluded that the Rho-Vol has an advantage as it can also analyse the actual size of the individual particles in one analyses, whereas analyses for these particles dimensions require traditional multi analyses tests. From the above graph, the size of the particles increases with the increase in density which supports the increase in mass and volume as indicated by their relationship with density of the particles.

Sample with same volume but different geometrical shapes can be compared by the following formula (Kolay and Kayabali 2006);

$$V_{sphere} = V_{cube} = V_{tri-pism} \leftrightarrow 4.19r^3 = b^3 = 0.43a^3 \qquad 1$$

Figure 3: A general view of angular (a), sub-angular (b), and rounded aggregates (c) used for the test (Kolay and Kayabali 2006).

Samples may fall within the same density range but have different orientations. It is important to understand such orientations as they affect the overall separation efficiency and classification of particles. From the research done (Kolay and Kayabali 2006), various formulation can be used to calculate the different dimensions of the particles. A simple and multiple regression analyses can be used to determine the relationship between the dependent and the independent variables. Particles size characterization parameters include the three axis diameter; shape coefficient, shape index and specific surface area (Zhao et al. 2015), which can be described as the height, length and breadth/width of the particle. From these, the index Zingg was defined to understand the particles of different shape and the motion equation of particles in the fluidised bed was derived as the ratio N/M (Zhao et al. 2015);

$$Fz = {}^{N}/_{M} = {}^{1h}/_{b^2} \qquad 2$$

Where N is the ductility of the particle and M is the flatness of the particle.

However, with the Rho-Vol where all measurements are done automatically by the camera-aided computer. And the results of the different particles orientations are indicated in the next graph.

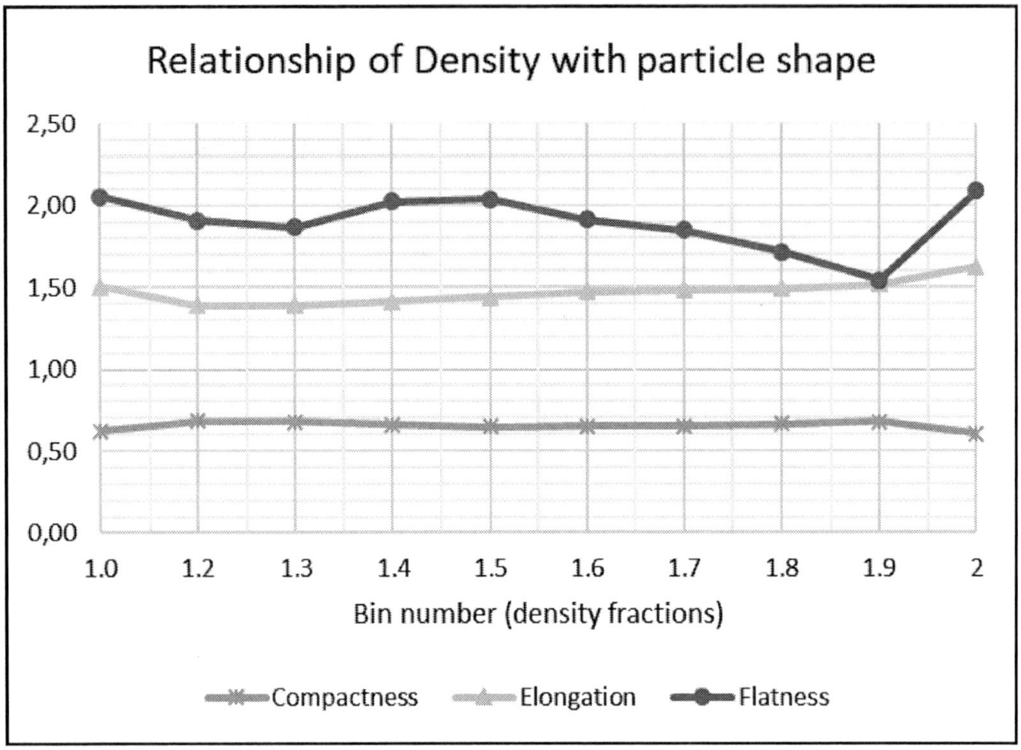

Figure 4: Results of particle Density, Compactness Elongation and Flatness as analyzed in the different bins.

Other useful considerations of each particle is the compactness, elongation and flatness of the particle and the Rho-Vol analyses for all the later concerning the density of the material. When using screening analyses, the particles are considered either undersized or oversized based on the orientation when it falls on the screen. The flat but elongated particle can be classified as a smaller particle based on the vertical fall orientation or alternatively as a larger particle based on the horizontal fall orientation, which in both ways of classification are not considerate of the actual size of the particle. From the results above, it can be seen that the flatness of the particles if higher than the compactness and elongation. The density separation does not consider these in the separation using the wet based coal beneficiation and the particles will be included in one separation bin. Doing this analyses can be beneficial for coal industry as it can be used to predict the abrasive and erosive behaviour of coal based on the orientation of the particles. Flat particles will tend to be abrasive and more erosive to the equipment's that encounter coal as compared to the elongated and compact particles.

4.4 Efficiency of Rho-Vol separation

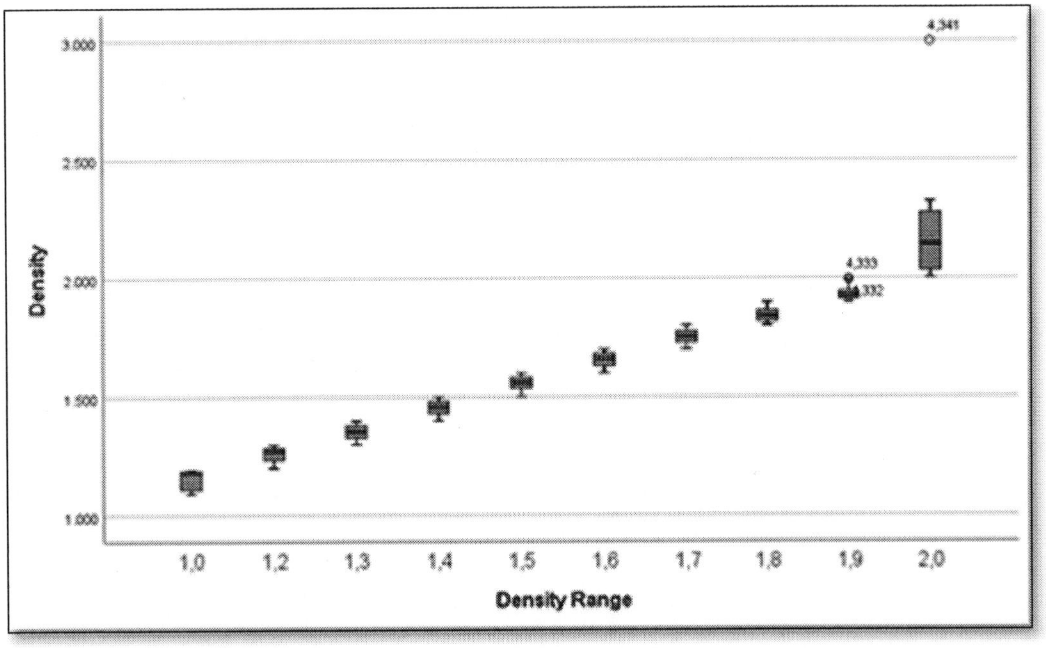

Figure 5: Normality and error plots of the different densities as separated by the Rho-Vol analyzer.

The figure above indicates the normality and error bars of the different particles as separated according to density. The Rho-Vol has higher efficiency of separation at lower densities like the float and sink analyses, where there were less errors and the distribution of material was smaller in each considered density. This is noted by the difference in the mean and the observed values as indicated by the box-whisker plot above. The efficiency of this method of analyses is higher at the densities of clean coal (between1.3-1.7), and the errors are increased at the other densities where there is less clean coal and more shales, rocks and other impurities and dominant. There was only two measures that were out of range at the density bin of 1.9 and one measure at a density higher than 2.0. This indicates a higher efficiency of separation for coal obtained in the different densities.

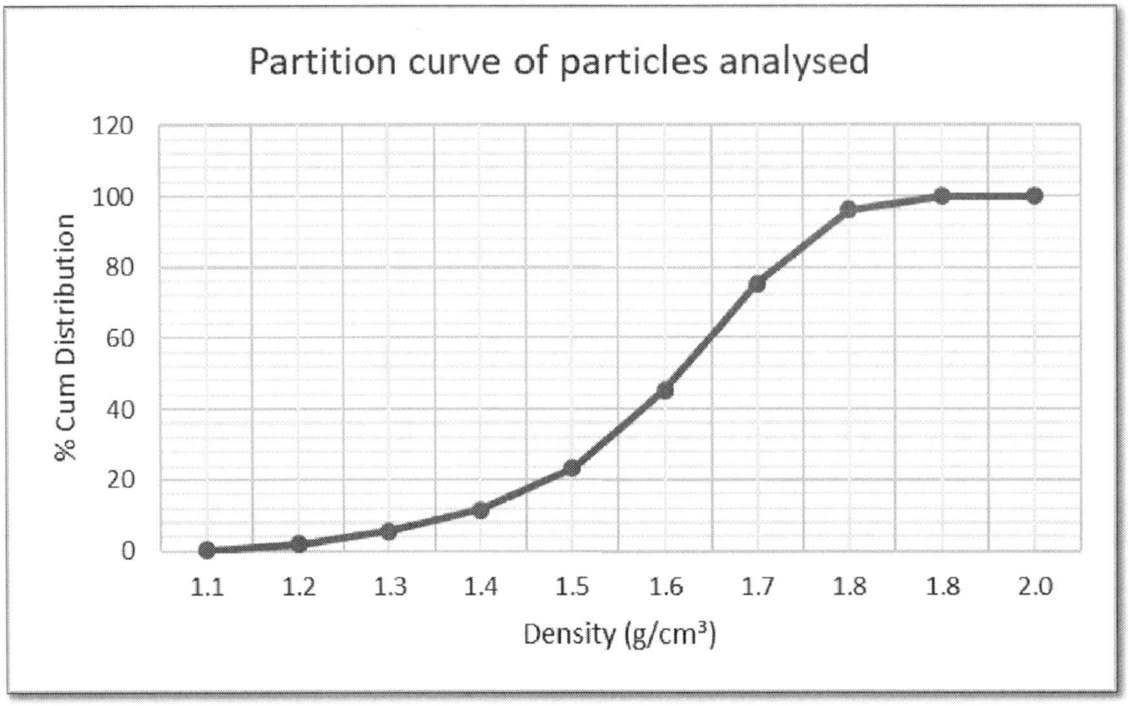

Figure 6: Particles distribution at different densities as obtained from Rho-Vol analyzer.

The figure above shows a good separation efficiency and distribution of particles observed while using the Rho-Vol dry material separator.

There are other properties of coal which are important but these require another separate method of analysis when wet coal beneficiation is utilized. The Rho-Vol mechanical particle separator can separate particles according to different properties and the separation can be done according to a preselected property. This method is a better method as compared to the float and sink washability as it gives real time results. The other advantages includes the;

- Cheaper operational cost as there is no need of a medium for density separation.
- Environmentally friendly as there is no need for water usage and this eliminates effluent production.
- Less labour intensive as the sorting is done automatically.
- Many parameters can be analysed at once.
- Less human error in data recording as the equipment is linked to a computer.

Based on the above, this equipment can be a useful tool in the coal industry to predict the yield and operational conditions of a plant.

6. CONCLUSIONS

The Rho-Vol material analyser is effective for coal dry beneficiation as it has an added advantage of analysing for many different parameters per particle. The relationship of these parameters can be studied and conclusions on the relationship can be made. The equipment is able to analyse for the shape of the different particles considered, which is a parameter that wet coal beneficiation does not include unless a different test can be considered. There is a higher efficiency of particle separation efficiency with less errors observed from particles sorted in density bins. The particles had a normal distribution as would be obtained in other beneficiation methods for coal. This equipment can be used as a predictive plant yield analyses as it is easier to use and less labour intensive than the wet process, and the results can be used to optimize the coal washing circuit. An added benefit to the equipment is that it has the ability to separate particles on multiple variable properties.

7. ACKNOWLEDGEMENTS

The author would like to thank the staff members at De-beers South Africa in Crown Mines who assisted on the analyses of the samples and everyone who has participated in the research work.

8. REFERENCES

Kolay, Ersin, and Kamil Kayabali. 2006. "Investigation of the Effect of Aggregate Shape and Surface Roughness on the Slake Durability Index Using the Fractal Dimension Approach." *Engineering Geology* 86 (4): 271–84. https://doi.org/10.1016/j.enggeo.2006.05.007.

Mishra, S. B., S. P. Langwenya, B. B. Mamba, and M. Balakrishnan. 2010. "Study on Surface Morphology and Physicochemical Properties of Raw and Activated South African Coal and Coal Fly Ash." *Physics and Chemistry of the Earth* 35 (13–14): 811–14. https://doi.org/10.1016/j.pce.2010.07.001.

Papanicolaou, C., T. Kotis, A. Foscolos, and F. Goodarzi. 2004. "Coals of Greece: A Review of Properties, Uses and Future Perspectives." *International Journal of Coal Geology* 58 (3): 147–69. https://doi.org/10.1016/j.coal.2003.10.006.

Yu, Dunxi, Minghou Xu, Jiancai Sui, Xiaowei Liu, Yun Yu, and Qian Cao. 2005. "Effect of Coal Particle Size on the Proximate Composition and Combustion Properties." *Thermochimica Acta* 439 (1–2): 103–9. https://doi.org/10.1016/j.tca.2005.09.005.

Zhao, Yuemin, Bo Zhang, Zhenfu Luo, Jingfeng He, Liang Dong, Liping Peng, and Luhui Cai. 2015. "Effect of Lump Coal Shape on Separation Efficiency of Gas-Solid Fluidized Bed for Dry Coal Beneficiation." *Procedia Engineering* 102: 1123–32. https://doi.org/10.1016/j.proeng.2015.01.235.

Kolay, Ersin, and Kamil Kayabali. 2006. "Investigation of the Effect of Aggregate Shape and Surface Roughness on the Slake Durability Index Using the Fractal Dimension Approach." *Engineering*

Geology 86 (4): 271–84. https://doi.org/10.1016/j.enggeo.2006.05.007.

Mishra, S. B., S. P. Langwenya, B. B. Mamba, and M. Balakrishnan. 2010. "Study on Surface Morphology and Physicochemical Properties of Raw and Activated South African Coal and Coal Fly Ash." *Physics and Chemistry of the Earth* 35 (13–14): 811–14. https://doi.org/10.1016/j.pce.2010.07.001.

Papanicolaou, C., T. Kotis, A. Foscolos, and F. Goodarzi. 2004. "Coals of Greece: A Review of Properties, Uses and Future Perspectives." *International Journal of Coal Geology* 58 (3): 147–69. https://doi.org/10.1016/j.coal.2003.10.006.

Yu, Dunxi, Minghou Xu, Jiancai Sui, Xiaowei Liu, Yun Yu, and Qian Cao. 2005. "Effect of Coal Particle Size on the Proximate Composition and Combustion Properties." *Thermochimica Acta* 439 (1–2): 103–9. https://doi.org/10.1016/j.tca.2005.09.005.

Zhao, Yuemin, Bo Zhang, Zhenfu Luo, Jingfeng He, Liang Dong, Liping Peng, and Luhui Cai. 2015. "Effect of Lump Coal Shape on Separation Efficiency of Gas-Solid Fluidized Bed for Dry Coal Beneficiation." *Procedia Engineering* 102: 1123–32. https://doi.org/10.1016/j.proeng.2015.01.235.

LIFE EXTENSION BY EXTRACTING VALUE FROM LOW-YIELDING COAL RESERVES

by J.F. Du Toit

Anglo American Coal – Khwezela Colliery

SUMMARY

Formerly known as Kleinkopje colliery, now Khwezela colliery – Bokgoni started in 1978 with opencast mining in two pits and expanded into four pits, five draglines & one double stage beneficiation plant with a design capacity of 9.2 Mt per annum. This respectable performance was followed by business threatening results and a need to seek innovative opportunities that will enable the mine to be profitable. Business focus shifted from mining high-yielding opencast coal to reclamation of low-yielding coal from mineral residue disposal facilities. Bokgoni now produces saleable product from previously rejected sources therefore reducing the environmental footprint and increasing Bokgoni's market competitiveness.

INTRODUCTION

Khwezela colliery - Bokgoni section, formerly known as Kleinkopje colliery, forms part of the South African Coal Estates (SACE) complex and is located 7km south west of eMalahleni and 120km east of Johannesburg, in the Mpumalanga province of South Africa. Bokgoni is an opencast operation that started mining in pits 3A / 4E during 1978, the beneficiation plant was constructed during the same period. During 1993 opencast operations expanded into 2A South and 5 West pits, followed by the opening of 2A North pit in 1998. The beneficiation plant was designed to process coal through a double stage circuit producing two products from a single feed. It had an installed capacity of 9.2Mt per annum with five Wemco drum modules, six dense medium cyclone modules and a fines processing plant at a total throughput rate of 2 100 tons per hour. In 2003 an expansion project was launched adding an additional run of mine (ROM) tipping bin and separating the high gravity and low gravity circuits into two single stage beneficiation circuits.

As ROM production from the opencast pits decreased, 9 out of the 11 modules in the beneficiation plant were mothballed reducing throughput capability to 2.4 Mt per annum. In 2008 Bokgoni started supplying coal into the domestic market as an additional product. A formal sales process to dispose of Bokgoni was initiated in February 2011 the asset included; four pits, five draglines & one beneficiation plant. During this time various business models were assessed to obtain a viable solution that will enable the mine to be profitable, these business models ranged from terminating ROM production, selling raw unprocessed coal and reprocessing previously discarded material. To suite the new business model Bokgoni was reconfigured during 2012 – 2014 to two ROM pits, two draglines and a partially mothballed beneficiation plant as indicated in Figure 1.

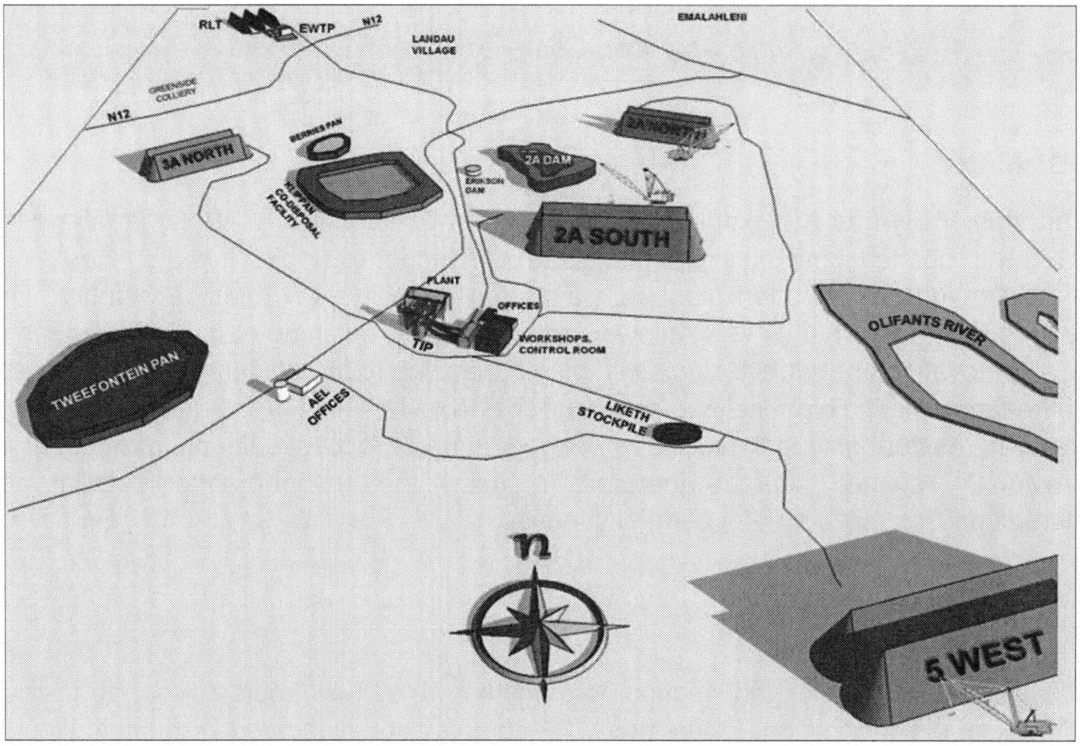

Figure 1: Bokgoni pit configuration

RECONFIGURED BUSINESS MODEL

5 West ROM pit was nearing its planned end of life, experiencing lower than predicted in-situ qualities and unfavourable mining conditions; production from 2A ROM pit was also looking grim as spontaneous combustion and aging equipment caused daily throughput constraints. During an initiative in January 2015 to seek opportunities that will enable the mine to be profitable, the focus shifted from mining high-yielding opencast coal to reclamation of low-yielding coal from previously disposed stockpiles and mineral residue disposal facilities (MRDF).

Historically 9.0 million tonnes of roof coal from previous ROM pits were stockpiled on top of the previous 'Kleinkopje' MRDF within a 0.7 km radius of the plant ROM tip. Borehole sample analyses, as seen in Table 1, indicated the raw calorific value (CV) of this coal was found to be nominal 12.6 MJ/Kg. This coal could be upgraded via a dense media separation process to a typical domestic market feed grade of 22.5 MJ/Kg at a theoretical yield of 36.4%. This lucrative opportunity could yield 3.3 million tonnes of product with virtually no mining cost.

Table 1: Combined roof coal borehole analysis

RD	Yield (%)		Ash (%)		CV (MJ/kg)		Volatile (%)		Sulphur (%)	
	Frac	Cum	Frac	Cum	Frac	Cum	Frac	Cum	Frac	Cum
1.40	3.1	3.1	9.2	9.2	30.1	30.1	27.4	27.4	0.5	0.5
1.45	2.2	5.3	12.6	10.6	28.2	29.3	23.0	25.5	0.5	0.5
1.50	3.4	8.7	17.5	13.3	26.4	28.2	22.3	24.3	0.4	0.5
1.60	9.2	17.9	24.3	18.9	23.6	25.8	20.5	22.4	0.4	0.5
1.70	10.7	28.6	31.8	23.7	20.5	23.8	18.9	21.1	0.4	0.4
1.80	10.5	39.1	40.7	28.3	17.1	22.0	17.9	20.2	0.6	0.5
2.25	60.9	100	68.7	52.9	6.5	12.6	15.0	17.1	4.7	3.1

MODULAR START-UP

During a trial wash of 3 900 tonnes in February 2015 a yield of 26.6%, air dried, was achieved at a CV of 22.8 MJ/Kg. The cut point density during the trial wash was 1.75 t/m^3. This being the nominal operating limit since the circuit is a gravity fed system. The roof coal material had a fine particle size distribution and existing infrastructure would not be able to manage the additional throughput, this required previously mothballed equipment to be recommissioned. Start-up of mothballed equipment was further emphasised by the business case requirement to beneficiate both the 1.6 Mt roof coal per annum as well as the annual ROM output from 2A pit.

Figure 2 depicts the layout of Bokgoni plant, equipment marked in orange and red were all mothballed during the first reconfiguration in 2012. Throughout 2015 equipment marked in orange were recommissioned and brought back into production.

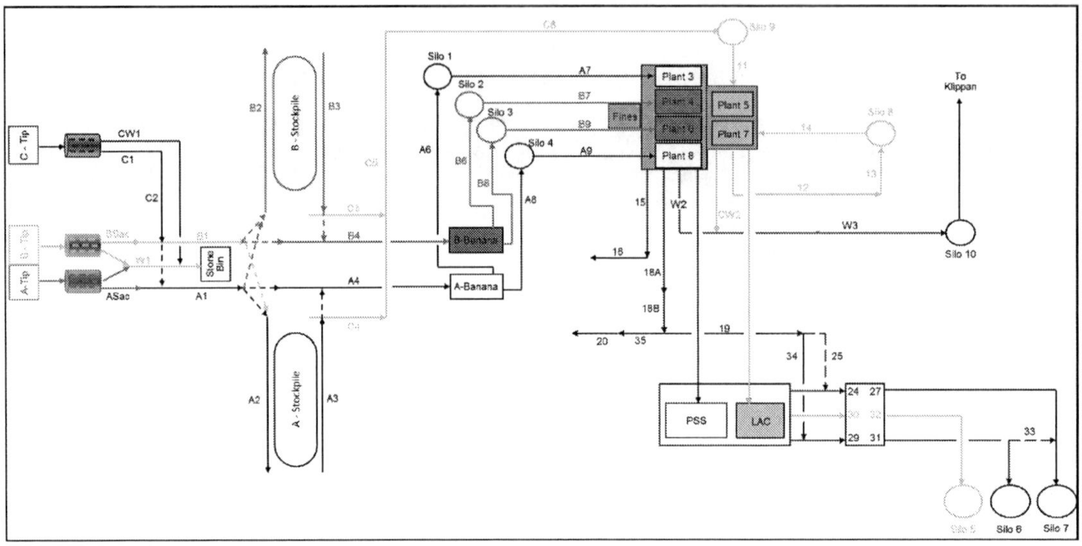

Figure 2: Bokgoni plant layout

Equipment recommissioned included: ROM crushing circuit, two Wemco drum modules, two dense medium cyclone modules, a fines processing plant, and thickener disposal equipment. Starting modules of a mothballed beneficiation plant and beneficiating roof coal was not without a challenge. The equipment had been on stop for nearly 4 years and parts were scavenged from standing equipment to keep the operational modules running without process interruptions. Equipment process interlocking was either not sequenced or in place and most of the process interlocks had to be programmed again.

Figure 3 shows the material characteristic variabilities observed in the roof coal feed; particles size variations, highly weathered surface layers, and horizontal and vertical inorganic intrusions. The weathered layer was also hard and compacted and demonstrated crust like properties to the roof coal below the surface. The weathered material was very sticky and frequently blocked conveyor transfer points, this was resolved by mixing the weathered material with material from outside the weathered zone deeper into the loading zone.

Figure 3: Photos indicating variability within the roof coal mining area

Calorific value from the head feed varied between 7 MJ/kg – 17 MJ/kg. Apart from the feed quality variation within the loading zone the particle size distribution also varied significantly with particle sizes smaller than 0.5 mm increasing as much as 30% within the feed. Fines circuits were under severe strain and thickener rake failures and thickener blockages was a frequent occurrence. Screen decks in the fines circuit were reconfigured to be more efficient at fines removal and cluster cyclone cut points were optimized. Product yields went as low as 23% overloading the existing discard drain and rinse screens and causing downstream process interruptions, these screens were replaced with higher throughput screens and beneficiation could be done at maximum rates.

Addressing the above-mentioned challenges accelerated roof coal production to such an extent that the stockpiled roof coal has been depleted and production has advanced into the 'Kleinkopje' MRDF below the roof coal surface with minimal production impact.

LANDAU 3 MRDF

With the formation of Khwezela colliery between Kleinkopje colliery and Landau colliery a chain reaction of new opportunities was created. Kromdraai pit was reaching its planned end of life threatening the coal supply to Navigation plant, a new coal supply had to be secured. The new Navigation life extension pit would supply coal to Navigation plant, however, this pit was divided in half by a 400kV overhead distribution line. Relocation of the line had to be done with the most feasible location going over the current Landau 3 MRDF.

Ore characterisation was done on Landau 3 MRDF with results ranging from low ash coal product rejects to burnt out and devolatilised rejects. Theoretical borehole yields varied between 28.8% and 75.9% with an average yield of 51.4% producing power station grade coal. Figure 4 shows various zones identified within the MRDF and Table 2 depicts raw qualities obtained. Zone 1 & 6 are the most undesirable zones in terms of ash content and calorific value, however this can be blended with material from zone 2 to create acceptable feed material. Blending these zones will alter ore characteristics and reduce material stickiness that causes transfer point blockages.

Table 2: Raw average qualities within Landau 3 MRDF

Zone	Description	Ash (%)	CV (MJ/kg)	Volatile (%)	Sulphur (%)
1	Top Zone	55.3	11.44	40.4	3.32
2	Main Zone	42.8	16.25	32.8	2.92
3	Bottom Zone	49.7	14.10	37.9	3.10
4	Devolatilised Zone	45.3	15.30	15.6	1.89
5	Old MRD Zone	34.7	18.70	30.6	3.29
6	Lens Zone	58.4	10.82	40.3	3.25

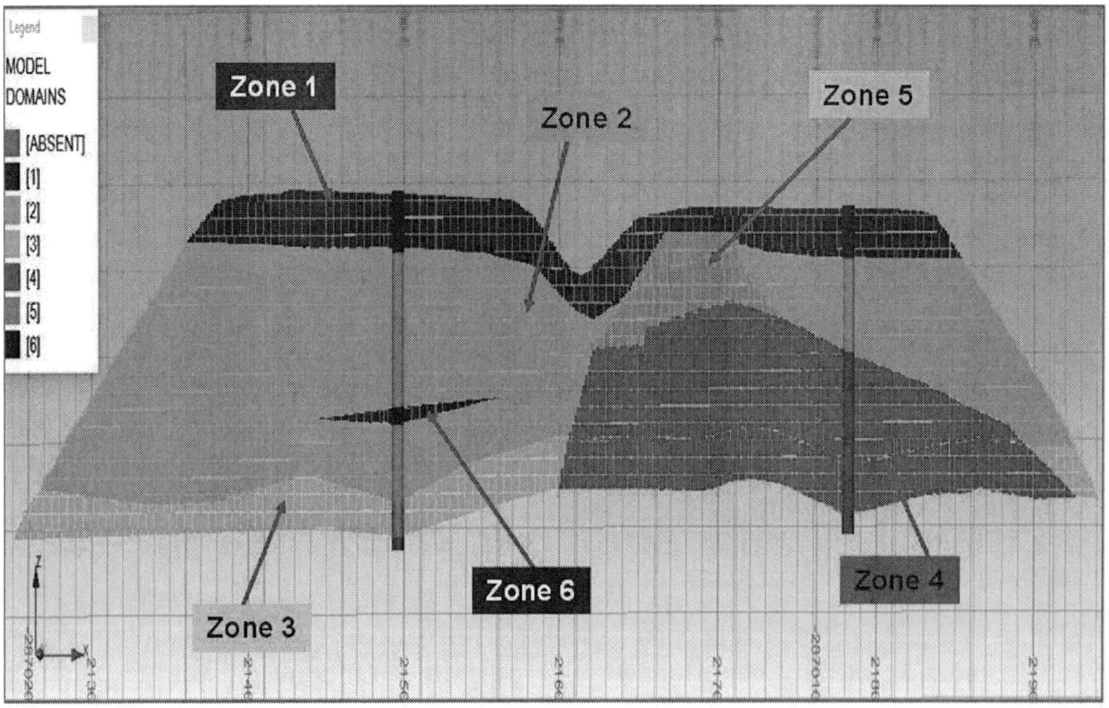

Figure 4: Cross section of Landau 3 MRDF

The journey of introducing and processing roof coal through Bokgoni beneficiation plant alleviated the strain of processing Landau 3 MRDF material as it had very similar ore characteristics. Bokgoni plant vehicle management plan was adjusted to accommodate merchant trucks delivering coal from Landau 3 MRDF to a designated stockpile area at Bokgoni tip from where it followed the same value chain as roof coal.

Washing roof coal, Kleinkopje MRD and Landau 3 MRD material through an aged coal handling beneficiation plant was not without a challenge but has certainly yielded positive results.

ACKNOWLEDGEMENTS

All involved, directly or indirectly, in all the above-mentioned work completed.

REFERENCES

- Dhlamini, E., 2016, Landau 3 mineral residue deposit reclamation project, Internal report
- Phali, M., 2015, Roof Coal Stockpile Rewash Opportunity, Internal report

Low relative density processing of coal with large diameter dense medium cyclones

Johan de Korte
Consultant
Pretoria, South Africa
dekorte.johan@gmail.com

Johann Claassen
Minopex
Johannesburg, South Africa
johann.claassen@minopex.co.mz

Prosper Rugube
Minopex
Johannesburg, South Africa
propser.rugube@minopex.co.mz

Andre La Grange
Consultant
Johannesburg, South Africa
andre.lagrange@minopex.co.mz

ABSTRACT

Large diameter dense medium cyclones are frequently employed to process coal larger than approximately 1 mm. When relatively low coking coal ash values are required, production of coking coal product often necessitates using relative densities below about 1.30. The paper suggests that it is not advisable to operate the large diameter dense medium cyclone studied at medium densities below about 1.25 t/m³. The organic efficiency of separation decreases from 90.2% at a medium RD of 1.30 to 73.9% at 1.25 with an associated coking coal yield loss to the thermal coal. Findings also support the notion that poor large diameter cyclone performance at low medium densities is more a function of factors such as high near-dense proportions in the coal, the percentage <8μm particles in the medium and the M:C ratio than the breakaway size phenomenon.

KEYWORDS: low density separation, large diameter cyclone, cyclone efficiency, EPM, near-density coal, organic efficiency

1 INTRODUCTION

Large diameter cyclones (Dc ≥ 1000 mm) is a common feature in most modern high-capacity coal processing plants today. These units not only allow smaller plant footprints with significant capital benefits, it also delivers good performance at circulating medium densities above 1.30 t/m³, a feed pressure of 9 Dc and medium:coal (M:C) ratios of at least 4:1 (Moorhead and Schutte, 2005, Meyers et al., 2014). The work of Chedgy et al. (1986) suggests that the good performance obtained from these units can be linked to increased medium stability due to lower shear stress at a given centrifugal acceleration compared to smaller diameter cyclones. This in principle implies that both coarse and fine particle separation efficiencies can be enhanced by increasing feed pressure to large diameter cyclones.

It is, however, not clear whether good large diameter cyclone performance levels are achievable at circulating medium densities below 1.30 t/m³. De Korte (2006) found that even though Ep values were good (Table 1), the organic efficiency deteriorated significantly below a cut-point density of about 1.40 t/m³ for Witbank coal when product ash levels below 10% was targeted (Figure 1). This decrease in cyclone performance levels can be linked to poor recovery of the lightest coal particles and an increase in the amount of near-density material (the density of liberated coal particles with an ash level below 10% is typically between 1.22 t/m^3 and 1.32 t/m^3).

Table 1 Efficiency results for cyclone performance in the Witbank coal field (De Korte, 2006)

Parameter	**Module A**	**Module B**
Feed ash %	13.0	13.0
Product ash %	5.5	5.2
Discard ash %	13.7	14.0
Product yield %	8.8	11.3
Cut-point density, d50* t/m^3	1.32	1.32
EPM	0.0121	0.0106
Organic efficiency %	63.4	79.4
Sink in float %	3.6	3.5
Float in sink %	1.7	2.2
Total misplaced %	5.3	5.7

*Note: circulating medium density 0.03-0.05 t/m^3 units below cut-point density

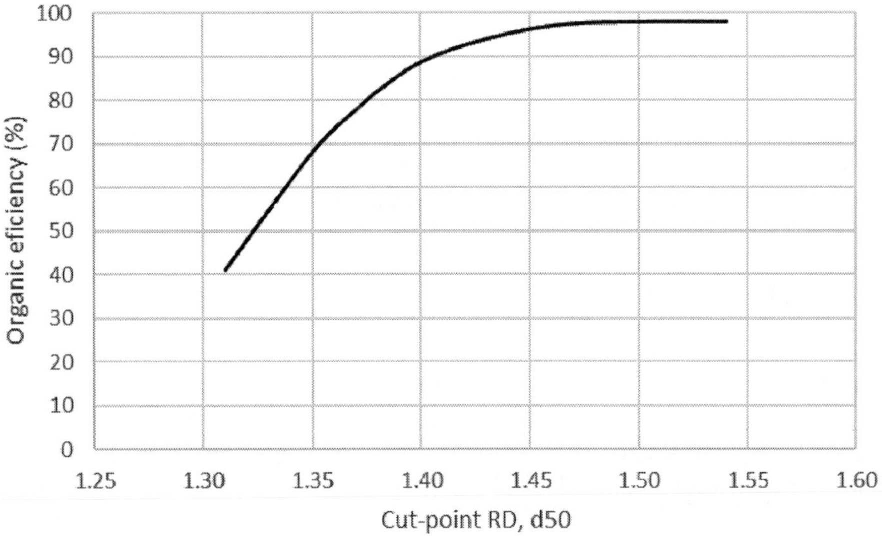

Figure 1: Cyclone organic efficiency as a function of cut-point density (De Korte, 2006)

If it is argued that the lightest coal particles tend to be misplaced at low medium densities in large diameter cyclones, then the same argument must hold for the smallest particles. Engelbrecht and Bosman (1995) indicated that a notable breakaway of Ep occurs at a particle size, the so-called breakaway size, below 4 mm. Meyers et al. (2014) however concluded from extensive studies for cyclone diameters between 700 mm and 1450 mm and feed pressures ranging from 6.9 Dc to 9.0 Dc that although the breakaway of Ep starts at a particle size of about 4mm, it only deteriorates significantly below approximately 1 mm as shown in Figure 2 (calculated using Equation 1).

$$Ep = 0.0149 + 0.054022 \times GMS^{-0.79655} \quad (1)$$

Where: GMS = geometric mean size of particle (mm)

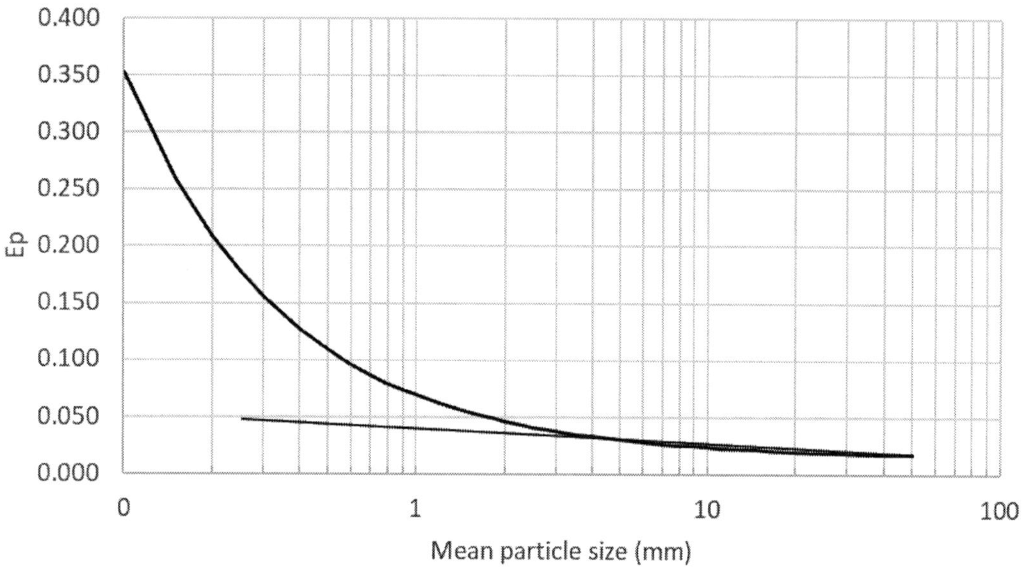

Figure 2: Ep vs mean particle size data for large diameter cyclones (Meyers et al., 2014)

This finding implies that a loss in (large diameter) cyclone efficiency witnessed when operating at low circulating medium densities may be caused by factors other than the breakaway size phenomenon in circuits where desliming occurs at a cut-point close to 1 mm.

2 LOW MEDIUM DENSITY PLANT TRIALS

The performance of a 1300 mm cyclone (Table 2) operated at 9 Dc was evaluated at circulating medium densities of 1.30 t/m^3, 1.25 t/m^3 and 1.20 t/m^3 to determine the:

- cyclone efficiencies at cut-point densities below 1.30 t/m^3
- Ep vs particle size relationship
- cut-off medium density beyond which the large diameter cyclone cannot stabilize the medium anymore

Table 2 Details of the large diameter cyclone evaluated

Parameter	Value	Unit
Size (diameter)	1300	mm
Inlet size	25	%diameter
Operating head	11.7	m (9 Dc)
Feed volume @ 9 Dc	1 917	m^3/hr
Medium:ore ratio	9.3:1(ave)	@ 9D and RD of 1.43
Cyclone cut range Max Min Average	 1.52 1.30 1.40	t/m^3
Cyclone yield to floats	43.9 (ave) 15.7 – 74.4	%
Cyclone yield to sinks	56.1 (ave) 25.6 – 84.3	%

2.1 Process flowsheet

Deslimed plant feed (> about 1 mm material) is processed in a double-stage large diameter cyclone arrangement as alluded to in Figure 3. The primary cyclone (Dc = 1450 mm) removes the bulk of the material as coarse rejects. The secondary/coking cyclone (Dc = 1300 mm) processes the primary cyclone overflow at density cut-points 0.2 – 0.3 units lower than the primary cyclone.

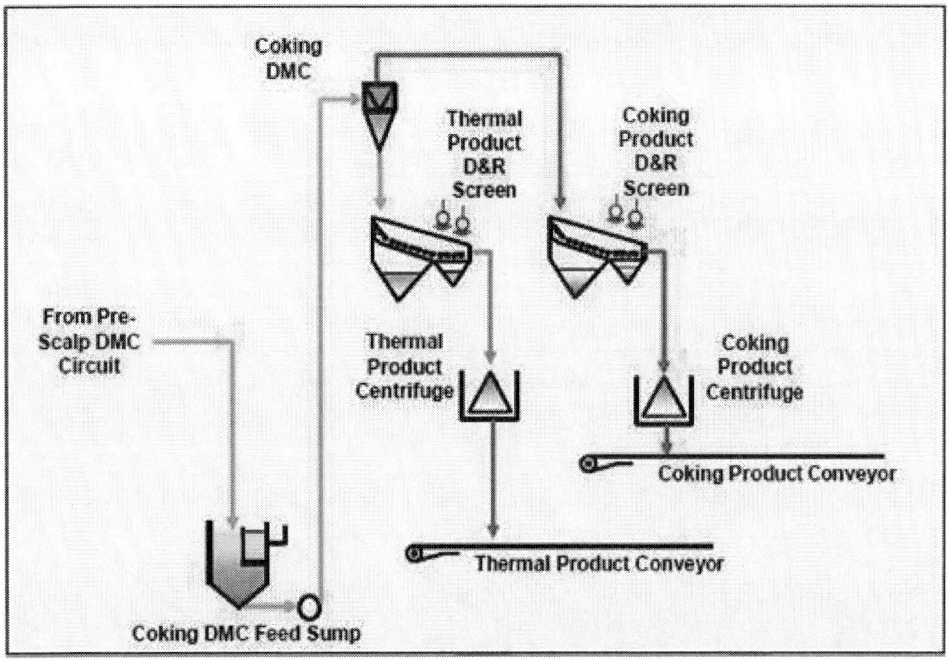

Figure 3: Simplified flowsheet of the coking cyclone circuit studied

2.2 Typical coal washability

The washability of the coal unit studied is characterised by increasing near-dense proportions as the cut-point density falls below about 1.35 t/m^3. A typical washability for the size fractions comprising the feed to the dense medium cyclones is shown in Table 3 and Figure 4. It clearly indicates that the plus 4 mm size fraction contains almost no coal with an ash content below 10% and most of the coking coal recovered in the low-density DMC arises from the minus 4 mm size fraction. This is further illustrated in Figures 5 and 6, which clearly show the difference in the size of coal on the primary feed drain-and-rinse (D&R) screen and the low-density cyclone product D&R screen, respectively. As the density of the medium to the DMC is lowered, the product becomes even finer.

Table 3 Typical washability of the coal unit studied

Relative density		Feed to DMC (-50 +4 mm)				Feed to DMC (-4 +1.2 mm)			
		Fractional		Cumulative		Fractional		Cumulative	
		Yield %	Ash %	Yield %	Ash %	Yield %	Ash %	Yield %	Ash %
F @	1.30	0.35	4.60	0.35	4.60	13.33	2.90	13.33	2.90
F @	1.35	2.21	10.80	2.56	9.95	10.12	9.60	23.46	8.68
F @	1.40	6.09	16.50	8.65	14.56	11.85	15.60	35.31	13.55
F @	1.45	5.99	21.20	14.64	17.28	4.94	19.10	40.25	15.82
F @	1.50	5.98	26.20	20.62	19.87	9.63	24.60	49.88	18.37
F @	1.55	5.77	32.10	26.39	22.54	3.95	25.40	53.83	19.90
F @	1.60	8.42	28.90	34.81	24.08	4.20	33.10	58.02	23.10
F @	1.80	15.60	43.80	50.41	30.18	9.88	42.00	67.90	28.95
F @	2.00	14.16	58.30	64.57	36.35	8.64	53.00	76.54	34.22
S @	2.00	35.43	72.70	100.00	49.23	23.46	70.20	100.00	46.97

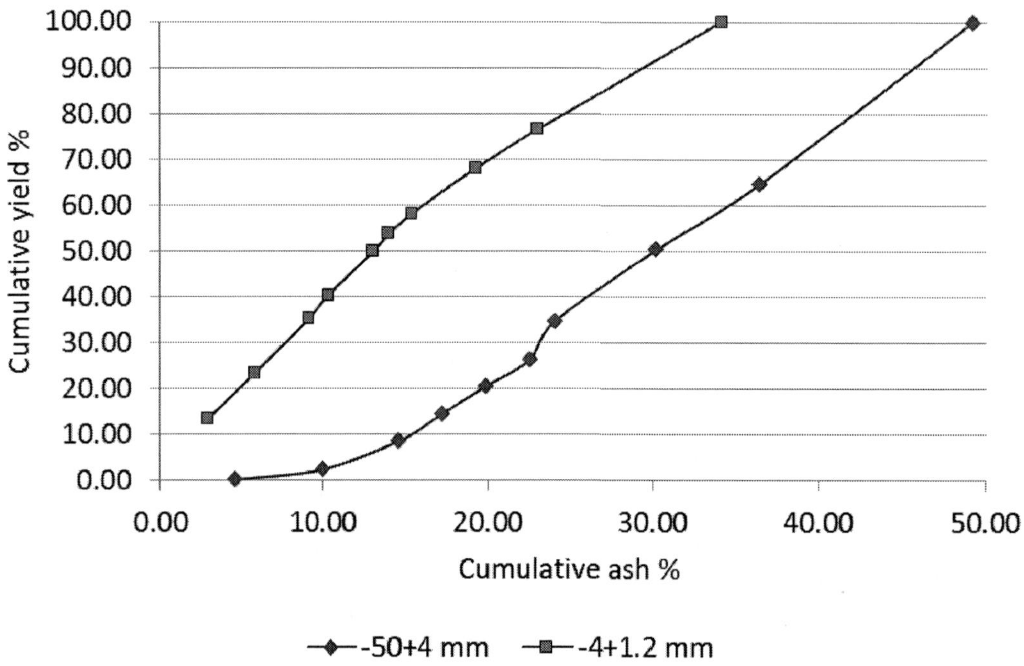

Figure 4: Typical washability of the coal unit studied

Figure 5: Feed to the primary DMC

Figure 6: Product from the low-density large diameter DMC

2.3 Test procedure

Three sampling campaigns at coking DMC medium densities of 1.30 t/m^3, 1.25 t/m^3 and 1.20 t/m^3 were conducted. Samples were taken simultaneously within a 5-minute period at the primary cyclone overflow screen discharge as well as the coking and thermal product D&R screen discharges. Approximately 100 kg of sample were collected at each point per campaign once the circuit was running stable at full capacity, at 9 Dc pressure and the desired circulating medium density. The SCADA density readings were confirmed using manual in-plant Marcy scale readings. Correct medium samples were also collected for each campaign.

The samples taken during the campaigns were subjected to float-and-sink tests to determine the washability characteristics of the feed to the cyclone, the coking coal product (floats) and the thermal product (sinks). The data obtained from the analyses were then used to determine the efficiency of the DMC when operating at the different circulating medium densities.

3 RESULTS AND DISCUSSIONS

3.1 Dense medium cyclone efficiency

The results obtained from the float-and-sink analyses are summarised in Table 4.

Table 4 Summary of efficiency test results

Parameter	**Circulating medium RD**		
	1.30	**1.25**	**1.20**
Feed ash %	19.1	19.9	19.8
Product ash %	14.2	12.2	10.9
Discard ash %	26.6	25.3	20.5
Product yield %	59.9	41.4	6.7
Cut-point density, d50 t/m^3	1.43	1.37	~ 1.27
EPM	0.0135	0.0184	-
Organic efficiency %	90.2	73.9	17.3
Sink in float %	8.34	10.85	-
Float in sink %	4.46	4.34	-
Total misplaced %	12.80	15.19	-
Near-dense material %	53.2	61.8	-

The partition curves obtained for the tests at the different medium densities are shown in Figures 7, 8 and 9.

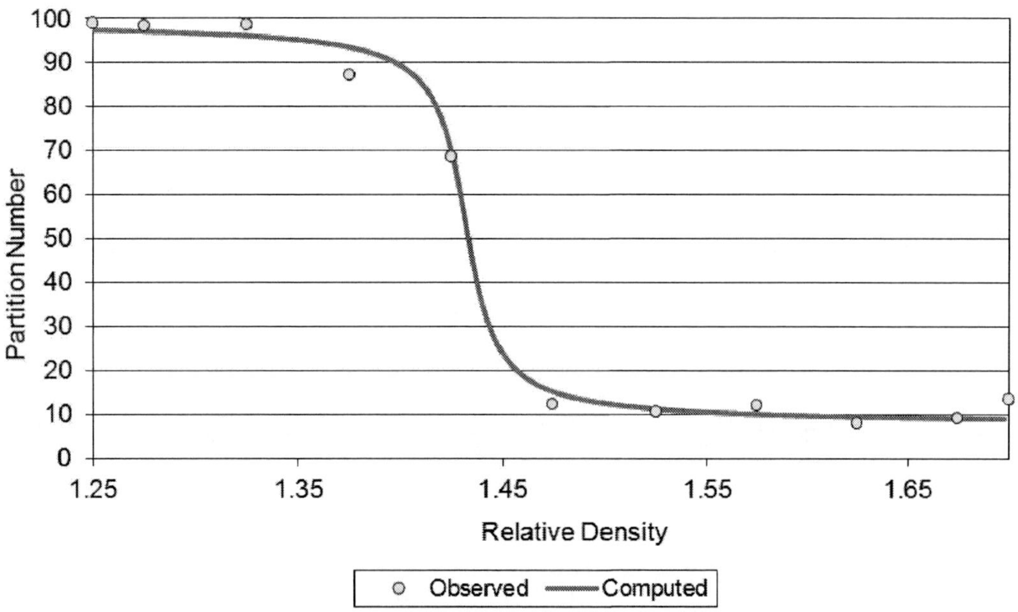

Figure 7: Partition curve for a 1300 mm DMC operated at a circulating medium density of 1.30 t/m^3

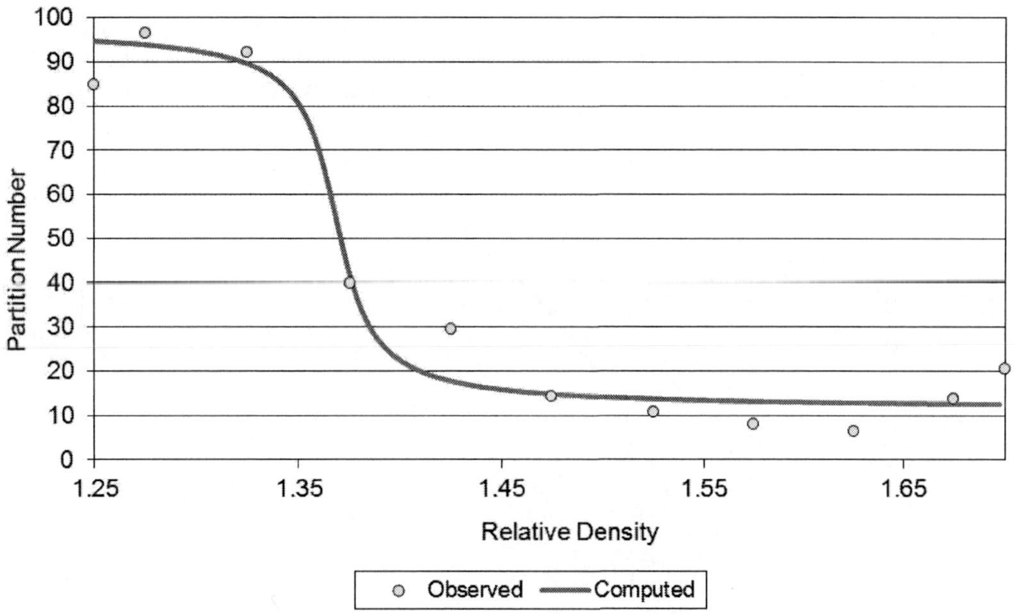

Figure 8: Partition curve for a 1300 mm DMC operated at a circulating medium density of 1.25 t/m^3

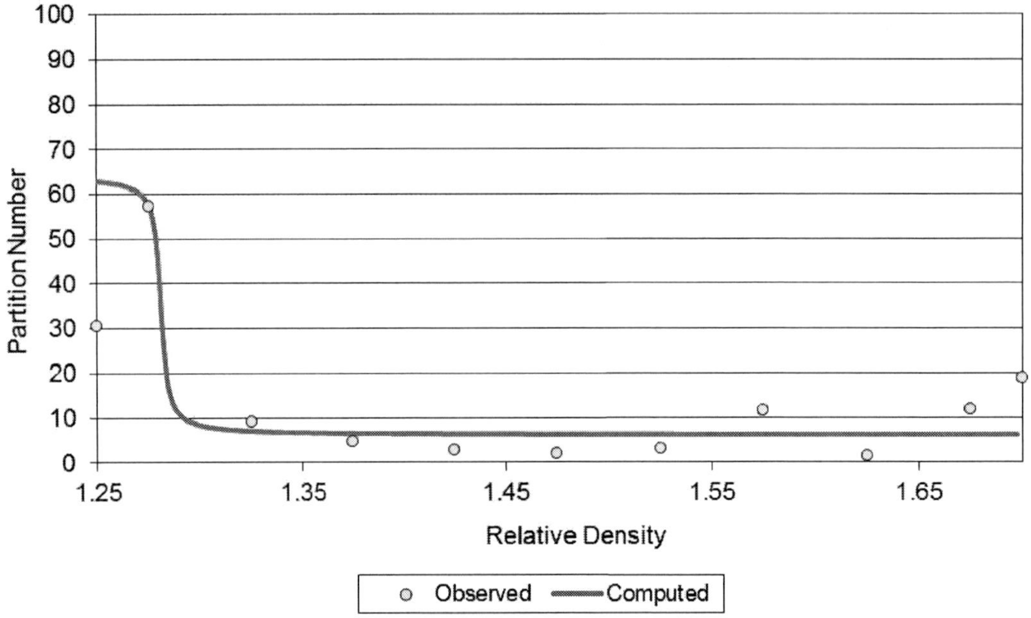

Figure 9: Partition curve for a 1300 mm DMC operated at a circulating medium density of 1.20 t/m^3

From Table 4 and Figures 7-9 it is evident that relatively good separations are obtained at circulating medium densities of 1.30 t/m^3 and 1.25 t/m^3, although the organic efficiency decreases from 90.2% at a density of 1.30 t/m^3 to 73.9% at a density of 1.25 t/m^3 (also refer to Figure 10). At these densities the high amount of near-density material in the feed to the cyclone is worth noting.

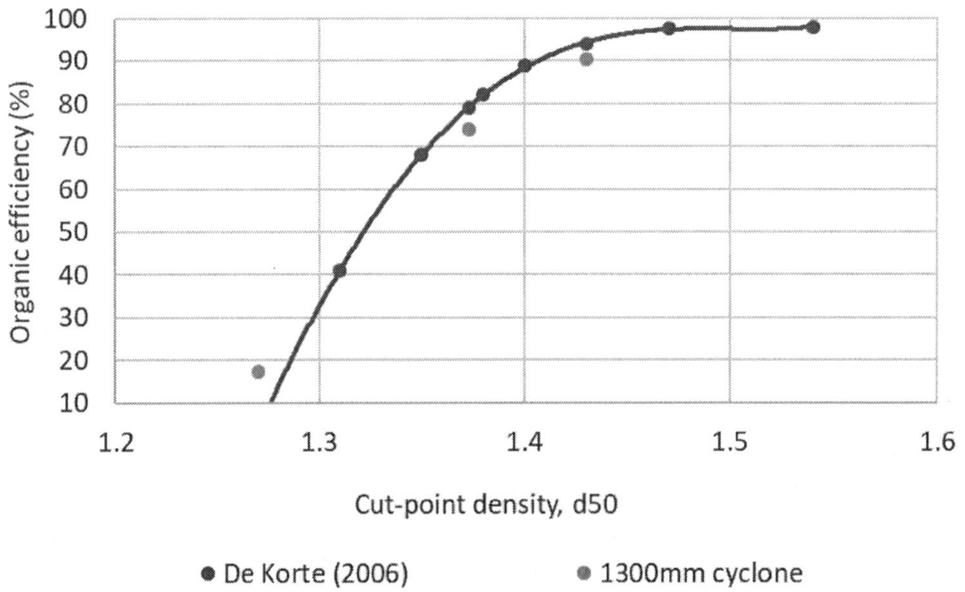

Figure 10: Organic efficiency for a 1300 mm DMC operated at circulating medium densities of 1.30 t/m^3, 1.25 t/m^3 and 1.20 t/m^3 using data from De Korte (2006) as reference (solid line)

When the relative density of the circulating medium is reduced to 1.20 t/m^3, the performance of the cyclone deteriorates completely and a very low coking coal yield (6.7%) is obtained. It is not possible to construct the complete partition curve for the 1.20 relative density test and therefore it is not possible to determine the cut-point density or EPM values for the cyclone when operating at such low density. It can be argued that the relative density of the medium in the overflow of the cyclone is too low (around 1.10 t/m^3) and only the smallest and lightest coal particles can be 'transported' to the product screen in the cyclone overflow.

3.2 Ep vs size

Figure 11 indicates that results obtained from this study agree with the work performed by Meyers et al. (2014). The poor performance of the 1300 mm DMC at medium densities below about 1.25 t/m^3 should therefore rather be linked to a significant increase in near-density material with a decrease in medium density (>60% at medium RD less than 1.25) and the inability of the cyclone to inherently self-adjust its discharges based on feed quality (Moorhead and Schutte, 2005). The self-regulating ability of the large diameter DMC is limited to high enough circulating medium densities, high enough levels of <8μm particles in the medium (actual about 12% vs 19-25% required), low levels of non-magnetics (actual >15$^v/_v$) and acceptable M:C ratios (actual M:C > 4:1) that enable transportation of coal particles to the cyclone overflow as mentioned earlier.

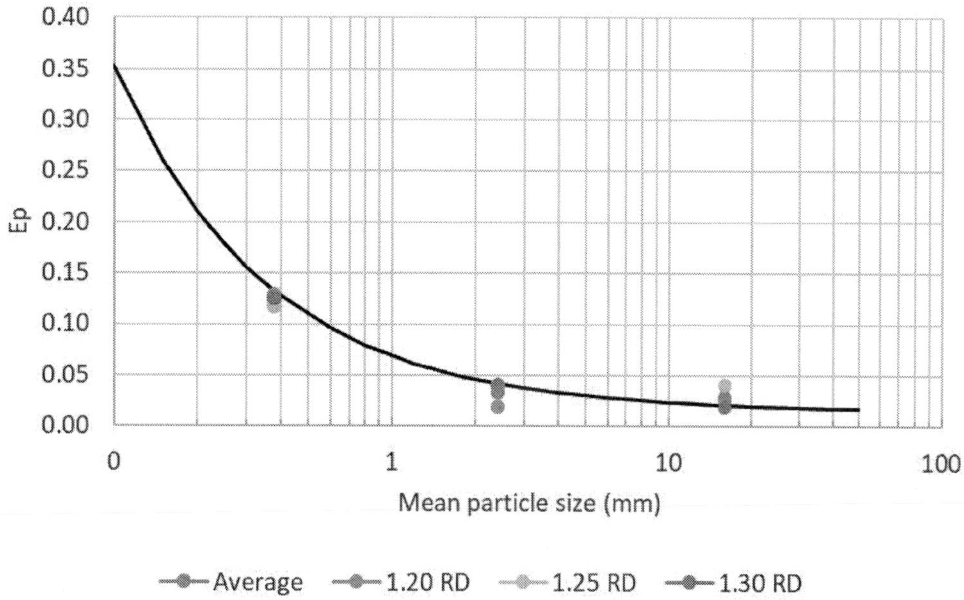

Figure 11: Ep vs mean particle size for a 1300 mm DMC operated at circulating medium densities of 1.30 t/m^3, 1.25 t/m^3 and 1.20 t/m^3 using Equation 1 (Meyers et al., 2014) as reference (solid line)

4 CONCLUSIONS

Tests conducted on a 1300 mm dense medium cyclone treating coal with high near-dense proportions at relative medium densities of 1.30 t/m^3, 1.25 t/m^3 and 1.20 t/m^3, suggest that it is not feasible to operate the circuit at medium densities below about 1.25 t/m^3. Results indicate a notable increase in the amount of near-dense material from 53.2% to 61.8% and a decrease in the organic efficiency of separation from 90.2% to 73.9% when reducing the medium RD from 1.30 to 1.25.

Data also suggests that poor cyclone performance experienced at medium densities below about 1.30 t/m^3 is probably more a function of the inability of the dense medium cyclone to self-regulate its discharges than the breakaway size phenomenon. It was shown that significant breakaway occurs only below a mean particle size of about 1mm. The study also highlights the necessity of a fine enough circulating medium as well as an acceptable M:C ratio when operating below an RD of about 1.30.

5 REFERENCES

Chedgy, D.G, Watters L.A. and S. T. Higgins. Heavy Media Cyclone Separations at Ultralow Specific Gravity. In Proceedings of the 10th International Coal Preparation Congress, Edmonton, Canada, pp 60-79, 1986.

De Korte, G.J. Low-density separations – the need and the problems. Paper presented at the FFF Indaba, Johannesburg, South Africa, 2006.

Engelbrecht, J. A. and Bosman, J. Design Criteria for an Improved Large Diameter Dense Medium Cyclone. In Proceedings of the Seventh Australian Coal Preparation Conference, edited by Smitham, J., Paper B3, 1995.

Meyers, A., Sherritt, G., Jones, A. and O'Keeffe, M. Large Diameter Dense Medium Cyclone Performance in Low Density/High Near Gravity Environment. International Journal of Coal Preparation and Utilization, Volume 34, pp 133-144, 2014.

Moorhead, R.G. and Schutte, A. Operating Guidelines for Low-Density Dense-Medium Cyclone Separations. Presented at the SA Coal Preparation Conference and Exhibition, Secunda, South Africa, 2005.

Southern African Coal Processing Society

"Coal Processing – Extracting value from low grade reserves" 20 to 22 August 2019

DETERMINING THE EXPECTED AND ACTUAL MEASUREMENT PERFORMANCE OF AN ONLINE ELEMENT ANALYSER ON A FINAL PRODUCT STREAM AT EXXARO's GROOTEGELUK 7 & 8 PLANTS

By H.N. Engelbrecht and M.V. Roper

Exxaro Resources Ltd

Quality control interventions emanates from process upsets and trends that moves product quality beyond natural variation expected in a stable process. The ability to detect these upset conditions and to react in an effective manner, protects both resource utilisation and compliance to product specifications.

Product quality control on coal processing plants traditionally rely on the combined sampler and laboratory (reference) system to measure and track product qualities. The introduction of online analyser (OLA) technology creates a second source of quality measurement and tracking information to be considered for quality control decision making.

The case study done on the final product stream at Exxaro's GG7&8 plants at the Exxaro Grootegeluk complex explores the OLA system's ability to meet the OEM performance guarantee, perform equal to the reference system without analytical delay time and assesses the effort to achieve and sustain this performance.

This paper demonstrates a conceptual comparison of the two measurement systems; the theoretical and practical comparison of measurement performance and shares an assessment of the effort to achieve and sustain the measurement performance of an online-analyser system on a final product stream.

The results demonstrate that the online analyser outperformed the OEM's performance guarantee and can achieve similar levels of performance to the reference system without the reporting delay.

INTRODUCTION

The GG7&8 plant is an integrated process plant producing power station (thermal) coal consisting of GG8 as a partial wet processing plant for the +2.5 mm Run of mine (ROM) fraction and GG7 as a crush and screen facility processing specific benches from the Grootegeluk pit. The final product from the plant combines the GG8 dense medium, GG8 unprocessed dry fines and dry GG7 product. The block diagram in

Figure *1* displays the combination of processing routes leading to final product. Also shown in Figure *1* are the entities that define the OLA and Reference measurement systems. A primary sampler creates the feed to both the OLA and Reference systems. The final product is stockpiled, with stockpiles closed off at target quality.

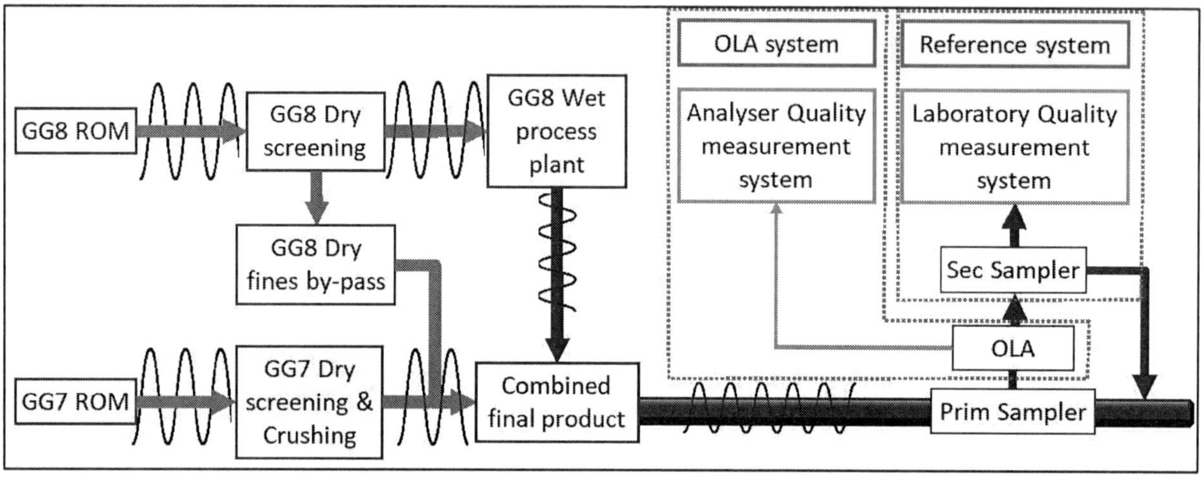

Figure 1: High level process flow of the GG7&8 plants including measurement systems.

Note:

Information source: Exxaro project work

As deduced from Figure *1*, there are various factors within the plant that can cause process upsets and trends e.g. a change in ROM quality or any processing equipment malfunction. Unattended upsets and trends will result in sub-consignments (hourly product packages) and ultimately consignments (completed stockpiles) that does not comply to product specifications. Compliant consignments could be made up of highly variable sub-consignments which normally indicates quality control compensation and results in sub-optimal resource utilisation.

Measurement system response demonstration

A simulated TV (True Value) for final product Ash% is used to demonstrate the measurement response of the reference and OLA systems graphically. The simulated TV is characterized by an excessive downward deviation in product quality that should be corrected immediately to avoid control system failure. The process upset is followed by a steady upward trend in product Ash%. The TV value is the reference in Figure *2* and Figure *3* against which the Reference and OLA systems response is demonstrated.

The reference system is simulated to have no sample preparation and analysis error and is only impacted by the averaging of increments.

Reference system
The Reference system includes a secondary sampler taking 60 increments every two hours. Samples are removed to an on-site laboratory executing preparation and analysis according to the relevant published standard methods, reporting a measurement result after a four-hour delay (response time).

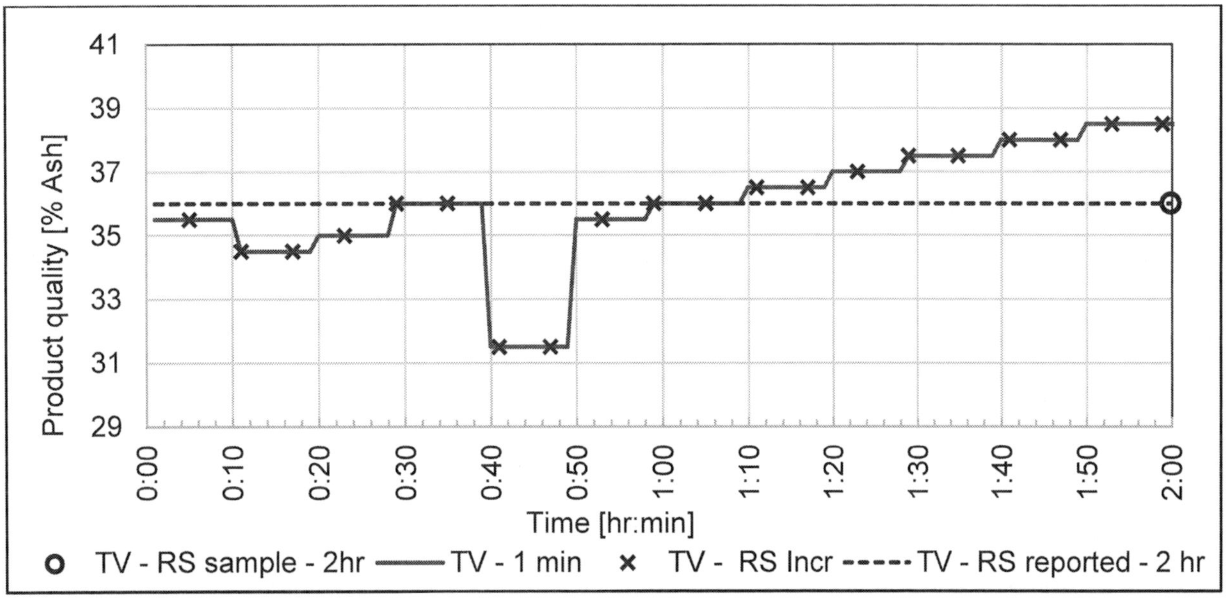

Figure 2 : Process Upset followed by upward trending of final quality

Note:

Information source: Exxaro project work
RS = Reference System
Incr = Increment

From Figure 2 the limited exposure of secondary sampler increments to the primary sample stream as well as the averaging effect is well demonstrated. Considering in addition the 4-hour response time, the reference system is not effective to identify process upsets. If such a scenario persisted, the process upset could have continued unchallenged for a period up to 6 hours. The reference system in Figure 2 does not highlight the upward trend but reports a decrease in total ash over the period due to the averaging effect of combining 60 sample increments over a two-hour production period.

OLA system
The OLA receives a 100% measurement exposure to the total primary sample stream. It reports the average value analysed for every minute. It is more sensitive to process upsets as can be seen in Figure *3*.

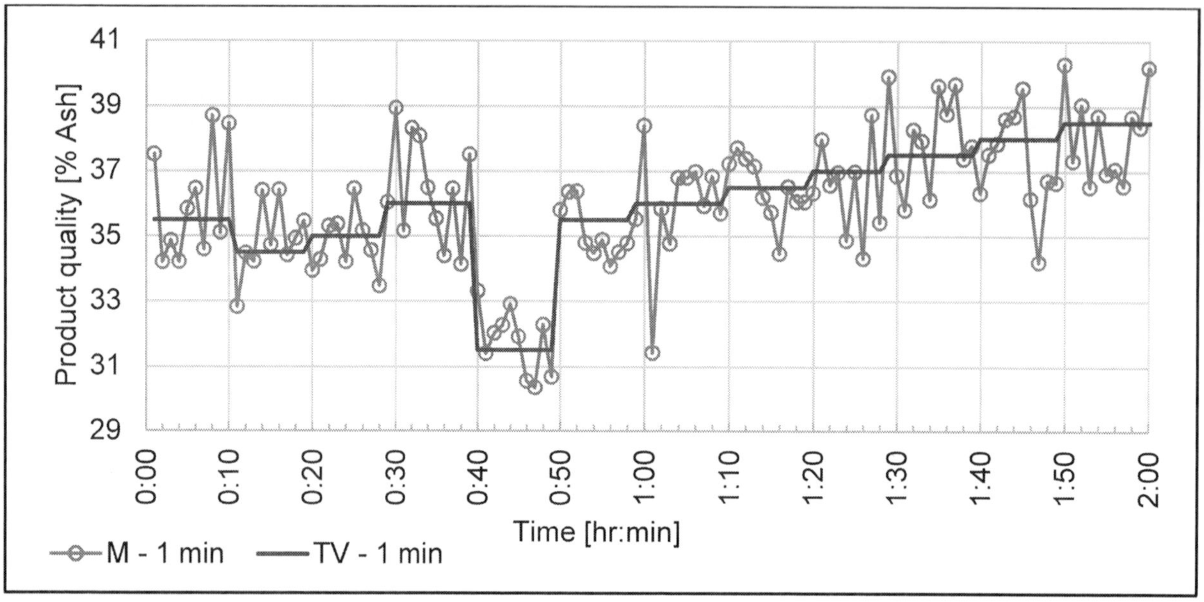

Figure 3: Process Upset analysed by OLA

Note:

Information source: Exxaro project work
TV = True Value
M = Measurement

The OLA, on a minute interval is affected by its own inherent noise, (stochastic measurement variation). For overall quality control a minute interval would be too erratic. In practise a balance must be found between measurement stability and measurement response.

This installation in the GG7&8 plant is non-typical. Most installations are done on sorting and blending applications, where the frequency of control decisions cannot accommodate the delay caused by the reference system. On a final product stream however, the operations team now has access to two independent and potentially conflicting sources of measurement information. To apply the OLA measurement system with confidence a level of trust had to be established through a comparative investigation.

Aim of this study

The aim of the study was to determine if the OLA system installed on the combined GG7&8 product, can achieve the same measurement performance as the Reference system, but eliminating the analytical reporting delay and to establish the procedure that will ensure that the system is reliable and maintains performance over time.

METHOD OF EVALUATION

A direct comparison of the two systems was done by executing the following evaluations:

- Literature survey to establish a factorised understanding;
- Confirmation of the expected operational performance;
- Dynamic Performance evaluation of the Reference and OLA systems; and
- Ongoing combined system performance management

The performance of the measurement system is evaluated by its ability to report product ash for a production interval with a required precision, within a required turn-around time, targeting zero bias.

For the OLA system a stochastic distribution of observations is assumed, and the concept of precision applied. Measurement precision (error, deviation, variation) is expressed as an estimate of the standard deviation and reported with a factor of one. This relates to $\pm ts$, where t is the value of Student's t (67% confidence level, two sided) and s is the standard deviation of the observations about the mean value.

For the reference system a standard deviation of the reported ash measurement is also applicable. Sampling, sample preparation and analysis activities each contribute to the measurement variance of the reference system.

Different factors determine the precision of the measurement systems. The composite impact of these factors is calculated by applying the following principles/assumptions:

1. Factors impacting measurement performance are independent.
2. Variance (V) of a system is an additive statistic:

   $V_{\text{measurement system}} = V_1 + V_2 + V_3 + \ldots + V_n$ (for n components)

   $V_{\text{OLA vs Reference}} = V_{\text{Reference}} + V_{\text{OLA}}$ thus $V_{\text{OLA}} = V_{\text{OLA vs Reference}} - V_{\text{Reference}}$
3. $s_i = \sqrt{V_i}$ *for i* = a component or system

Literature Survey to establish a factorised understanding:

Table 1 below summarises the factors considered in this paper for the two measurement systems:

Table 1 : Factors impacting measurement performance applicable to the Reference and OLA systems

	Factor	Cause of variation:	Reference system	OLA system
1	Exposure	Sample representativeness of the measured process stream.	Sampling is guided by ISO 13909-8:2001.	OLA covers 100% of measurement process stream, but nuclear radiation interacts with about 1% of nuclei. (Cierpisz S, 2005)
2	Sample preparation & analysis	Execution of materials handling and processing procedures and analytical methods.	Preparation and analysis are guided by the relevant ISO / SANS standards	Not relevant

	Factor	Cause of variation:	Reference system	OLA system
3	Ash variability	Changes in coal characteristics: PSD, moisture, ash composition, ash range & level etc.	Not relevant	PGNAA technology is superior to other radiometric methods, (Cierpisz S, 2005)
4	Measurement time	Stochastic nature of OLA measurements.	Does not impact measurement variation. Benchmark time for the reference system is 2 hours, the sample removal interval.	Longer measurement times reduces variance.
5	Response time	Delay between the actual and the reported ash.	Has a major impact, is determined by the laboratory processing time.	Mainly determined by the measurement time. Signal processing very fast compared to measurement time.

Note:

Information source: Exxaro project work

From literature and project work a quantitative assessment gives perspective to the relative contribution of factors 1 to 4 as listed in Table 2 below for a two-hour measurement time.

Table 2: Quantitative assessment of factors impacting measurement performance for the Reference and OLA systems

	Factor	Reference system		OLA system	
		s	*V*	*s*	*V*
		[%Ash]	[%Ash^2]	[%Ash]	[%Ash^2]
1	Exposure	0.50*	0.250	0.027***	0.001
2	Sample preparation & analysis	0.50**	0.250	0.00	0.000
3	Ash variability	0.00	0.000	0.7***	0.490
4	Measurement time	0.00	0.000	0.118****	0.014
	System	0.71	0.500	0.71	0.505

Note: * CPSA, 2015

** Sowerby B.D., Lim C.S., Abernethy D.A., Liu Y, Maguire P.A., 1998

*** Cierpisz S, 2005

**** Exxaro project work

For the factors observed in Table 2, the two systems perform similar. To achieve the initial and sustainable performance of each system requires its own specific effort to control the relevant causes of measurement variation.

The OLA system is flexible regarding the duration of measurement time as it only requires the averaging of minute to minute observations. Consideration is required for the stochastic nature of observations, yielding more stable measurements at longer measurement times.

The comparison of response time between the systems is difficult due to the long delay of the reference system. The response time for the reference system is 4 hours compared to the OLA system of less than a minute. This is one of the key differences between the two systems determining its ability to support the tracking of cumulative product qualities and providing timeous information for process decision making. The reference system is more suited for cumulative quality tracking over longer production periods. No comparative analysis was conducted on this factor between the systems.

Confirmation of the expected operational performance:

The reference system is expected to perform at published estimates (refer Table 2) as the sampling system was designed and runs according to ISO standards. Laboratory preparation and analysis practices are executed according to relevant standards.

The OLA system is expected to perform to published estimates (refer Table 2) depending on the measurement time setup. For dynamic performance testing the measurement time was an hour and during normal operations it is set to 30 observations, equivalent to 30 minutes under full production.

The OEM performance guarantee for the By-line PGNAA analyser is included in Table 3 and was used as a reference during dynamic testing:

Table 3: GG7&8 OLA Performance Guarantee

COAL QUALITY PARAMETER	ASSUMED RANGE OF VALUES	RMSD GUARANTEE
As Received Ash	9.21 – 42%	2.588

*Performance guaranteed on a 1-hour, 1-sigma "as received" basis.

Dynamic Performance evaluation of the Reference and OLA systems:
ASTM 6543 and ISO 15239 standards were applied to assess the performance of the of the two systems.

ASTM 6543 standard applies the RMSD (Root Mean Squared Difference) statistic to describe measurement performance of the two systems in comparison. It does not isolate one system from the other and produces a combined system performance. Its application to rate the performance of the OLA system requires an unbiased and precise reference system.

The ISO 15239 standard includes a technique to test the analyser dynamic precision and uses the *s* (sample standard deviation) statistic to describe the measurement performance. It isolates the performance of the OLA system and is thus able to evaluate its performance. The statistical two-instrument test, applied in this study, requires a bias test to assess potential bias and remove it from the sampling system. The additive principle of system variance is applied to isolate the OLA system performance.

During the study the following evaluation protocol was agreed with the OEM and executed:

- Bias evaluation on the secondary sampler;
- Dynamic validation;
- Dynamic performance testing with application of both ASTM and ISO methods; and
- Direct comparison of RMSD performance achieved to performance guaranteed.

ISO 15239 method enabled the project team to calculate the OLA system and Reference system variation for comparison and verification.

Ongoing combined system performance management:
Table 4 shares some of the management actions to ensure ongoing measurement performance:

Table 4: Ongoing system performance management actions

	Factor	Reference system	OLA system
1	Exposure	Sampling health monitor	OLA feeder control setup
2	Sample preparation & analysis	SPC charts and Round Robins	-
3	Ash variability	-	3 Monthly CRS checks
4	Measurement time	PLC control program for sampling.	OLA system setup
5	Combined system	Daily comparison check	

Note:

Information source: Exxaro project work
CRS – Calibration Reference Standard

Sampling Health monitor
The overall quality monitoring system is divided into five sub-systems. A condition-based algorithm (Sampling Health Monitor) is utilised to track all the components daily in each of the five sub-systems:

- *Peripheral – production rates, feeder conveyors, discharge conveyors;*
- *Primary Sampler – Feed to the OLA, hydraulic power pack;*
- *OLA – Instrument internal health;*
- *Secondary Sampler – Feed to the reference system, hydraulic power pack; and*
- *Reference System – Sample size.*

A sub-system is highlighted as defective if one of its factors monitored, using the condition-based algorithms, are found to be defective. Fault conditions are highlighted and corrected to ensure system health. An example of a sub-system and its components tracked can be seen below for the Reference System.

The following factors can adversely affect the Reference System performance. They are the result of sampling system health and are monitored continuously.

1. Belt speed (OLA feeder)
   The analyser must run at above 8Hz belt speed to ensure continuous material flow to the secondary sampler. If the belt speed is insufficient, shearing of the analyser discharge (sampler feed) will occur.
2. Sample mass
   To do the analyses required, a minimum sample mass of 20kg is required by the reference system. If the sample mass is insufficient, it may be due to a problem with the sampling system.
3. Feeder Hopper running time
   The OLA is required to run 80% of the time, above the minimum belt speed, to ensure the secondary sampling system receives enough feed to be analysed by the reference system.
4. Number of secondary cuts
   To meet the ASTM 2234 standard, it is required that a minimum of 30 sample cuts of material is taken per sample interval (one hour).

Daily Production Ash Comparison check

The OLA and the reference system do not measure the same time intervals. The OLA reports a Rolling 30 measurement average and the reference system reports the average over a 2-hour interval. It is a complicated and time-consuming process to compare the two analysis systems. To ensure the comparability of the OLA to the Reference System, both systems' qualities are used to calculate an ash-cumulative product stockpile based on the tons produced per day. If there is a deviation of greater than 1%Ash, fault finding is immediately initiated.

Utilising the Sampling System Health Monitor, it can be established whether the deviation is due to the sampling system. If not, the correct intervals are pulled from the OLA to match the Reference System intervals for the day. If the deviation remains for the aligned intervals, joint fault-finding is done with the OEM. The system can be re-calibrated using the CRS and parameter adjustments are made by the OEM if required.

3 Monthly CRS checks

On a quarterly basis, 3 CRS's are used to verify the OLA measurement stability by applying the relevant procedures as prescribed by the ISO 15239 standard. These standards are created to represent the range of production qualities that can be expected during normal operation. This process culminates in confirming the instrument's linear response.

The CRS data is analysed by completing the three tests below.

F-test:

- Verify that the random measurement variability (Static repeatability / Measurement precision) remains acceptable
- Testing if the Variance ratio from test "f_n to f_{n+1}" is not significantly higher than a critical value for each standard.

T-test:

- Verify that the level of measurement response (Measurement accuracy) remains acceptable
- Testing if the test statistic tm in the comparison of test t_n to t_{n+1} is significantly greater than a critical value for each standard.

Linear Regression:

- Verify that the linear response (Correlation stability) remains acceptable.
- Comparison of off-set, slope and R^2 of the straight line when plotting the average measurement result for t_n to the certified value, per standard, over time.

For this article, only the linearity of the data is investigated and further discussed.

EXPERIMENTAL PROCEDURE

Dynamic Performance evaluation of the Reference and OLA systems:

Sampler bias testing

A bias test campaign was done on the secondary sampling system used to collect physical samples.

Dynamic validation

Dynamic validation entails comparative testing to determines the reliability of the sampling systems and laboratory analyses. It enables the OEM to make final adjustments to the OLA setup and calibration before dynamic testing. 30 one-hour composite production samples are collected of appropriate mass and submitted for analyses.

Analyses done include:

- Inherent Moisture
- Surface moisture
- Total moisture
- %Ash (As Received)
- %Sulphur (As Received)
- XRF analyses determine the ash constituents on 50% of samples.

Dynamic performance testing

Dynamic performance testing consisted of a minimum of 30 one-hour comparisons taken in duplicate, resulting in 60 measurement values from the reference system. These samples were analysed in duplicate by an independent laboratory.

For statistical analysis of results and to correctly calculate the analyser performance, it is important to test the range of operation. At least 5 samples close to the minimum operating range and 5 samples at the maximum operating range should be taken. Plant operating parameters should be adjusted to achieve this range.
Submit the samples to the external laboratory for the following analyses:

- Inherent Moisture
- Surface moisture
- Total moisture
- % Ash.

The prescribed data analysis according to ASTM 6543 and ISO 15239 was followed to determine the outcome of relevant tests:

- ISO 15239: Determining the OLA and reference system performance by implementing the two-instrument test.
- ASTM 6543: Comparison of actual RMSD to the performance guarantee

Ongoing combined system performance management:

Sampling Health monitor

Daily, historian data of all the components of the overall system is extracted and loaded into the condition-based algorithm file. Each sub-system is analysed in terms of a robot table. Analysis is done per hour. Each fault-condition has a pre-defined corrective action which will be initiated upon completion of the System Health Analysis.

Daily Production Ash Comparison check

An automated ash-simulation is done utilising Reference System data as well as Historian data. The simulation is updated daily. If the deviation between the Reference System and the OLA ash cumulative is greater than 1%, fault finding is initiated. The first step is to determine the Sampling System Health.

If the sampling system is healthy the second step is to manually align the OLA and Reference System intervals. The corresponding data is downloaded into the cumulative-ash-simulator to determine if the deviation is due to the interval impact. If the interval impact is minimal the OEM is contacted to assist in further fault-finding.

3 Monthly CRS checks

Instrument stability is determined by inserting the three reference standards into the OLA analysis zone for calibration validation.

The results of the test are captured and compared to historic data. If there is a deviation, the OEM is required to complete fault finding. The OEM can adjust the system parameters to correct equipment setup.

RESULTS AND DISCUSSION

Dynamic Performance evaluation of the Reference and OLA systems:
The application of the ASTM 6543 and ISO 15239 protocols on the Reference and OLA systems for a 1-hour measurement times is summarised in Table 5:

Table 5: Results of the dynamic performance test conducted on OLA

	Reference system		OLA system		Combined system			
	s [%Ash]	*V* [$\%Ash^2$]	*s* [%Ash]	*V* [$\%Ash^2$]	Bias (Ref vs OLA)	*V* [$\%Ash^2$]	*s* [%Ash]	RMSD [%Ash]
OLA - assumed actual (1 hr. interval)	0,71	0.500	0,72	0.518	0	1.018	1.009	1.009
OLA Performance test (1 hr. interval)	0,49	0.242	0,57	0.324	0,014	0.566	0,752	0,739

Note:
Information source: Exxaro project work

The dynamic performance test was conducted over a range of 8,8 %Ash from 29.1 to 37.8 %Ash.

ISO 15239: Determining the OLA and reference system performance:

- The measurement variation of the OLA system in terms of *s* reports 15 % higher than the Reference system.
- The Reference system *s* of 0,49 %Ash is calculated on the standard deviation of the differences between duplicate samples taken by the secondary sampler. This approach is simpler but inferior to the more complete three instrument test. It benefits the reference system and transfer an undetermined portion of its variation to the OLA system.
- Both systems outperformed the assumed performance
- The observed bias between the two systems is negligible.

ASTM 6543: Comparison of RMSD to performance guarantee

- The tested RMSD outperformed the OEM guarantee by 71%.
- Although the %Ash range during the performance test was a lot smaller than the OEM guarantee, hinting towards an expected improved performance, the achieved improved performance was unexpected.

Based on the test work outcome, the operations team recognises the OLA system as a complementary source of measurement information complementing the quality control process.

Ongoing combined system performance management:
An OLA in a quality control application needs regular monitoring to ensure that no drifting has taken place and accuracy is maintained.

Sampling Health monitor
Table 6 below is an example of the results of the condition-based monitor used to establish overall system health and identify sub-systems causing sub-standard measurement performance.

Table 6: Overall Sampling System Health Record – (20 hour extract)

	Sub-systems				
Interval	Peripheral	Primary Sampler	OLA	Secondary Sampler	Reference System (Sample Check)
Hour 1					
Hour 2					
Hour 3					
Hour 4					
Hour 5					
Hour 6					
Hour 7					
Hour 8					
Hour 9					
Hour 10					
Hour 11					
Hour 12					
Hour 13					
Hour 14					
Hour 15					
Hour 16					
Hour 17					
Hour 18					
Hour 19					
Hour 20					

Note:
Information source: Exxaro project work

The primary cause is highlighted in red, fault condition occurring because of the primary fault is orange and all other factors monitored but in a healthy state are highlighted in green.

During the 24-hour period shown in Table *6* the first major impact was due to the peripheral area, in this case the feed to the sampling plant. This is noted, but no fault-finding is required. Feed to the sampling plant is driven by the GG 7&8 plant production.

The second major impact was the secondary sampler taking insufficient sample cuts. The operations team investigated for possible blockages or material build-up. The maintenance team was contacted to ensure all the limits and actuator valves were in working condition. This resolved the issue.

Daily Production Ash Comparison check

Daily monitoring is done by simulating a cumulative stockpile built for the day for both OLA and the reference system as shown in Figure 4.

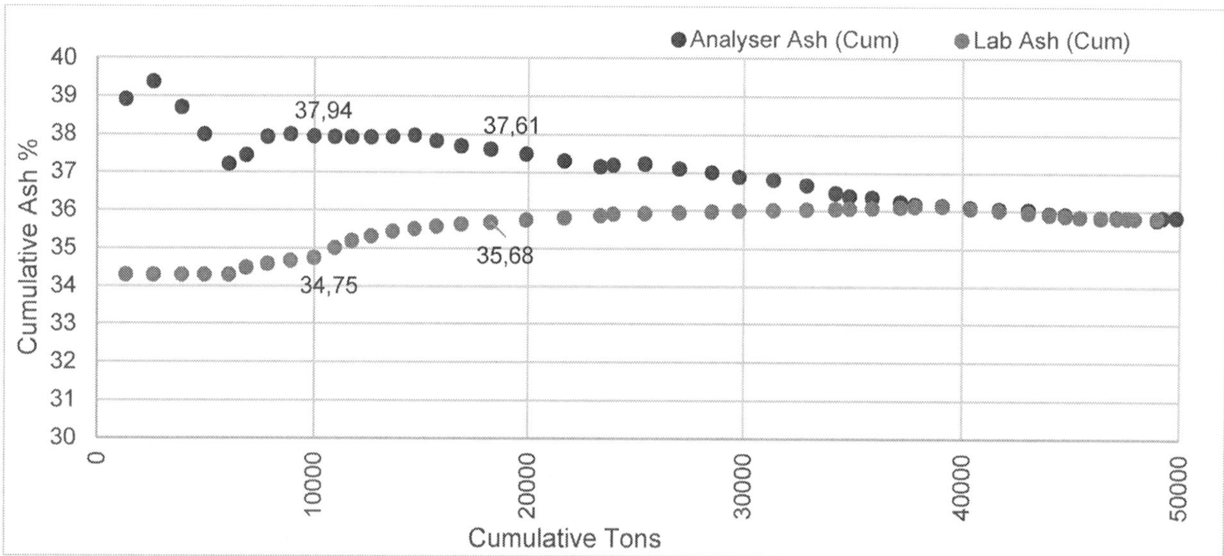

Figure 4: Comparison of the Reference and OLA systems for building a 50000t stockpile at a target of 35% Ash

Note:

Information source: Exxaro project work

The daily deviation is logged as displayed in Figure 5. If a constant bias is identified that is not due to the sampling system, the OEM can make parameter adjustments in the OLA operating system to correct it.

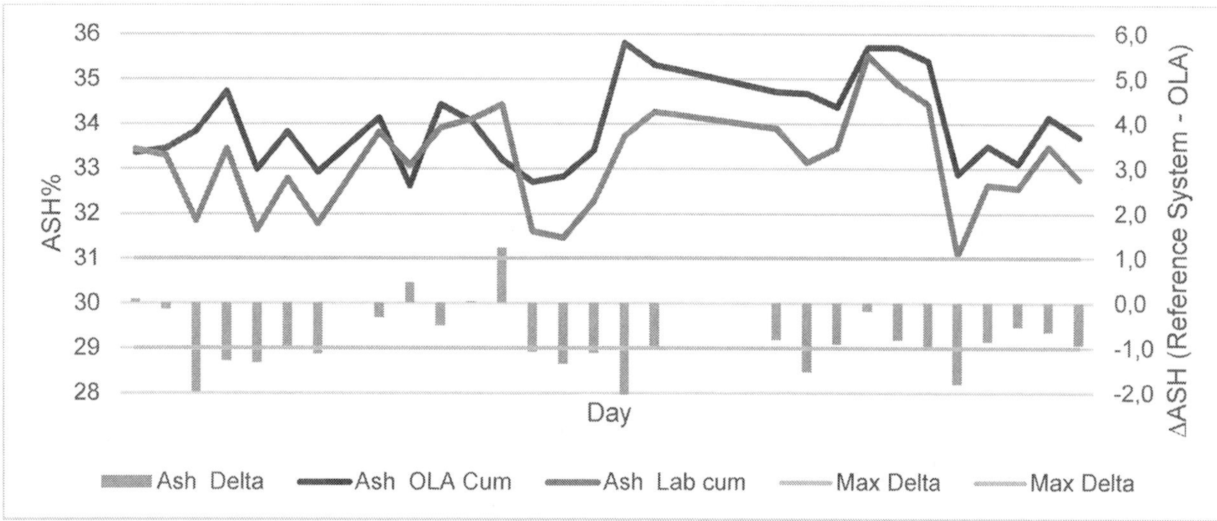

Figure 5: Cumulative Ash comparison for 30 days for OLA and Reference System

Note:

Information source: Exxaro project work

3 Monthly CRS checks

In Figure 6 below the results of evaluating the linear response for the reference standards are shown. The slope is required to be close to 1, the intercept should approach zero and the R^2 value should be close to 1. As can be seen in the results, this remains true for the tests done. Based on the linear regression test, there is an acceptable linear response.

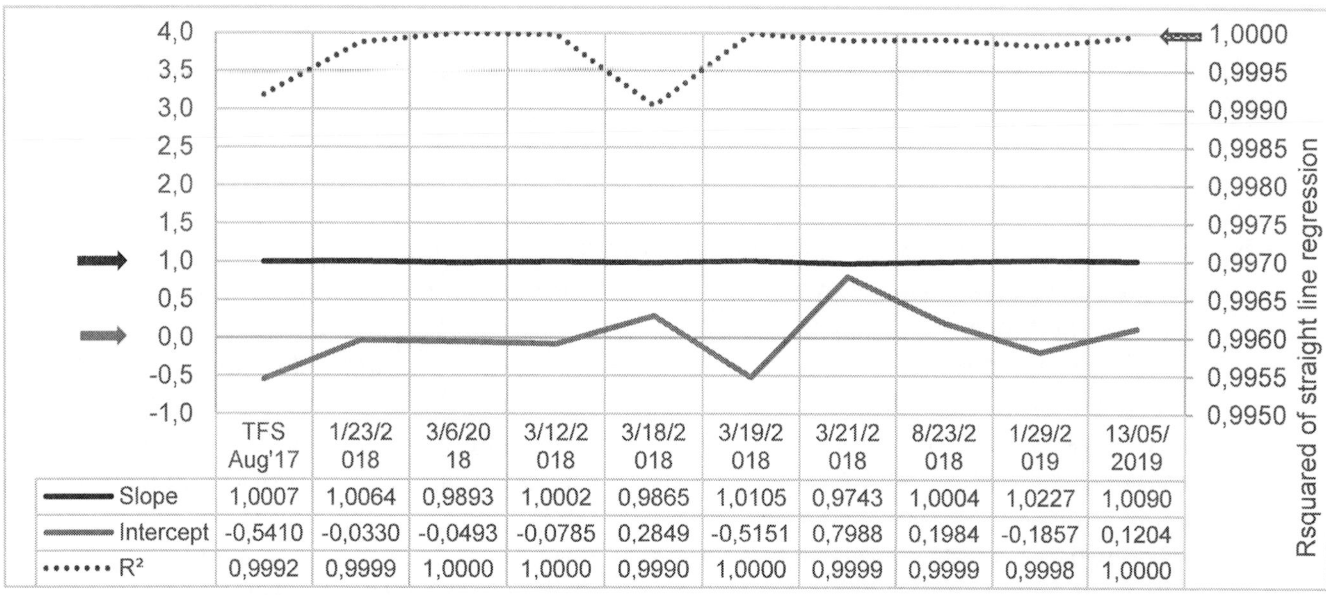

	TFS Aug'17	1/23/2018	3/6/2018	3/12/2018	3/18/2018	3/19/2018	3/21/2018	8/23/2018	1/29/2019	13/05/2019
Slope	1,0007	1,0064	0,9893	1,0002	0,9865	1,0105	0,9743	1,0004	1,0227	1,0090
Intercept	-0,5410	-0,0330	-0,0493	-0,0785	0,2849	-0,5151	0,7988	0,1984	-0,1857	0,1204
R^2	0,9992	0,9999	1,0000	1,0000	0,9990	1,0000	0,9999	0,9999	0,9998	1,0000

Figure 6: Confirming linear response of OLA data per 3 monthly checks

Note:

Information source: Exxaro project work

Operational application of OLA system:
At GG7&8 a rolling 30 measurement average is trended, updating every minute, used for immediate intervention in the case of a process upset or drifting in overall quality. In Figure 7 the OLA and reference system are compared over 24hours during an actual production day.

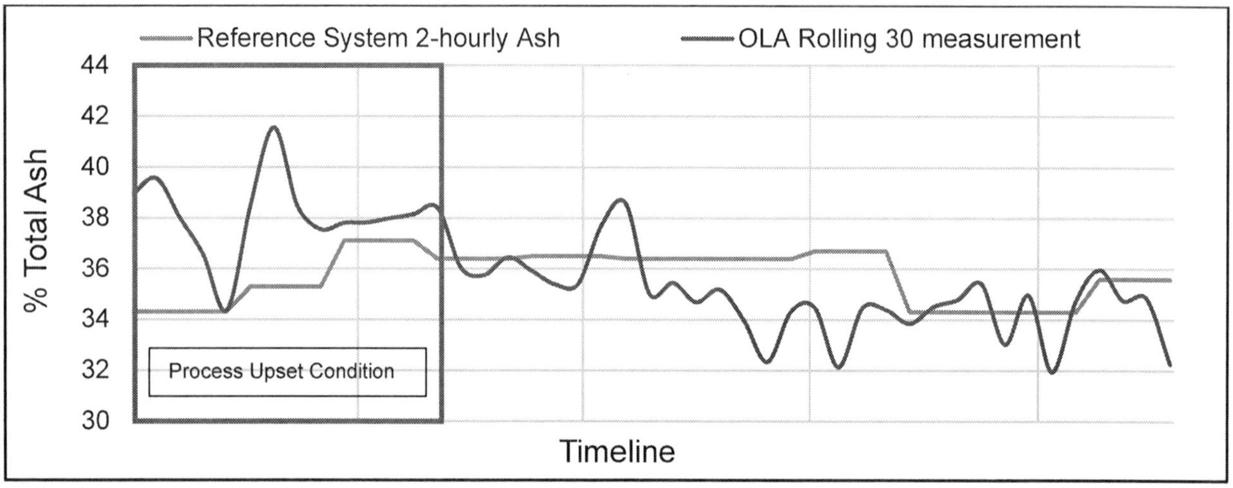

Figure 7: OLA vs Reference System actual production Ash comparison

Note:

Information source: Exxaro project work

As indicated in Figure 7, during this time one process upset occurred. Due to the more sensitive nature of the OLA, the process upset was identified and corrective measures were initiated after immediate trouble shooting. Be reminded that reference system Ash is reported with a 4-hour response time and that it did not identify the process event.

CONCLUSIONS

The following conclusions have been drawn from the completed study:

- OLA system has been demonstrated to perform like the Reference system at hourly measurement intervals
- OLA system is more flexible ito measurement times requiring only and programmatic adjustment.
- OLA system complements the reference system and enables faster decision making, reducing the impact of the 4-hour response time of the Reference system to a 1-minute calculation cycle.
- Reference system is best suited for cumulative ash tracking compared to application for control decision making due to its insensitivity to process variation and delayed response time.
- OLA system has gained the trust of operations teams and is applied in quality control decision making.

- Combined and separate system performance must be maintained through monitoring and intervention when required.
- Health monitoring algorithms executed manually leads to delayed responses by operations and maintenance teams and will benefit from the planned real time monitoring system being developed.

ACKNOWLEDGEMENTS

The authors acknowledge

- Exxaro management enabling the project and approving this publication
- the support of colleagues in the execution of the OLA project
- the trust of the operations team in the application of OLA for decision making.

REFERENCES

1. ASTM, 2000, Standard Guide to the Evaluation of Measurements Made by On-Line Coal Analyzers, ASTM D 6543-00, Volume **05.06**
2. Cierpisz S, 2005, Analysis of errors of coal quality monitors, IFAC Proceedings Volumes, Volume **38** (Issue **1)**, Pages 69-73
3. CPSA, 2015, Coal Preparation in Southern Africa, The Southern African Coal Processing Society, Section 3
4. SABS, 2012, Solid mineral fuels – Evaluation of the measurement performance of on-line analysers, SANS 15239:2006 Edition 1, ISO 15239:2005 Edition 1
5. Sowerby B.D., Lim C.S., Abernethy D.A., Liu Y, Maguire P.A., 1998, On-Conveyor Belt Determination of Ash in Coal, CSIRO, Pages 113 to 119

Medium Density Measurement without the Need for a Gamma Ray Source

T.M. Powell *Ultra-Dynamics Pty Ltd,*
PO Box 420 Ipswich Queensland 4305 Australia
Keith Buttemer, *Scanmin Africa (Pty) Ltd*

ABSTRACT

This paper describes the development, testing and operational experience of a density gauge for online measurement of magnetite medium density in pipes. Unlike industry standard, gamma ray based density gauges, no ionising radiation is used. The measurement is based on an inductive technique and the sensor consists of a pair of coils which are supplied in, either, a spool piece, or, probe which can be installed in process pipe work for direct measurement of either Rd or magnetite content.

The measurement principle involves the application of a highly sensitive magnetic susceptibility measurement. The resulting magnetic susceptibility distribution is used to identify the components of the medium and hence allows the summation of the mass fractions of these components to give overall density of the medium. Readings are made at 25mSec intervals which can be smoothed as required by the user. System update times are not affected by pipe size or density range resulting in faster tracking of density changes.

The cost of the gauge is comparable to convention gamma ray based systems without the significant additional cost imposts of Occupational Health and Safety regulatory requirements. The choice of either spool piece or probe are decided based on individual application eg wear, pipe size etc.

The system has been supplied to a number of process measurement applications and operational results are presented.

Further technical developments are underway to extend the application of the technology to low magnetic susceptibility applications such as tailings thickener underflow measurements. Early results are most encouraging.

1.0 Introduction – Basis of the Technology

densiMAG™ measures the magnetic susceptibility of materials by measuring the changes that these materials cause to an alternating current (AC) magnetic field. Changes in the AC magnetic field are caused by a complex mix of eddy currents and the action of magnetic domains. The magnitude of these changes are dependent on the material's magnetic and electrical properties, sample geometry and quantity of material presented for measurement.

Eddy currents are induced in a conductor when it enters the magnetic field core (inside the coil). Figure 1 shows that a current is produced when a magnet is moved over a copper coil. Figure 2 shows that this process can be reversed i.e. a current in a coil will produce a magnetic field.

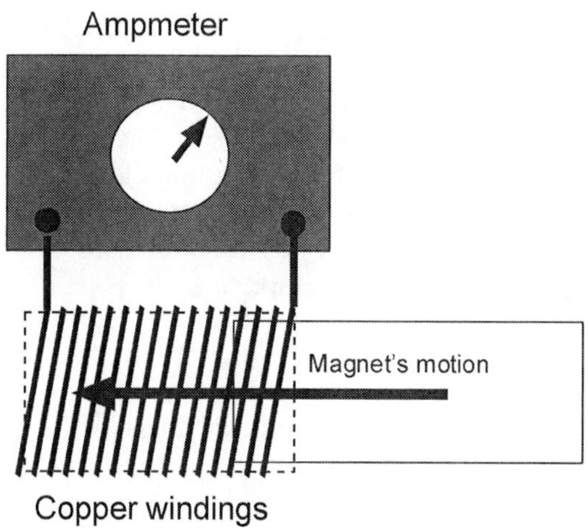

Figure 1

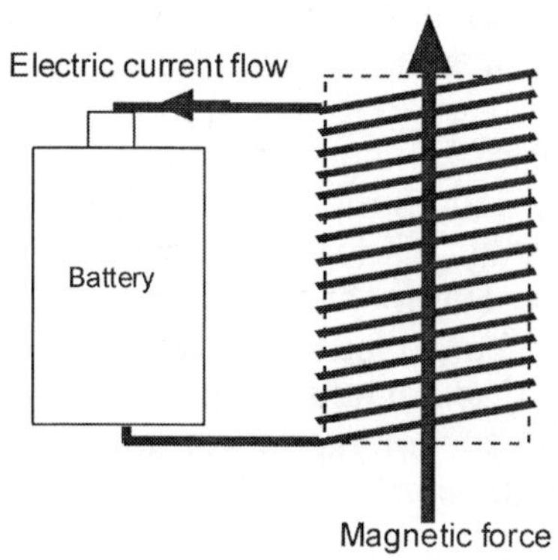

Figure 2

Magnetic domains are used to describe magnetic characteristics of ferrous materials. A ferromagnetic material consists of magnetic domains, which are small units that align themselves in an applied magnetic field. Figure 3 shows the movement of domains induced by a bar magnet.

Figure 3

When a ferromagnetic material is subjected to an applied magnetic field, the domains align. The alignment of domains cause the material to have considerable magnetic flux density. If the applied magnetic field is changing (i.e. an AC magnetic field) the flux density of the material and the moving of the domains have a significant effect on the applied AC magnetic field.

MAGNASAT subjects the material to be tested to an AC magnetic field. The strength of the eddy currents produced is a function of the magnetic properties of the material in the AC field. The eddy currents produced generate their own magnetic fields which have the effect of altering the electrical properties of the coil applying the magnetic field.
Ultra-Dynamics Pty Ltd's development of densiMAG™ has included investigations into the effects of all the factors that influence changes in the magnetic field. Studies included the effect of sample presentation geometry, as well as the development of a measurement coil/reference coil combination for the precise and rapid analysis of samples.

The initial development has been concentrated on the measurement of highly magnetic materials such as magnetite and ferrosilicon. The technique is well suited to the measurement of highly magnetic materials. Future software development will focus on less magnetic materials such as those found in tailings thickener applications.

2.0 System Description

The On-Pipe densiMAG™ system is designed to provide accurate measurements of process media as they are pumped around the process. The system utilises a magnetic technique for the measurement of magnetic susceptibility which can be converted to other process parameters of interest such as density, weight percent magnetite, % magnetics etc.
The system consists of a Measurement Spool Piece Assembly and an Electrical Cabinet. The Spool Piece Assembly is designed so that the measurement is made without contact with the process medium and consists of a pair of balanced Measurement and Reference Coil. These two coils are electrically matched so that there is a continuous removal of any extraneous magnetic field influences. The Electronics Control Cabinet houses the electronics for data processing as well as the electrical terminations for interfacing with the plant. A 5 metre length of interconnection cable is also supplied.

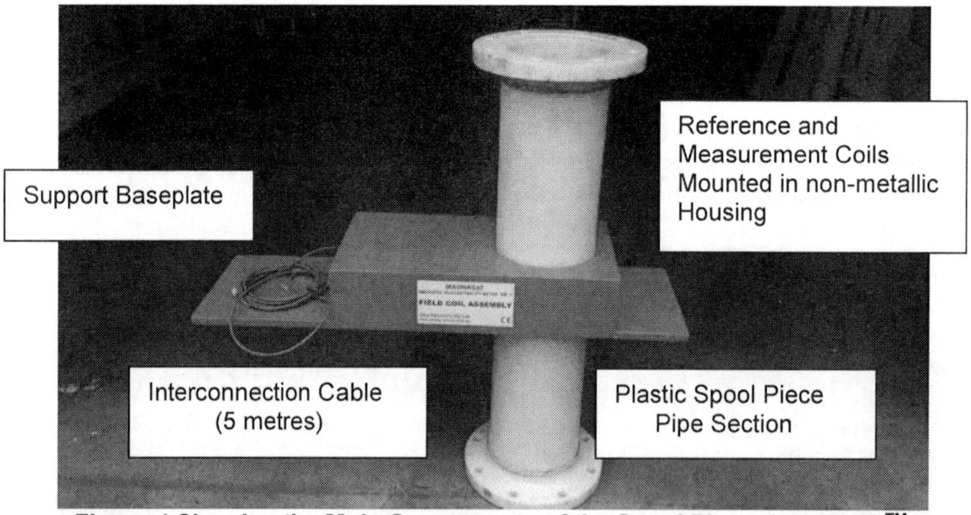

Figure 4 Showing the Main Components of the Spool Piece densiMAG™

3.0 Applications

The main application of interest explored in this paper is the measurement of the "correct medium" circuit. densiMAG™ has been found to be suitably accurate for this application. The conventional method for control of the dense medium density is to use a gamma ray based on-pipe density gauge. These gamma ray gauges suffer from some commercial and technical drawbacks such as the need for a site radiation officer, regular safety checks including wipe tests etc. Gamma ray gauges rely on the absorption of the gamma ray energy as it passes through the pipe and process medium to provide density measurement. Since the radioactive disintegrations are random the system update time has to be increased with increasing pipe diameter and/or density. densiMAG does not suffer from this problem and so large pipe applications can have the same system update time as smaller diameter.

4.0 Results

Results have been obtained from a number of installations and these are shown in graphical form below.

Results obtained from a test programme carried out to validate the technology are as shown in Figures and the ollowing shows the results obtained. One each of the Spool Piece and Probe type densiMAG options were installed in a dense medium circuit, the density was progressively altered and the results from three systems observed:

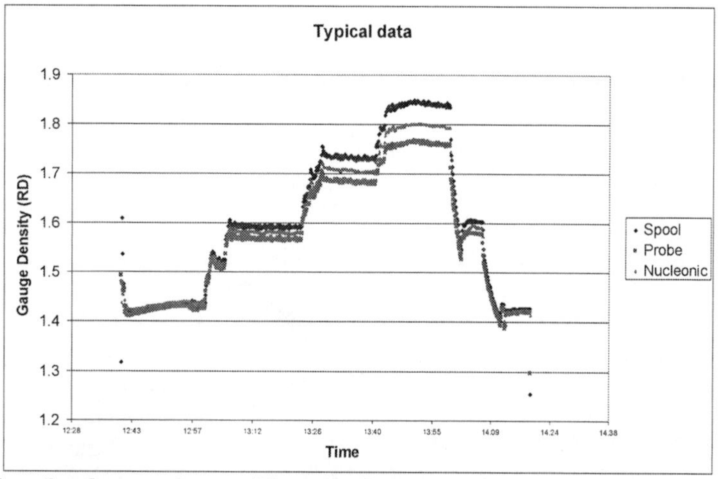

Figure 5 Showing the Comparison of Results Between densIMAG and a Gamma Ray Gauge

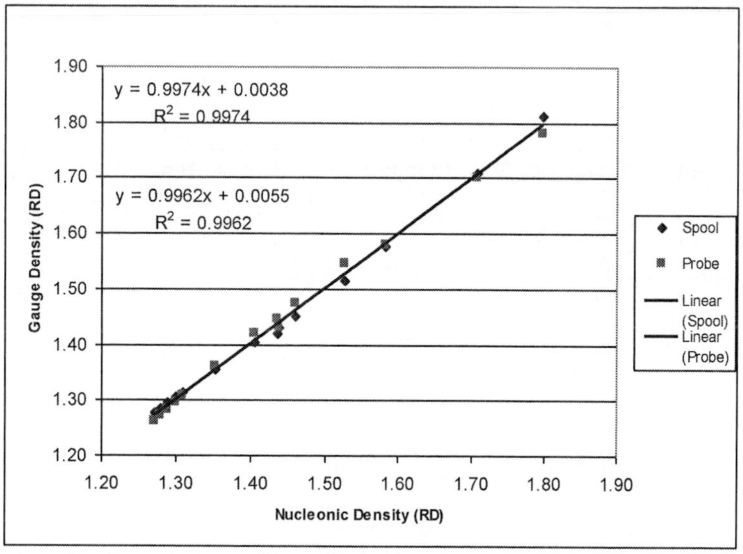

Figure 6 Showing the Correlation of Results Between densiMAG and a Gamma Ray Gauge

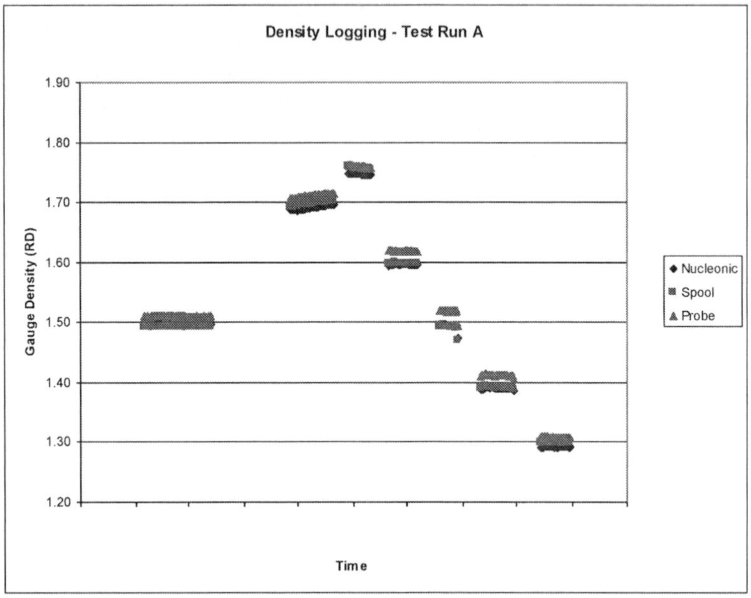

Figure 7 Showing How densiMAG and a Gamma Ray Gauge Tracked Changes in Density

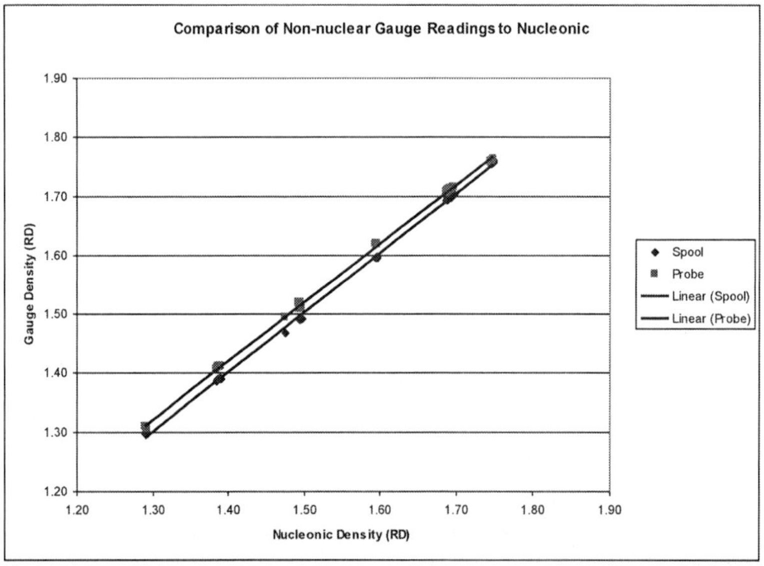

Figure 8 Showing densiMAG and Gamma Ray Gauge Results as a Scatter Plot

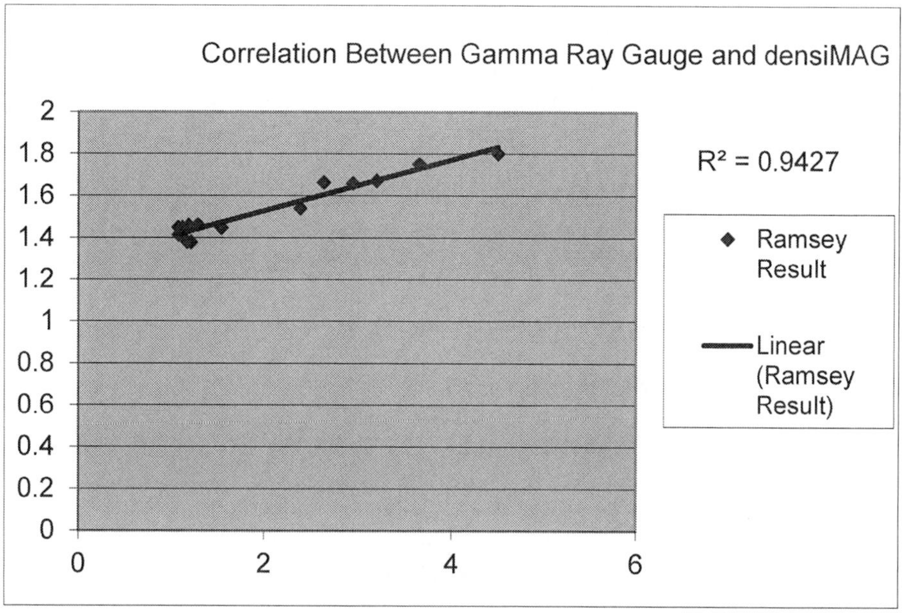

Figure 9 Showing the Results Obtained at a Coal Washery for densiMAG and a Gamma Ray Gauge

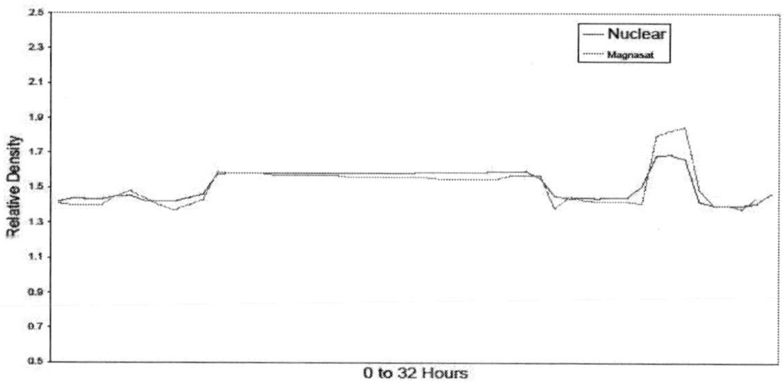

Figure 10 Showing a Tracking Plot for Both densiMAG^{TM} and a Gamma Ray Gauge Installed at a Coal Washery

5.0 Discussion of Results

The results shown were obtained in coal processing facilities and illustrate a good correlation between the industry standard gamma ray based density gauges and densiMAG^{TM}.
One important aspect of the densiMAG^{TM} technique that should be made clear is the effect of clays on the measured result. densiMAG^{TM} does not "see" non magnetic fractions of the slurry such as clays which may be present. It is our understood that these clays are desirable and even essential for the efficient processing of coal and therefore will be present during normal operation. If the clays are present in the slurry at a constant level then their effect on density can be compensated for and removed by a simple offset in the calibration parameter set. If

they are present at varying levels then they could introduce a significant error to the densiMAG™ reading.

6.0 Operational Experience

Several commercial systems have already been supplied including the following:

Australia

Wilpinjong Coal Preparation Plant – Correct Medium Circuit

BMA Peak Downs Mine – correct medium circuit

Anglo Coal Moranbah North – Magnetite make up circuit

New Hope Coal – correct medium circuit

Iluka Mineral Resources – mineral sands plant feed monitoring

JKTech University of Queensland – research project

New Zealand

Taharoa Sands – iron sands plant, plant feed monitoring

Chile

CAP Iron Ore Processing Plant – tailings loss monitoring

Mexico
A.H.A.M.S.A. Iron Ore Pellet Plant – pellet plant feed monitoring

Holland

Enerco, Coal Preparation Plant – correct medium circuit

7.0 Conclusion

Ultra-Dynamics Pty Ltd densiMAG™ Non Nuclear Density Gauge can be utilised in coal preparation plant control in place of the conventional gamma ray based technologies. The one parameter that can affect the performance is the presence of varying quantities of clays. Provided the non magnetic clays content of the slurry remains relatively constant then this parameter can be calibrated out. Further work is to be carried out to extend the application of the technology to other applications such as thickener underflow control.

Southern African Coal Processing Society

OPTIMISING A REFLUX CLASSIFIER CIRCUIT

By P. Ramsaywok, G Legodi and S Ndanduleni

Exxaro Resources Ltd

The objective of this paper is to share the optimisation journey of a pilot Reflux classifier (RC) circuit in the Exxaro GG1 plant, at Grootegeluk beneficiation complex in Lephalale, South Africa. The RC was commissioned in 2016, however the stability of operation as well as the quality control of the coal product was not fully achieved. Various test campaigns and optimisation studies were conducted to understand and optimise the RC circuit at GG1. The studies revealed circuit design inefficiencies, improved technical understanding of the current pilot circuit and the circuits capability to reach a stable operation. The operation was stabilised with its current inefficiencies to produce an average 12.5% ash product for prolonged periods.

INTRODUCTION

Traditionally with gravity separation, as finer particle sizes are processed, efficiency tends to decrease. Separation of coal in the finer fraction size (1x0.5mm) particle sizes has gained much publication in the last decade with the deployment of Reflux classifiers (RC), low-cut spirals and fine dense medium cyclones (dmc). Low-cut spirals are extending the cut density capabilities of spirals and RC as well as fine coal dmc addressing the traditional lower separation efficiencies when processing fine coal (-1mm). Various studies have confirmed the low-cut spirals' ability to cut lower than traditional spirals. The RC technology has also been proved to have high efficiencies provided the feed is introduced in a narrow size range (typically 3:1 or 4:1; top size: bottom size). Fine dmc is an extension of dmc technology treating fine (-1mm) coal.

The RC utilises teeter bed separator and lamella settling principles. Its main advantages over spirals is higher throughput for a given footprint and the ability to dynamically change product qualities as required. The main disadvantages are increased water utilisation and a high degree of dependence on the feed preparation circuits. There is a perceived complexity associated with the technology.

GG1 RC flowsheet

A RC2000 was installed at Exxaro GG1 plant and commissioned in 2016. The design intent of this plant was to treat 1x0.212mm particles to produce a Semi soft coking coal product of 10.3% ash.

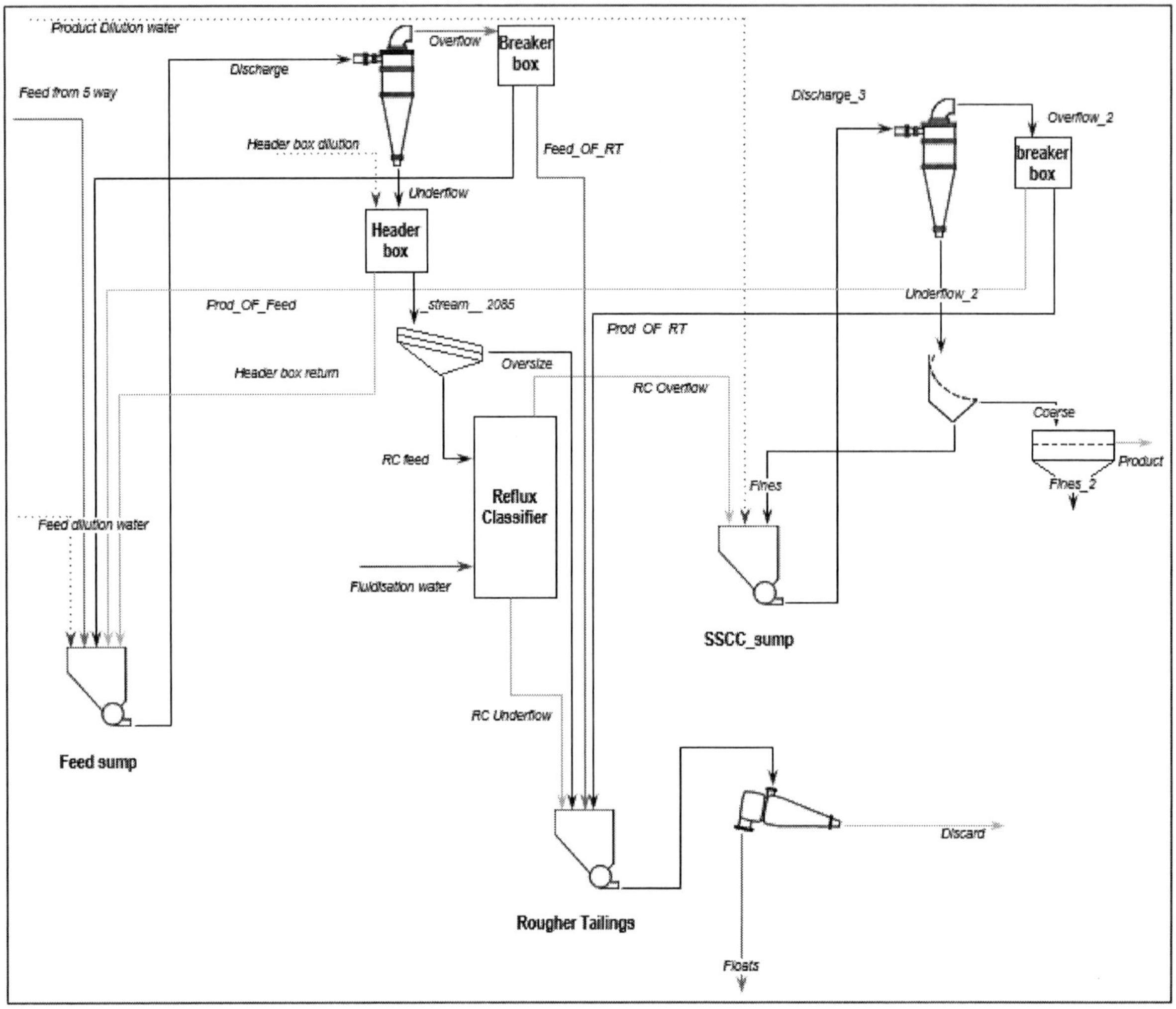

Figure 1: Schematic representation of GG1 RC flowsheet

Fines (-1mm) from the main plant reports to a thickener from which it is pumped to the fine circuit consisting of spirals. One of the spiral modules have been removed to make room for the RC plant with its flowlines shown in **Figure 1**. Slurry feed is pumped to classification cyclones to remove ultrafines (-90µm). Cyclone underflow is passed over a static oversize protection screen to remove +1mm particles. The -1mm from the oversize protection screen enters the RC. RC is pumped to another set of classification cyclones and the underflow from the cyclones dewatered through a sieve bend and a vibrating dewatering screen and then reports to the Semi-soft product conveyors. Raw water is used to fluidise the RC.

Typical RC feed properties at GG1

The RC plant processes the fine coal from Bench's 2,3 and 4. Products from these benches are for power generation and export coking coal markets . **Table 1** provides a typical particle size distribution as well as the fractional ash for the GG1 RC feed. The particle sizes outside the design envelope are typically 6% by mass +1mm and 36.5% -212µm. The ash quality deteriorates as the particle sizes become smaller.

Table 1 : Particle size distribution and fractional ash for typical RC Feed at GG1

Typical RC feed particle size distribution			Fractional Ash	
Particle size µm	Cumulative passing (mass %)		Particle size µm	Ash (mass%)
Feed	100.0		Feed	32.9
1000	94.0		+1000	22.9
850	89.7		+850	23.4
710	84.5		+710	23.0
600	77.4		+600	24.1
500	67.9		+500	25.2
450	58.6		+450	26.8
355	53.5		+355	29.0
212	36.5		+212	30.1
180	33.2		+180	33.5
125	25.7		+125	36.0
90	21.0		+90	36.3
63	17.9		+63	38.1
45	15.6		+45	39.9
-45	n/a		-45	51.8

Figure 2 shows a typical grade recovery curve based on washability analyses for GG1 RC feed with -212µm removed. This highlights the need to be able to dynamically adjust product qualities as market demands change.

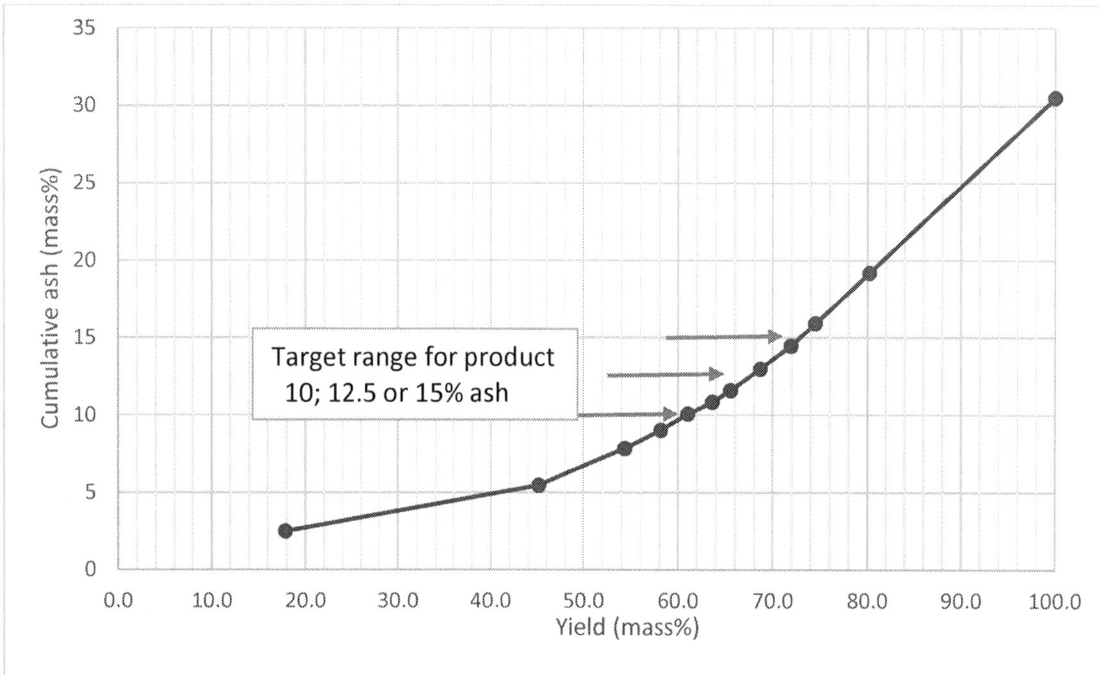

Figure 2: Typical grade recovery curve (based on washability) for GG1 RC feed with target product qualities indicated.

CHALLENGES IN IMPLEMENTING A RC CIRCUIT INTO THE EXISTING GG1 SPIRAL CIRCUIT

Retrofitting a RC into an existing spiral circuit

Retrofitting a RC into a spiral circuit comes with its own set of challenges. One of the main challenges being the changes in volumes of water required. The additional water in the circuit impacts pipe sizes, tank and sump sizes. Any spare capacity in the existing circuit is quickly used up. This then limits the extent of parameter variation, such as fluidisation water rate in the RC, that can be explored.

Circuit inefficiencies

Upon evaluation of the GG1 RC circuit, the main process inefficiencies relate to:

Oversize +1mm removal; and

Undersize-212µm removal.

Oversize is removed via a static oversize protection screen. In operation, slurry runs over the surface of the screen reporting to the RC discard causing a portion of the feed bypassing to discards. To overcome this issue operationally, mixed screen panels with square apertures between 10-15mm are used. Although this minimised feed bypass, oversize in the form of +1mm reports as RC feed. The +1mm in the RC feed was normally around 6% by mass but in some occasions as high as 10%.

The ultrafines (-212µm targeted) in the RC circuit is removed by 900mm diameter classification cyclones. The cut-point on these cyclones was measured and was found to range between 90µm-105µm. The -212µm in the RC feed varied between 20-40% and averaged 36.5% by mass during the RC campaigns. The ash quality of the -212µm fraction was above 42%. To demonstrate the impact of -212µm on RC product quality, the RC was run on erratic upset conditions for a week. Shiftly changes affected the amount of fluidized water and fluctuating feed to the classification cyclones also meant that there was some inconsistency in the feed volumes to the RC during this period. **Figure 3** shows the variation of the RC product quality during this period with qualities between 10 and 35% ash. All production samples for the period were retained and randomly selected samples over the period were screened in the laboratory removing the -212µm and the +212µm analysed for its ash content. The screened RC product (with the -212µm fraction removed) show periods of less than 10% ash and periods of 10-15% ash.

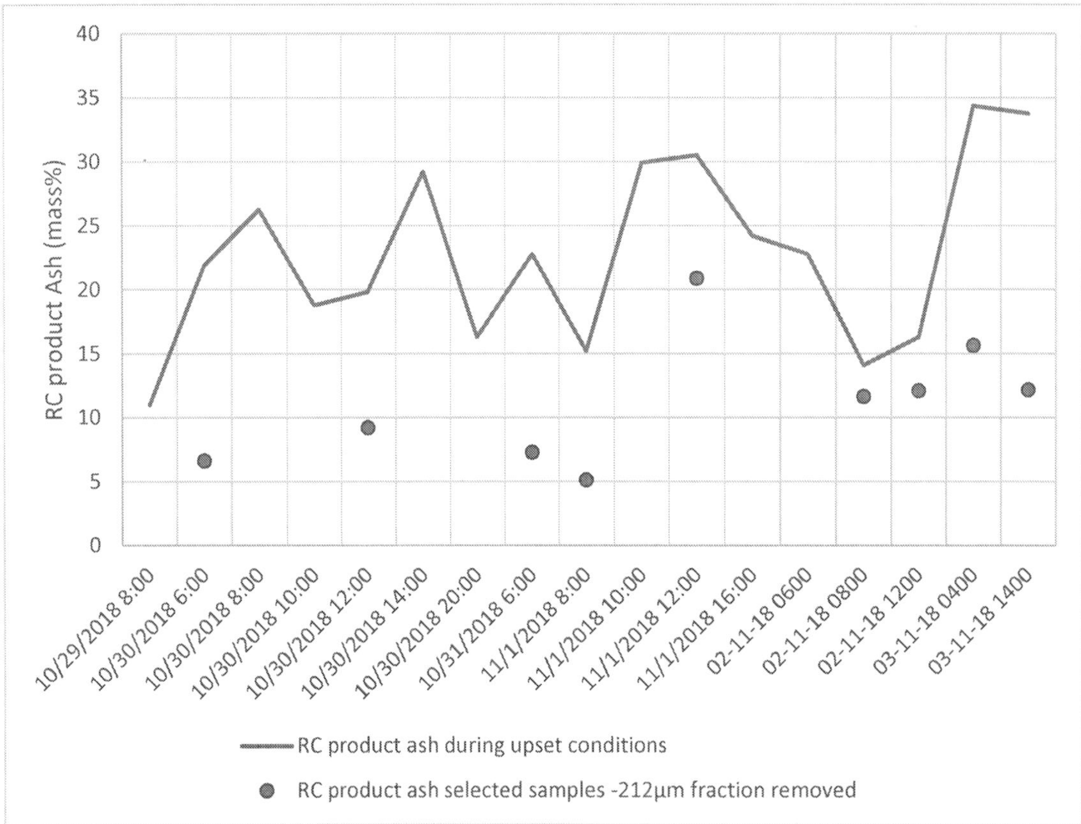

Figure 3: Effect of -212μm on RC product quality. Conditions in the RC were deliberately upset highlighting production variations and feed preparation variations significantly impacts product quality.

PROJECT WORK TO OPTIMISING THE RC CIRCUIT

Circuit stabilisation

With an understanding of the circuit deficiencies, it was decided to evaluate stabilisation of the RC circuit for a prolonged period to produce a 12.5% ash product instead of the 10.3% design intent. The reason for the change in target from 10.3 to 12.5% ash was to allow the unit to operate in normal production as the plant was producing a 12.5% ash at the time. **Figure 4** summarises the effect of RD set-point and fluidisation rates to vary product ash quality. Also shown in **Figure 4** is the settings selected for the stabilisation test (RD set-point of 1250 kg/m^3 and fluidisation water rate of 35m^3/h). The set-point chosen was on the conservative side to account for circuit inefficiencies expected during production. The lower fluidization set-point of 35m^3/h was chosen to manage water in the existing spiral plant. Fluidisation rates below 35m^3/h were not evaluated during the optimisation as the range was unknown at the time when the circuit was stabilised.

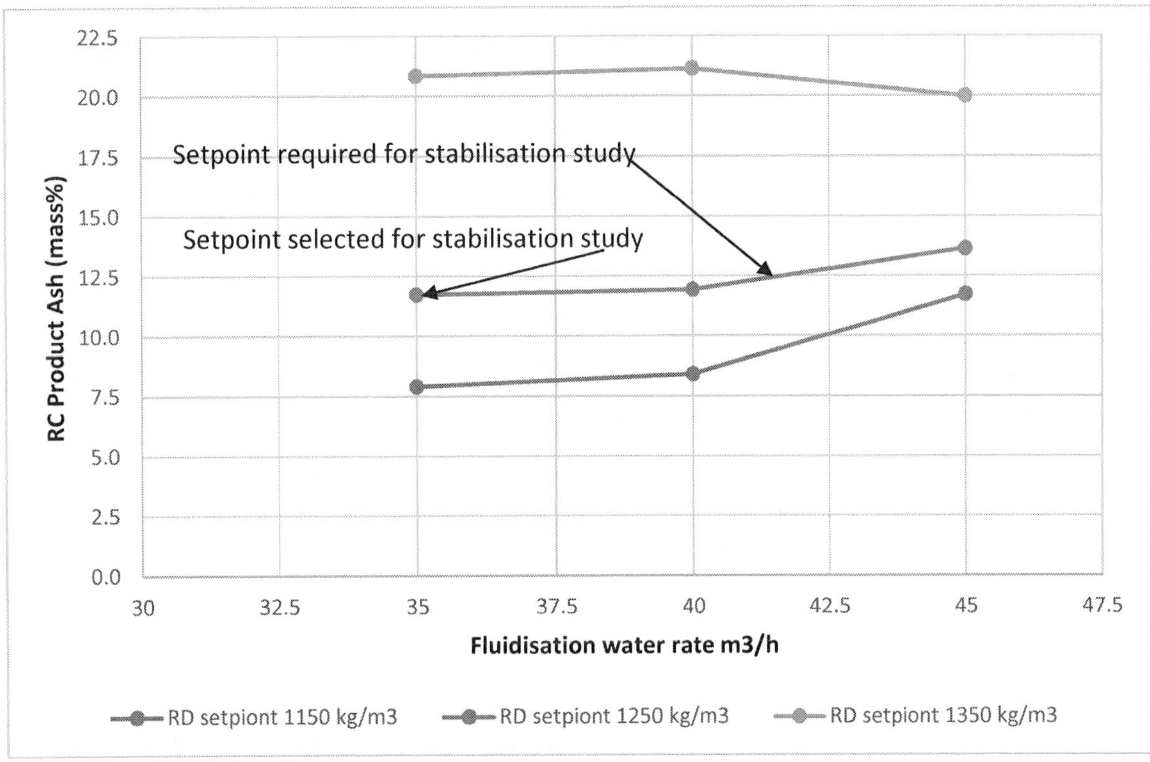

Figure 4:RD set-point and fluidisation water rates evaluated to produce a 12.5% ash product. Set-points selected for evaluation of a stable operation is also indicated.

Figure 5 summarises the production data after set-point intervention for a stable run. Periods of stable operation between 10 and 15% ash was observed for a period of approximately 2 months. Considering the inefficiencies of the GG1 circuit, the stability tests were deemed as successful.

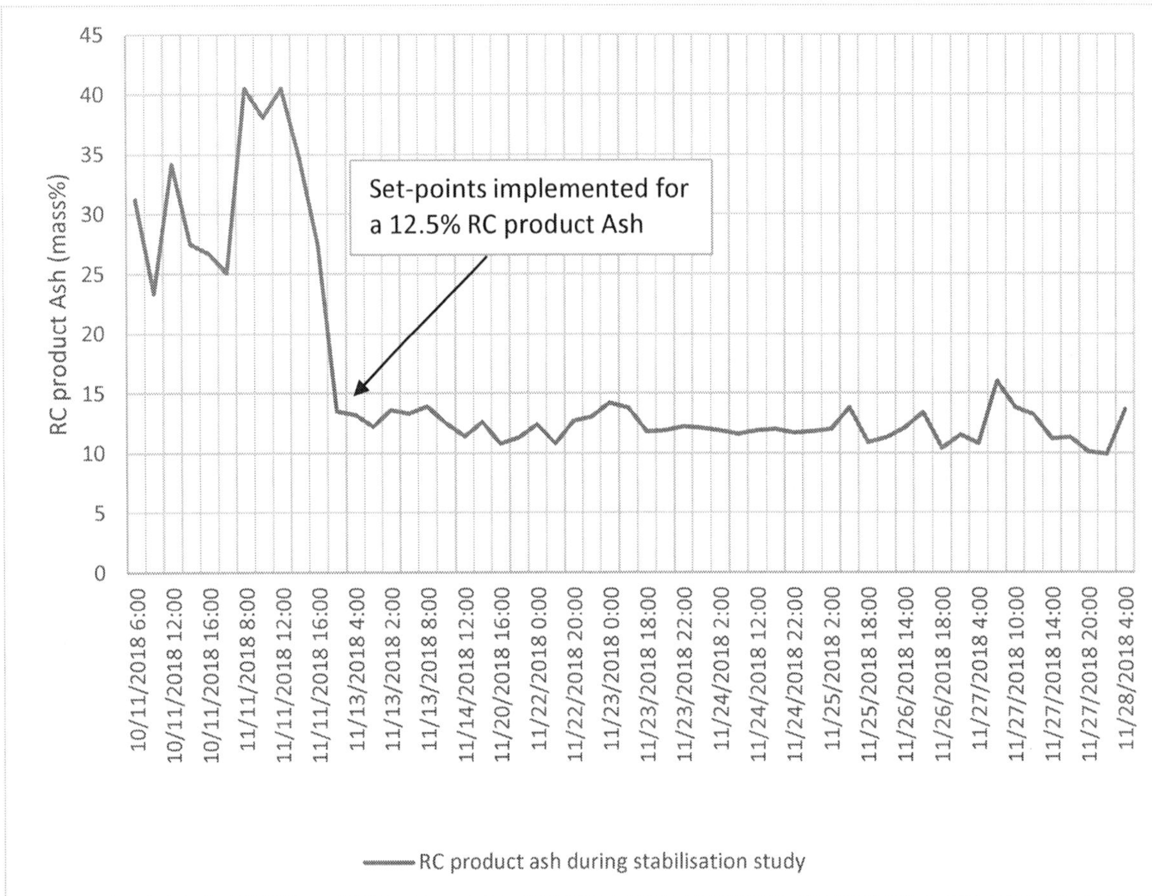

Figure 5: RC product quality observed from production data at a RD set-point of 1250 kg/m³ for the stabilisation study to targeting a 12.5% product Ash.

Changing product qualities on the fly (on demand)

One of the advantages of having a RC in you plant is the ability to adjust to varying product qualities by simply applying new set-points. As an example, **Table 2** shows the impact of RD set-point, demonstrating the ability to shift the cut-point produce a 10 to 15% ash on demand.

Table 2 : Impact of RD set-point to vary product quality

Set-point kg/m³	1175	1275	1375
Feed Ash mass%	30.5	31.5	32.1
Product Ash mass%	10.9	11.8	15.3
Cut-point D50 measured	1.35	1.45	1.56

CONCLUSIONS AND RECOMMENDATIONS

From this study to get a stable RC circuit at GG1 the following conclusions are made:

- Production variation and upsets impacts are predominantly seen in the RC product with almost all cases leading to poorer product quality. In a coal preparation plant, the RC is not a forgiving piece of technology and is heavily reliant of proper feed preparation in terms of oversize and undersize removal. This adds complexity to a RC circuit in terms of operation and maintenance.
- When stabilising a RC operation in setting target product qualities, consideration must be given to the entire circuit and not just the RC itself.
- The RC can be stabilised and supports dynamically changing cut-points to produce varying product qualities.

The following recommendations are made for improving existing or new RC circuits:

- Change oversize protection to vibratory screens; and
- Implement high frequency screening on either the feed or product streams of the RC to remove -212μm as applicable to this feed.

REFERENCES

n/a

TAKING CARE OF YOUR PLANT'S CYCLONES IN ORDER TO TAKE CARE OF YOUR PLANT

F.H. Enslin; E. Bekker

Multotec Process Equipment.

ABSTRACT

With a very rich and colourful history in South Africa, the coal industry has reached a point where the quality of coal resources is declining and differs quite significantly from the coal that was mined 30, 40 and 50 years ago. On top of that it also seems however that the level of experience in the industry may be declining. To maintain and improve the efficiency of the process and to operate the Coal Processing Plant (CPP) at a more economically rewarding level, it is necessary to refresh operational personnel's understanding of how an integral piece of CPP equipment, like the cyclone (both dense medium and classification) work, the factors impacting on its performance and common mistakes that are observed in the industry on a frequent basis.

This paper will briefly describe the impact that poor dense medium cyclone performance has on material with various levels of near density material and its economic impact. It will also discuss the effect of varying feed conditions and the influence on cyclone efficiency for hydro cyclones on a high level. Although cyclones have a fairly simple design with no moving parts, the effect they have on a plant's overall performance is staggering, and we therefore cannot afford them to be operated at a substandard level with the modern coal reserves.

Furthermore, common field observations will be discussed and the impact it can have on the operation of the cyclone unit as well as the overall process. Where relevant, the observations will include other pieces of equipment as well. Most of these observations are preventable and can be quite easily corrected if plant personnel understand the equipment's working principles, respect their function in the CPP and know what to look out for.

The paper concludes by looking at some of the tools used by Multotec to do cyclone inspections onsite, specifically the use of a 360° VR camera which reduces inspection time drastically and gives detailed imagery of the internal wear surfaces.

With a focussed and disciplined approached it is still possible to extract sufficient value from modern low-grade coal reserves.

BACKGROUND

From the very beginning humans investigated natural phenomena and the manipulation thereof for our own advantage. It is therefore no surprise that the manipulation of centrifugal forces in classification applications by means of hydrocyclones dates back to even before 1890. In 1891 a 43-year old Eugene Bretney filed a patent for a water purifier (Bretney, 1891) which would turn out to be an early ancestor of the cyclones (hydro- and DMS) we still use to this day in our processing plants. As can be seen from the following drawing, taken from Bretney's patent, the cyclone as we know it today has evolved over the past 130 odd years.

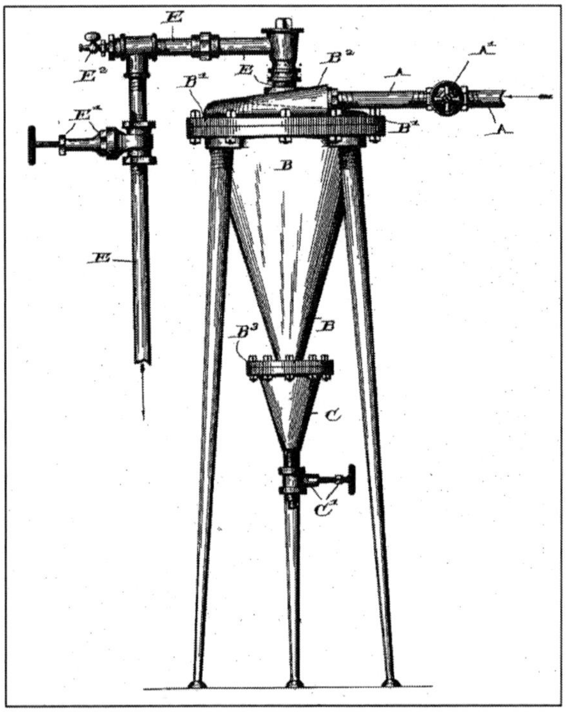

Figure 1: Bretney's design for a water purifier (Bretney, 1891)

Between 1900 and 1937, 30 additional patents were issued for the use of devices resembling the hydrocyclone of today. According to Smith (2000:3) the first commercial application of the hydrocyclone came about in 1937, when hydrocyclones were sold for cleaning sand and dirt particles from fibre suspensions in the pulp and paper industry. By 1940, hydrocyclones as we know them today, made an appearance in the paper and pulp as well as coal washing industries. At the same time, Maximiliaan Driessen, of the Dutch State Mines (DSM) was working on hydrocyclones for thickening and classification of slurry streams (Bradley, 1965). The Dutch State Mines hydrocyclone was first commercially implemented in the US in 1948.

The DSM continued with their ground-breaking work in the development and application of cyclones in specifically coal processing plants and in 1946 Driessen *et al.* (1951) filed the very first patent for a dense medium separation (DMS) cyclone. Although the patent describes using loess (a type of sand), instead of the magnetite we use today in DMS applications for coal, the principles developed by the DSM still form the basis of for all modern DMS cyclone designs (Singleton, 2013:27).

Cyclones (and specifically the DMS cyclone) have become the key processing unit in coal processing plants since its development more than 70 years ago (de Korte, 2012:2). With its capability of efficient separations, ease of operation and accurate control, it has proven itself as a workhorse in the coal processing industry. Unfortunately, a lot of processing plants neglect the operation, maintenance and sizing of their installed cyclones. This may be due to their relative simplistic design, the fact that they don't have any moving parts or even complacency or a lack of proper understanding from those who operate them. Nonetheless, with South African coal reserves which get more difficult to separate, yielding less product and at lower product qualities (Enslin, 2017:3), it is of utmost importance to ensure that the heart of the coal processing plant (the cyclones) is performing optimally to improve the plant's economic viability.

This paper focuses on some of the common mistakes and practices seen on a regular basis at various coal processing plants, how it affects the plant overall and what can be done to fix the situation.

THE DENSE MEDIUM CYCLONE'S ROLE IN THE COAL PROCESSING PLANT

Below is a simplified flow diagram of a coal processing plant commonly found in the Witbank coalfields. This plant consists of a Coarse and Smalls Low Gravity circuit (producing the higher quality coal, 6000kcal/kg or 5500kcal/kg) from which the sinks are collected and fed to the High Gravity circuit (producing the lower quality coal, 5500kcal/kg or 4800kcal/kg). There are also however examples where the floats (and not the sinks) of the primary DMS circuit is fed to the secondary DMS circuit so that the higher quality coal is produced in the secondary circuit. Choosing this configuration will depend on the specific plant's washabilities and the cyclones' underflow capacities. The fines are beneficiated by means of spiral classifiers and with the assistance of hydrocyclones. There are also plants that may cut a bit coarser on their desliming screens than the 1mm used in this example, but the fact remains that the +1mm (or even sometimes +2mm) material is normally processed by means of dense medium separation and then specifically using DMS cyclones. Although there is a magnitude of different coal processing plant configurations available, the basic principles remain the same and since the plant described here and depicted below is very common for the Witbank coalfields specifically, the examples given in this paper will be based on this plant configuration. It is however possible, without great effort, to adapt the examples and calculations given in this paper for any specific coal processing plant under consideration.

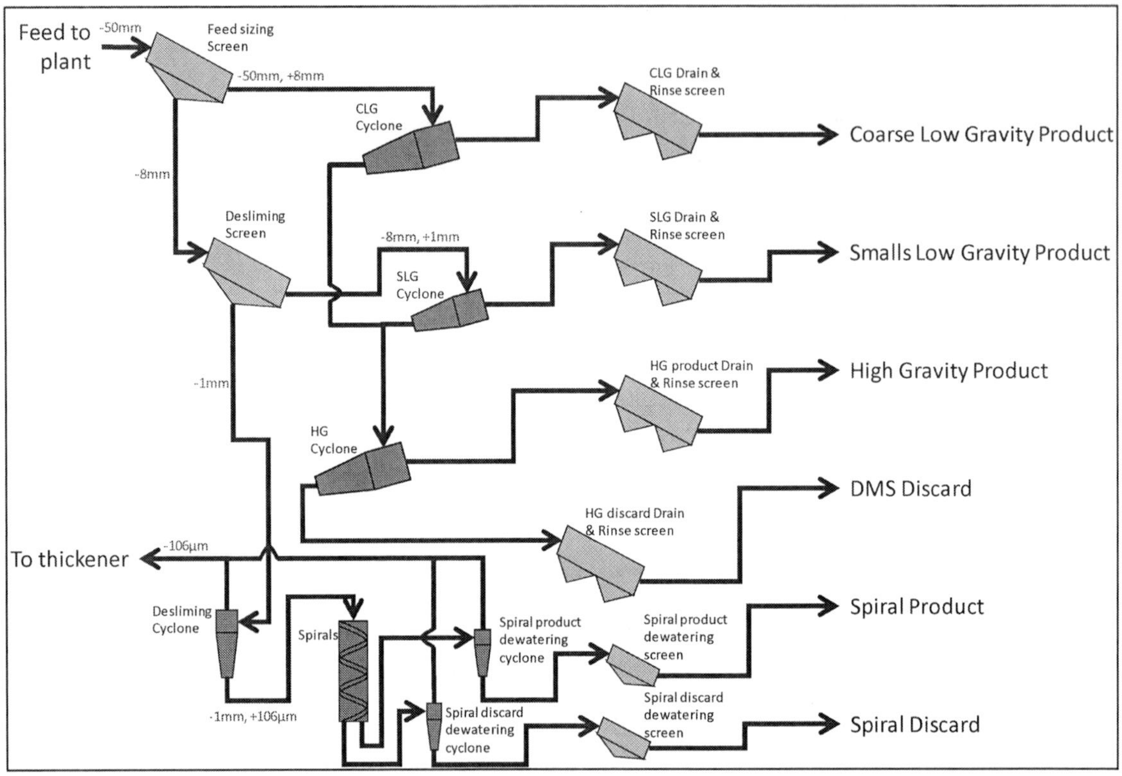

Figure 2: Simplified flow diagram of a common coal processing plant in the Witbank coalfields.

Based on several samples taken of the feed to various coal processing plants, the particle size distributions indicate that the fines (-1mm material) represents anything from 2% to 33% of the total feed to the plant. This means that 67% to 98% of the total feed to the plant will be beneficiated in the plant's DMS circuit. For the sake of the example and calculations to follow, a figure of 80% will be assumed, but as can be seen from the following graph, this figure may vary significantly.

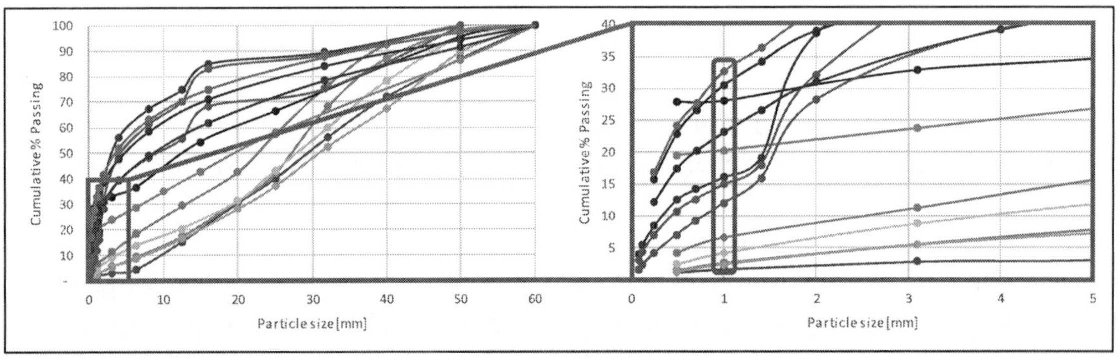

Figure 3: Particle size distributions of the feed to various coal processing plants.

With a few basic assumptions (based on what is happening in reality) it becomes an easy calculation to quantify exactly how important dense medium cyclones are in the economic viability of a coal processing plant. The following assumptions are made for the calculation to follow:

Table 1: Assumptions made for profit creation calculation.

Feed to plant (FTP)	1,000	tph
Portion fed to DMS cyclones	-50, +1	mm
Portion of FTP fed to DMS cyclones	80	%
Expected DMS cyclone product yield	70	%
Revenue per saleable ton (4800kcal/kg)	R850	

Based on these assumptions, one can expect 560tph of product passing through the cyclones (1000 x 80% x 70% = 560tph). At a selling price of R850/t the potential revenue created by the dense medium cyclones is in the order of R132/sec (R850 x 560tph ÷ 3600sec/hr = R132.22). Should we assume that you would require 2 x 1000mm cyclones to beneficiate this coal and that the profit for selling the coal is only a mere 15%, the profit generated by a single 1000mm cyclone under these conditions is enough to pay for that cyclone in just more than a single shift's time (R300,000 ÷ (R132.22 ÷ 2 x 15%) = 8hrs 25min). Generating a profit in the order of R10/sec it becomes clear why it is worthwhile to ensure the cyclone is operating optimally.

As mentioned previously, if 80% of the total feed to plant (-50, +1mm) is processed by means of Dense Medium Separation (and then specifically DMS Cyclones), 1% misplacement of floats to the underflow for a 1000tph CPP (producing a 4800kcal/kg product) relates to a potential loss of revenue in the order of R104,000 per day. These are only typical values for a coal processing plant in the Witbank-coalfields, so for more variance in the %Fines (-1mm) in the FTP, product yield achieved by the DMC and misplacement of floats to the cyclones' U/F, consider the following graphs.

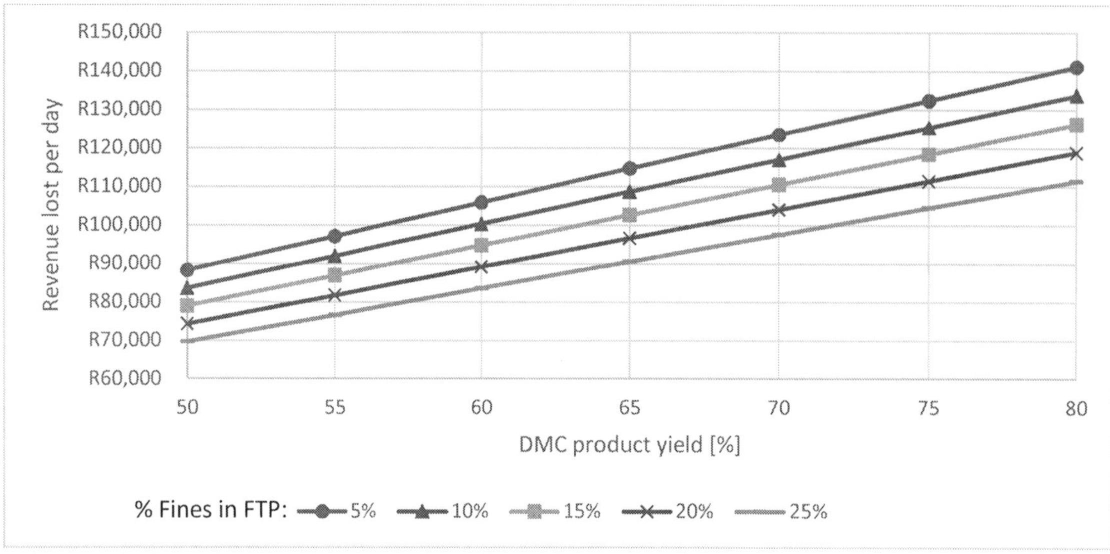

Figure 4: Possible revenue loss for a 1000tph plant with a 1% misplacement of floats to the U/F at various DMC yields and fines (-1mm) in the FTP.

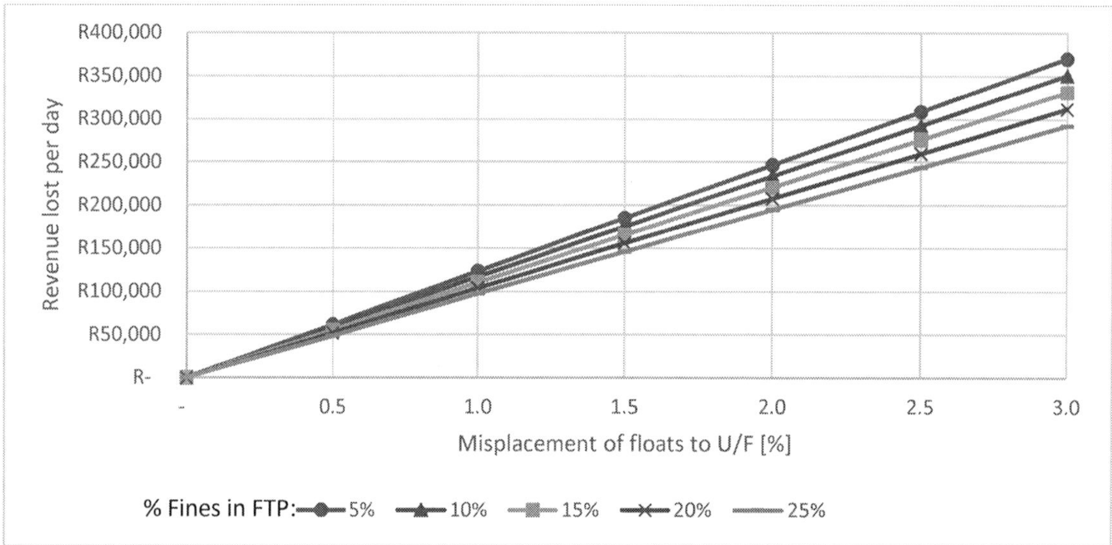

Figure 5:Possible revenue loss for a 1000tph plant with a 70% DMC product yield at various floats to U/F misplacements and fines (-1mm) in the FTP.

Now that it is clear what the potential gains are for operating a cyclone optimally, it is important to identify and address situations and conditions that can contribute to the misplacement of product to the cyclone's underflow and other aspects that can have a negative impact on the profitability of the specific plant.

Near density material is probably one of the biggest contributors to misplacement in a DMS cyclone. In fact, Claassen (2013:767) stated that the amount of near density material in the feed to a DMS cyclone is the single biggest variable that impacts DMS cyclone efficiency (in the region of 98%). Near density material is typically defined in

South Africa as those particles whose densities are within 0.1g/cm^3 from the specific separation density (Magwai & Claassen, 2013). There is however an alternative definition which choose to define the amount of near density material as those particles whose densities are within 0.05g/cm^3 from the specific separation density. For the sake of this paper the 0.1g/cm^3 variance definition is assumed. The following graph is a simplified indication of how near density material can be determined from washability data.

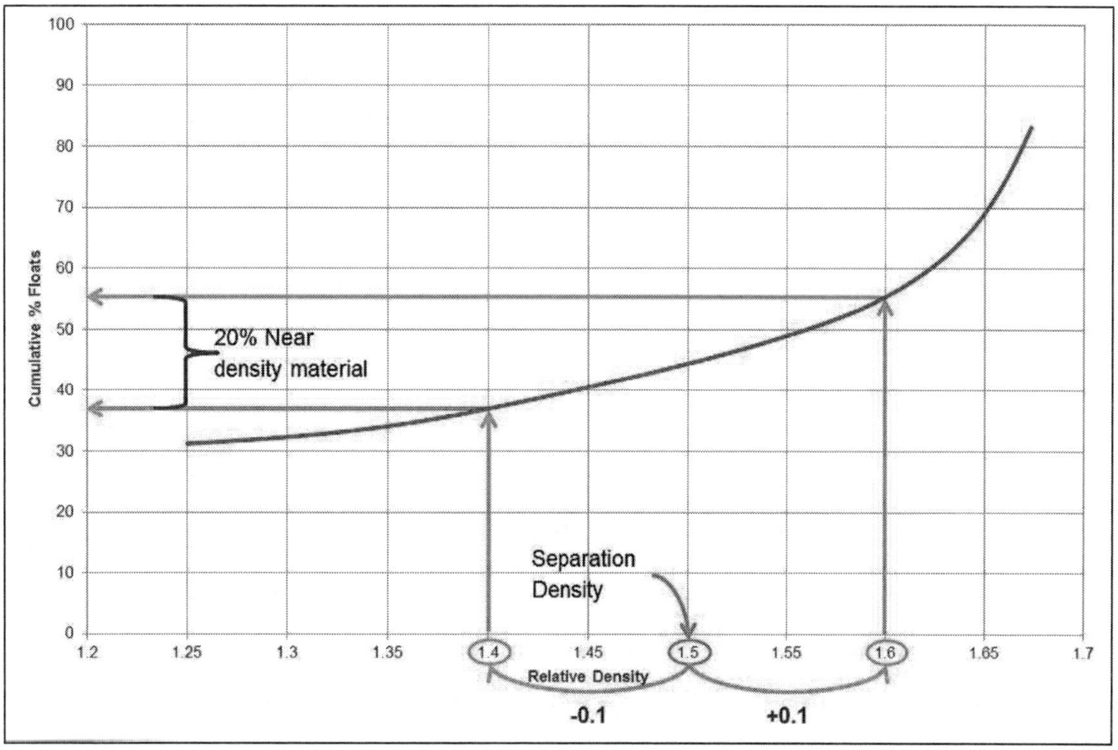

Figure 6: Simplified approach in determining near density material.

The impact of near density material on the DMS process has previously been reported to be mainly through a lowered organic efficiency and increased misplacement of particles in the DMS cyclone (Horsfall, 1987; de Korte, 2008). There is a direct relationship between the amount of near dense material in the feed to the DMS cyclone and the cyclone's ability to efficiently beneficiate the coal with minimal misplacement of particles, this relationship is quantified in the following table:

Table 2: Near density material with associated degree of difficulty of separation.

Near Density Material	Degree of Difficulty
0% - 7%	Simple
7% - 10%	Moderate
10% - 15%	Difficult
15% - 20%	Very difficult
20% - 25%	Exceedingly difficult
> 25%	Formidable

The following example illustrates the impact of near dense material on the amount of misplacement taking place in the cyclone. Below are two samples of washabilities to be processed in the same plant, through the same equipment and both washing for a product with 18.6% ash.

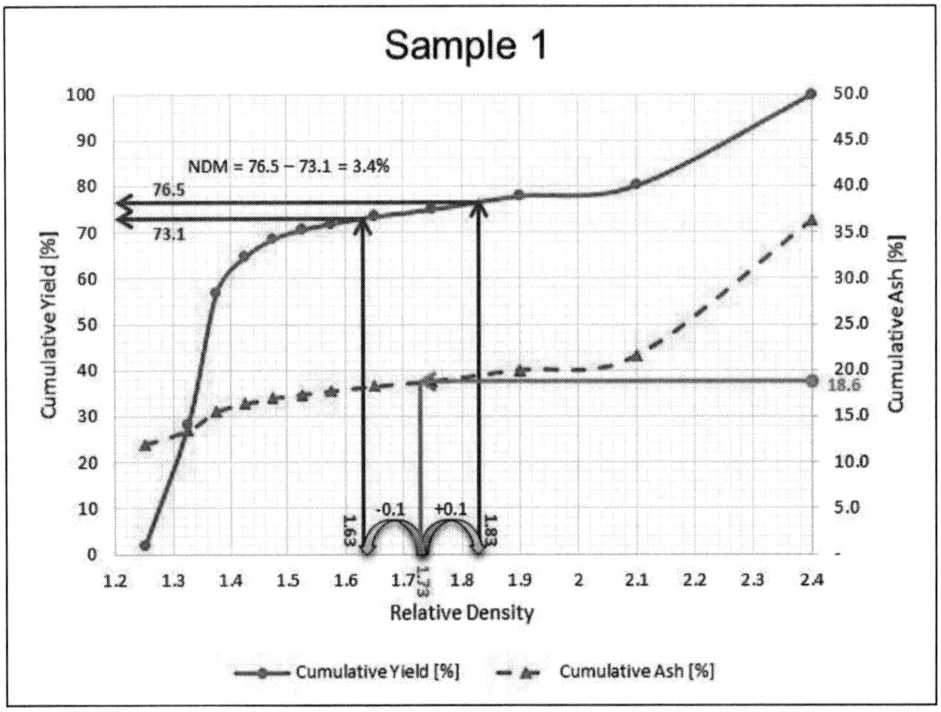

Figure 7: Washability data with low levels of near density material.

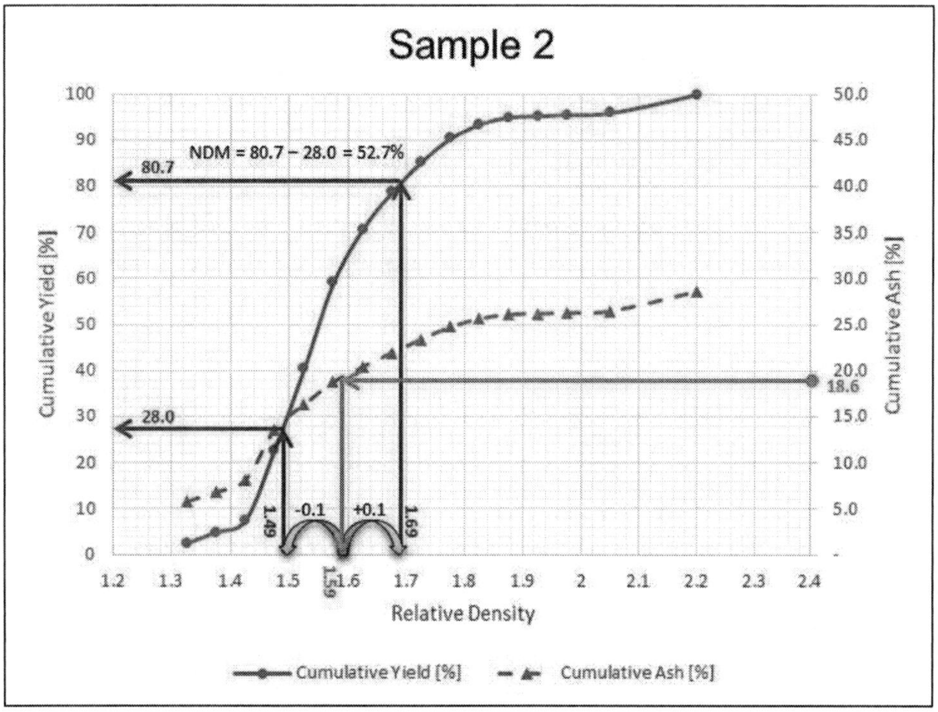

Figure 8: Washability data with high levels of near density material

Even though the DMS cyclone will be able to produce a 18.6% ash product with both samples and the product yield will be relatively similar (74% vs. 64%), the fact that Sample 2's near dense material is almost 50% more than that of Sample 1 (52.7% vs. 3.4%) results in the misplacement of floats to the spigot being almost 20 times higher for Sample 2. The misplacement of product to the cyclone's spigot for Sample 1 is a mere 0.1% whilst the same misplacement under similar operating conditions for Sample 2 is in the order of 2%. Note that near density material is a function of the specific cut density and is therefore directly related to the required product qualities and ore washability. Remembering that every 1% misplacement (in a 1000tph plant) of product to the cyclone's underflow relates to approximately R100,000 of potential revenue lost per day. These values quickly add up and over a month's time the effect on the plant's bottom line is tremendous.

With the amount of near dense material in the feed to the plant largely being a function of the site's specific geology and what the plant receives from Mining, there are fortunately still some opportunities the CPP can consider minimising the effect of high values of near dense material. According to Magwai & Claasen (2013:7) the measures to deal with excessive near dense material can be divided into two groups namely: changes around the DMS cyclone and then changes involving the feed to the cyclone. The four options under changes around the DMS cyclones include:

- Adding a barrel to the cyclone.
  A barrel is a straight cylindrical section added between the inlet head and top cone of a cyclone as indicated in the following schematic:

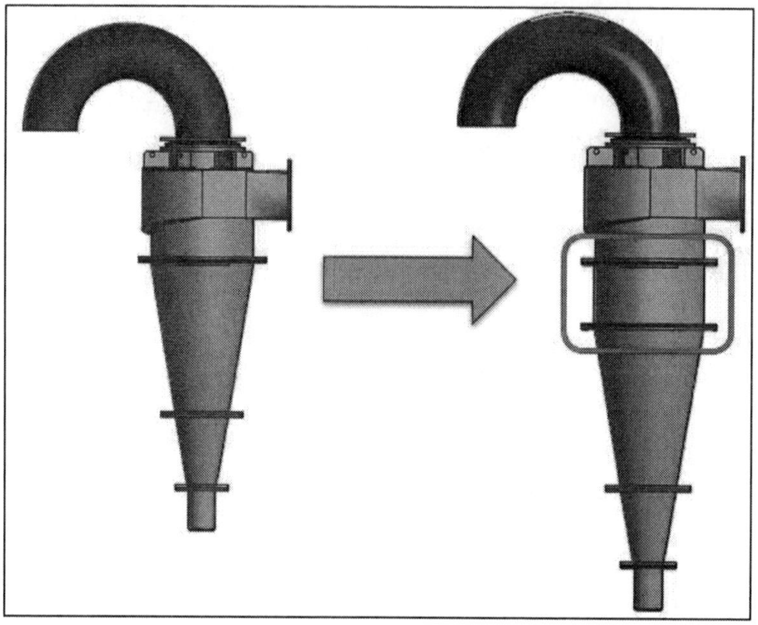

Figure 9: Addition of a barrel section to a cyclone.

By adding a barrel to a cyclone, one increases the cyclone's volume and therefor also the residence time of each particle being processed by the cyclone. Keeping in mind that the residence time of each particle in the cyclone is very limited, one can expect a greater probability that a particle will classify correctly if it gets a longer opportunity to do so.

Special care must be taken when considering adding a barrel to a cyclone however, because it could increase the cyclone's density differentials as well. The increase in density differentials will lead to an increase in the density cutpoint and this needs to be taken in account.

- Medium-to-ore ratio in the feed to the cyclone.
  The recommended medium-to-ore ratio (on a volumetric basis) when treating South African coal is 3.5:1 (Peatfield, 2003:363). Washing coal with relatively high near dense material at medium-to-ore ratios lower than the recommended 3.5:1 will result in inefficient cyclone operation and excessive misplacement. According to Magwai & Claassen (2013:7) there is however an upper limit beyond which further increase in the medium-to-ore ratio no longer makes a difference to the cyclone's efficiency but only reduces the cyclone's capacity. This upper limit for coal processing is in the order of 5:1.

- Spigot diameter.
  The two factors that has the biggest impact on a cyclone's capacity is the medium-to-ore ratio in the feed to the cyclone and the cyclone's "discard capacity" determined by it's spigot diameter. After a DMS cyclone has been sized by an OEM, great care should be taken before any changes are made to

a cyclone's spigot size since there are two opposing views on how the spigot size influences a cyclone's ability to effectively beneficiate ore with a lot of near dense material.

On the one hand, Sripriya *et al.* (2001) mentions that near dense material can build up inside the cyclone which can become a problem if the spigot is not large enough to discharge all the discard. In this case some of the discard particles can be misplaced to the overflow. Therefore, by increasing the spigot diameter, one can alleviate the load on the spigot and reduce misplacement of particles.

In contrast to this view, He and Laskowski (1995) believes that better separation efficiency can be achieved with smaller spigot diameters at the expense of lower cyclone throughput. The increased residence time of the feed particles in the cyclone due to the cyclone's lower throughput is said to improve their chance of reporting to the correct product streams.

When sizing a spigot it is important to take the following factors into account:

- Expected yields based on washability data;
- Volumetric capacity of a certain spigot at a certain feed head;
- The DSM principle of spigot ID to vortex finder ID ratio ≤ 0.8:1

- Cyclone diameter.

  By increasing the cyclone's diameter, one also increases the unit's volumetric capacity and by keeping the solids throughput to the cyclone similar (to what was fed to the smaller diameter cyclone) you can operate the cyclone at a higher medium to ore ratio which will counter the effect of near dense material as previously discussed.

  One also needs to keep in mind that although dense medium cyclones separate on a density basis they are still inherently classification cyclones with both the ore and medium being subjected to centrifugal and drag forces. The following formula is used to calculate the centrifugal forces acting on each particle in the cyclone, pulling the particle to the outside diameter of the cyclone:

$$F_C = \frac{m.v^2}{r}$$

Where: F_c = Centrifugal force [N]
m = mass of the particle [kg]
v = velocity of the particle [m/sec]
r = radius of the circular motion [m]

As can be seen from the formula, by increasing the diameter or radius of the cyclone, the centrifugal force acting in on each particle will reduce. This will cause less magnetite particles to migrate to the sides of the cyclone, causing a lower density differential and therefor also a lower cut density.

Furthermore, with the increase in the cyclone's diameter and the associated decrease in centrifugal forces as well as the limited residence time of a particle in the cyclone, the efficient separation of smaller particles will be influenced. The smaller centrifugal force acting in on the smaller particles, might not be sufficient to move them into the correct density zone in time before exiting the cyclone. The result is that bigger diameter cyclones will not separate smaller particles as effectively as the bigger particles in the same slurry stream. For each cyclone diameter there is a certain particle size (known as the breakaway size) at which separation efficiency starts to deteriorate. Please note that the breakaway size doesn't mean that the cyclone won't be able to separate particles smaller than that size, it simply means that the efficiency of separation will start to deteriorate for particles smaller than that cyclone's specific breakaway size. Breakaway size can be calculated with the following formula:

$$d_B = 6 \times \emptyset^{1.64} \times 10^{-5}$$

Where: d_B = Breakaway size [mm]

Ø = Cyclone diameter [mm]

It therefor becomes clear that if you are able to increase your cyclone's volumetric capacity (by means of the addition of a barrel section or utilising a bigger diameter cyclone), you will be able to operate the unit at higher medium-to-ore ratios which will have an impact on the amount of misplaced material. The higher medium-to-ore ratio comes at the possible price of lower cut densities, lower efficiencies for smaller particles or a change in cyclone throughput (depending on what measure is incorporated). These are all factors that needs to be taken into consideration before any changes are made to the physical cyclone.

By simulating the washabilities represented by Sample 2 (from the above example), the following comparison was made before and after making the following modifications:

- Bigger cyclone diameter (1000mm to 1150mm)
- Including a barrel in the cyclone configuration
- Increasing the M:O ratio from 3.1:1 to 4:1

Table 3: Results obtained from optimising aspects around the cyclone.

		Base Case	Optimised case
Feed	tph	800	800
Cyclone Diameter	mm	1000	1150
Barrel		No	Yes
M:O ratio		3.1:1	4:1
Near Dense Material	%	25.9%	25.9%
Product yield	%	54.8%	55.4%
Product	tph	438.7	442.8
Product in U/F	%	2.4%	2.1%
Misplaced product	tph	8.80	7.54

As can be seen from the results summarised in the table above, by making the suggested changes, the plant managed to misplace 1.26tph (8.80 – 7.54) less product to the cyclone's underflow. 1.26tph may not seem like much at first, but this amounts to a possible gain in revenue in the order of R26,000 per day.

The second group of measures that can be taken to address the negative impact of near dense material on the cyclone's efficiency has to do with changes to the feed to the cyclone. By blending the feed with the high value of near density material (Sample 2) with feed containing less near density material (Sample 1), the near density material is actually "diluted" and the negative effect on cyclone efficiency is alleviated correspondingly. Applying this rationale to the two washability samples given earlier in this paper and blending them 1:1, before beneficiating them through the original 1000mm cyclone (without a barrel), generates the following results:

Table 4: Results obtained from mixing low and high near density material prior to beneficiation.

		Base Case	Mixed Sample case
Feed	tph	800	800
Cyclone Diameter	mm	1000	1000
Barrel		No	No
M:O ratio		3.1:1	3.1:1
Near Dense Material	%	25.9%	6.4%
Product yield	%	54.8%	67.1%
Product	tph	438.7	537.0
Product in U/F	%	2.4%	1.68%
Misplaced product	tph	8.80	4.43

From the table it is clear that mixing the high near dense material with sufficiently lower near dense material has a much bigger effect on the misplaced product than what was achieved by making physical changes around the cyclone. This gives some insights on the importance or "luxury" of having a plant feed blending stockpile in front of the coal processing plant. Mixing of coal ores are however only a viable option if the two (or more) ores have similar cut densities for the required product qualities. If the required cut density for the mixed ore is lower than the required cut density for the ore with the less near dense material (Sample 1 in this case), running at the lower cut density will result in a yield loss from the one portion of the mix (Sample 1). One then has to weigh out the gain in cyclone product from the less misplaced material in the mixed feed versus the loss of cyclone product due to operating at a too low cut density for one portion of the feed.

Combining the two suggested approaches gives even better results, even though it is only slightly better than what was achieved with the mixed sample case.

Table 5: Summary of results obtained after attempts of optimisation.

		Base Case	Optimised case	Mixed Sample case	Optimised, mixed sample case
Feed	tph	800	800	800	800
Cyclone Diameter	mm	1000	1150	1000	1150
Barrel		No	Yes	No	Yes
M:O ratio		3.1:1	4:1	3.1:1	3.9:1
Near Dense Material	%	25.9%	25.9%	6.4%	6.3%
Product yield	%	54.8%	55.4%	67.1%	67.3%
Product	tph	438.7	442.8	537.0	538.4
Product in U/F	%	2.44%	2.11%	1.68%	1.57%
Misplaced product	tph	8.80	7.54	4.43	4.11

Comparing the initial base case with the optimised, mixed sample case, an additional R42,000 can be made every day by incorporating these changes. At the time of doing these calculations, this amount is sufficient to cancel out the effect of a 22% drop in the current coal selling price (API-4).

Although near density material is the single biggest contributor to misplacement in a DMS cyclone, there are also other situations that will contribute to a cyclone's inefficiencies (for both DMS and hydrocyclones). These situations and observations made in the field will be discussed later on in this paper.

THE HYDROCYCLONE'S ROLE IN THE COAL PROCESSING PLANT

Whether it is to deslime, dewater, densify or degritt, the hydrocyclone in a coal plant doesn't necessarily beneficiate the coal as such, but rather assist or improve downstream processes. Without a desliming cyclone in front of the spirals, the spirals won't be able to beneficiate the plant's fines effectively. Much of the plant's water is also recycled from the fines circuit back from the thickener and to the DMS circuit and therefor issues in the fines circuit will correspondingly have a direct impact on the plant's DMS circuit. Keep in mind that the fines can be in the order of 20% (based on the earlier assumption made in the example) of the total feed to the plant and inefficiencies in this circuit will also have a significant impact on the plant's overall yield.

South Africa has about a billion tonnes of discarded thermal-grade coal fines (Enslin, 2017.3). Redick *et al.* (2007) mentions that the South African coal mining industry is currently disposing of about 10 million tons of ultra-fine coal (<150 μm) per year. The quality of these fines is very similar to the Run-of-Mine (ROM) coal qualities and in some cases even better. More and more coal beneficiation plants are starting to wash, add or mix this material with their product since there is still substantial potential energy locked up in these fines. The question arises however whether our plants are first of all running with optimised fines circuits and secondly have enough capacity for the additional fines. With frequent visits to "older generation" coal processing plants washing dump material or weathered coal, it becomes clear that their fines circuit is the bottleneck for their entire process because of capacity constraints. The fines circuit simply wasn't designed initially to handle much more than 10% fines in the total feed to the plant. Below is a table with design specifications for 4 coal processing plants in the Witbank coalfields proving that they were not designed for fines in the order of 20% of total feed to plant:

Table 6: Design specifications of various plants in the Witbank coalfields.

	Plant 1	Plant 2	Plant 3	Plant 4
FTP [tph]	1010	600	2000	200
Fines [tph]	114	60	133	25
% Fines in FTP	11%	10%	7%	13%

Care should be taken to avoid increasing the feed to the hydrocyclones to such an extent that their efficiency is compromised. For effective operation it is suggested that the %Solids$_{v/v}$ in the feed to the cyclone should be in the order of 10% to 15%. Values more than 15% is still acceptable under certain specific circumstances, but as soon as the %Solids$_{v/v}$ in the feed to the cyclone exceeds 30%, inefficiencies become too high and it is not worthwhile to operate that specific cyclone anymore. Since it isn't always as easy or quick to determine the %Solids$_{v/v}$ in the feed to the cyclone, the following graph was created to correlate the slurry density (which is much easier to determine onsite) with the %Solids$_{v/v}$ at various solids SG's. As can be seen from the graph, if you work with a solids SG of 1.55, you can't really feed a slurry much thicker than 1.08g/cm^3 to a hydrocyclone and still expect sufficient separation.

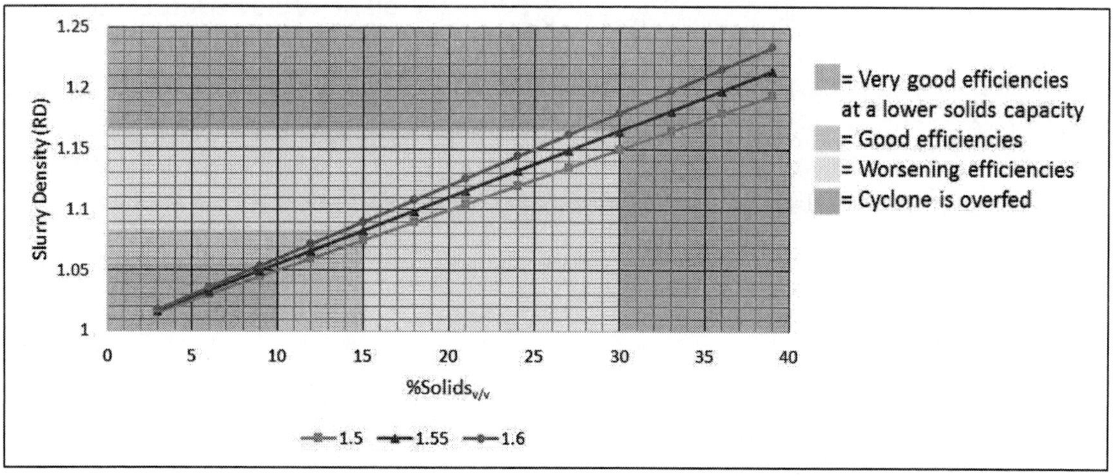

Figure 10: Correlation between %Solids$_{v/v}$ and slurry density in the feed to a cyclone at different solids SG values.

The following scenarios describes the potential consequences of having a fines circuit that doesn't have sufficient capacity for the increase of fines in the feed to the coal processing plant. Scenario 1 is based on a plant with a feed rate of 1000tph, a 70% product yield in the DMS circuit and initially 10% fines in the feed to the plant:

Table 7:Different scenarios depicting the effect of increased % Fines in the feed to plant

		Scenario 1	Scenario 2	Scenario 3
FTP	tph	1000	1000	500
Solids SG	g/cm^3	1.55	1.55	1.55
Fines in FTP	%	10	20	20
Feed to DMS	tph	900	800	400
DMS product yield	%	70	70	70
DMS product	tph	630	560	280
Feed to Fines circuit	tph	100	200	100
Desliming cyclone volumetric capacity	m^3/hr	430	430	430
%Solids$_{v/v}$ in feed to desliming cyclone	%	15	30	15
Expected desliming cyclone efficiency		Good	Bad	Good

Should the % fines in the feed to the plant increase from 10% to 20% and the feed rate to the plant is kept constant at 1000tph, the %Solids$_{v/v}$ in the feed to the desliming cyclone will increase from 15% to 30% and the cyclone's efficiency will drop significantly (Scenario 2). This will have a negative impact on the recoveries made by the downstream equipment and the water recycled back to the DMS circuit if the plant's thickener is already operating at its limits. To get the %Solids$_{v/v}$ in the feed to the desliming cyclone back to an acceptable level, the plant will have to cut back on the total feed to plant and thereby reducing the amount of clean coal produced via the DMS circuit. In this specific (and very extreme) example, if the fines in the feed to the plant was to increase with 10% (from 10% to 20%), the plant would be forced to reduce the total feed to plant to ensure that the fines circuit still runs within operating parameters. The knock-on effect will be a decrease of 56% (from 630tph to 280tph) of the DMS product. Since the biggest portion of the plant's product is produced via the DMS circuit, many plants in this situation would opt to run their fines circuit at a lower efficiency and keep the total feed to plant at a maximum.

COMMON FIELD OBSERVATIONS CAUSING CYCLONE INEFFICIENCIES

The following observations are made on a regular basis in the coal processing plants in the Witbank coalfields. The examples given in this paper will have a negative impact on DMS as well as Hydro-cyclones and should be avoided as far as possible:

- Bends in the feed pipe to the cyclone.

The first prize would be to have a straight pipe feeding into the cyclone (or at least straight for nine pipe diameters right before the cyclone) in an attempt to get close to a laminar flow into the cyclone. If this is not possible at all, the pipe should at least turn in the same direction the material will follow inside the cyclone.

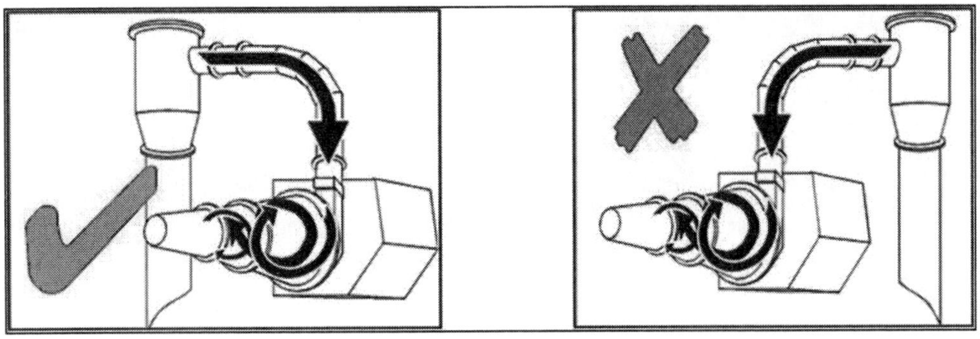

Figure 11: Installation of bends in front of a cyclone.

In the following examples, the bends right before the cyclones' inlets will cause "pre-classification" of the feed particles resulting in unnecessary turbulence and particles being in the wrong positions entering the cyclone. Together with the drop in efficiency, it will also have a negative impact on the expected wear life of the cyclones. Keeping in mind that the residence time for each particle in the cyclone is very limited, situations like these should be avoided at all cost.

Figure 12: Non-optimal installations of bends prior to cyclones.

- "Home-made" feed distributors and manifolds.
  The idea behind a feed distributor is to divide the feed equally between the cyclones with no bias. Should you get this wrong, not only will it influence wear life, but also the product produced by each cyclone. If the cyclones in the same cluster don't get the same feed, you cannot expect them to deliver the same product.

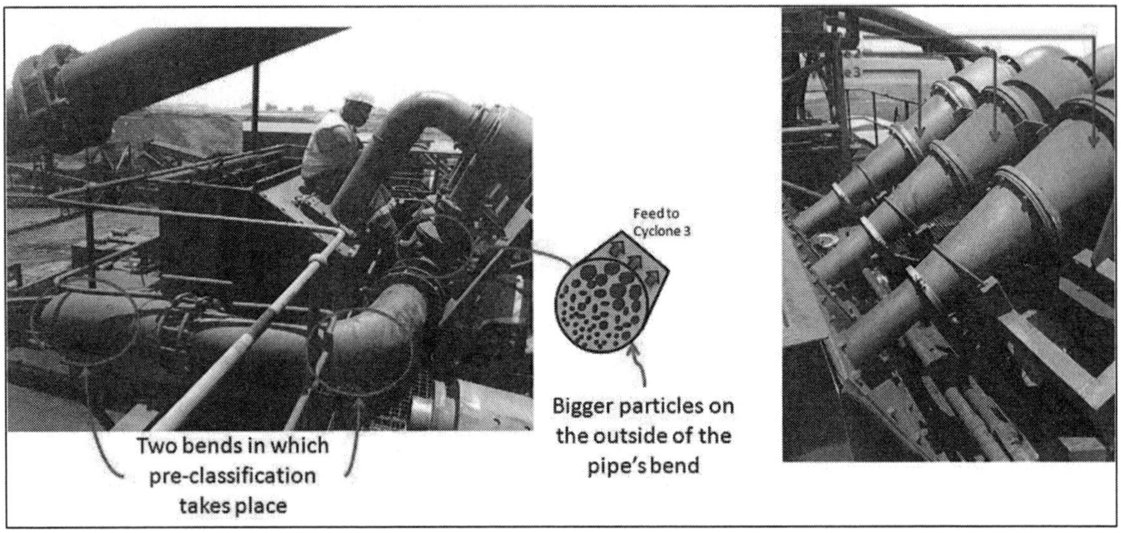

Figure 13: Effect of feeding a cluster of cyclones from a "home-made" manifold.

- Siphoning created in the cyclones' overflows.
  The cyclone's overflow should discharge at atmospheric pressure at a level equal to or higher than the feed inlet. If for instance the overflow discharges at atmospheric pressure at a level far lower than the spigot discharge, a siphon will be created in the overflow and a high misplacement to overflow can be expected. This can be detrimental for equipment, like a thickener, that is very sensitive to oversize particles in the feed and is fed by the overflow from a dewatering or desliming cyclone. If the overflow discharges into a tank or body of water, also ensure that the overflow doesn't discharge below the water level.

Figure 14: Siphoning created by inappropriate overflow installation.

- Cyclone installation angle.
  The installation angle of DMS cyclones is chosen to ensure that the contents of the cyclone drains out when the feed to the cyclone is stopped. Should there be any flat surfaces (relevant to the horizontal), material will build up and may cause blockages during start-up.

Figure 15: Cyclone installation with horizontal surfaces that is prone to cause blockages.

A cyclone with a cone angle of 20° (typical for a DMS cyclone) will normally then be installed at a 15° angle.

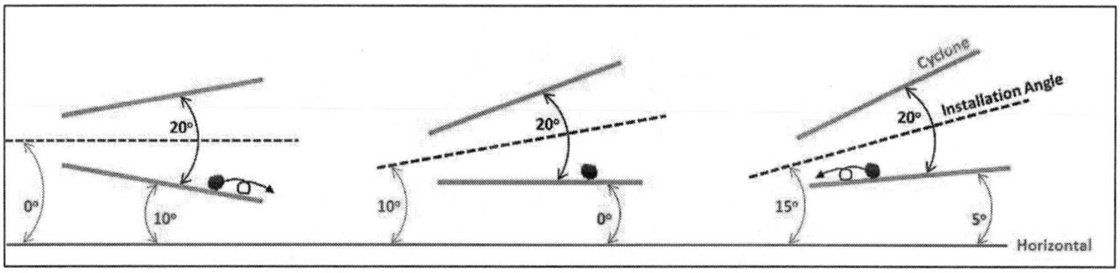

Figure 16: Rationale behind deciding on the installation angle of a cyclone.

- Pipe ID vs. inlet ID
  The cyclone's inlet inner diameter is designed specifically to affect the entering particles' velocity as it enters the cyclone. The inner diameter of the pipe feeding the cyclone therefor shouldn't be smaller than the inner diameter of the cyclone's inlet.

Figure 17: Cyclone feed pipe with a smaller ID than that of the cyclone's inlet.

- Using flexible hoses around the cyclone.
  When flexible hoses are connected to the cyclone, care should be taken to ensure that they are always open and allow free flow of material through them.

Figure 18: Flexible hose blocking free flow of material.

Making sure that your cyclones' physical installation is up to standard is a relative easy way of ensuring that inefficiencies are addressed.

OPERATING CYCLONES EFFICIENTLY FOR LONGER.

Now that it is known how important cyclones are in the coal processing plant and factors that can influence their efficient operation has been addressed, the aim is to keep them in this state for as long as possible. This is the point where maintenance management around the cyclones comes into play.

Normally there are three levels of maintenance that can occur in and around a cyclone in the plant. The very first level is where we react on a certain event – corrective maintenance. This is not ideal since there is only reaction after an event already took place. The time between the event taking place and it being corrected means that there is either no production or poor efficiencies. Nonetheless, for that period of time, the cyclone is not performing at the level it should and time wasted is money lost. This is also commonly referred to as a "break-down" and this is typically when a cyclone is worn through and leaking. Although this is an opportunity for plant personnel to showcase their creativity in patching holes, the cyclone is still not performing optimally and if possible, care needs to be taken to ensure that the plant doesn't operate in this maintenance space.

Figure 19: Patched cyclones in operation.

The second level of maintenance (and this is where OEMs also need to play a big role) has to do with preventing certain catastrophic events from taking place. Here special attention needs to be given to the material of construction of the cyclone (make sure it is ideal for the specific situation) and ensure that the cyclone is operated within the predetermined design specifications. The concept of preventive maintenance requires that the relevant stakeholders monitor facilities, systems and equipment so closely that they know where to look to find small problems before they can become larger, costlier headaches. One also must take care not to go overboard with the preventative maintenance as excessive preventative maintenance will increase the plant's overall maintenance cost. As can be seen from the following graph (TÜVRheinland, 2019) there is an optimal maintenance zone where the plant's total maintenance cost is at a minimum.

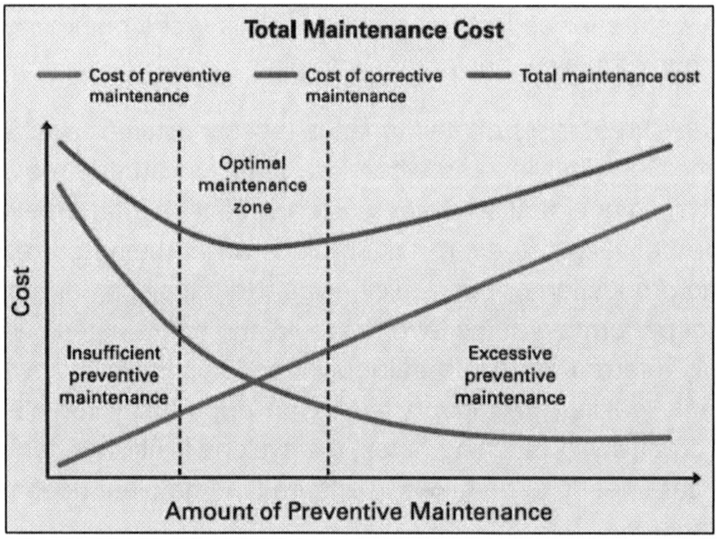

Figure 20: Balance between preventative and corrective maintenance.

The top level of maintenance that one needs to strive to achieve is predictive maintenance. This goes further than preventative maintenance, requiring that the relevant stakeholders know not only where a problem will arise but when, and then take steps to avoid it. Some of the tools that can be used to assist in this regard are:

- Multotec's Green dot tiles.
  The idea behind this concept is to imbed the ceramic tiles used to line cyclones with a green ceramic disk of which the thickness can be predetermined. This side of the tile will then be the side that gets fixed to the cyclone wall. With time, as the tile wears through, the green dot will become visible indicating that the tile is wearing thin. At this stage plans need to be in place to replace the cyclone or that specific part, before the cyclone gets a catastrophic failure. With time one can also build up history to determine what the specific wear rates (mm/month or mm/t) are for the different sections of the cyclone which will allow for better planning.

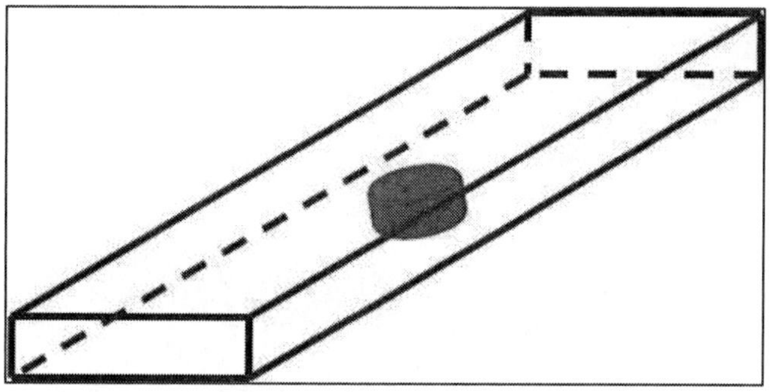

Figure 21: Simplified schematic of Multotec's greendot tile.

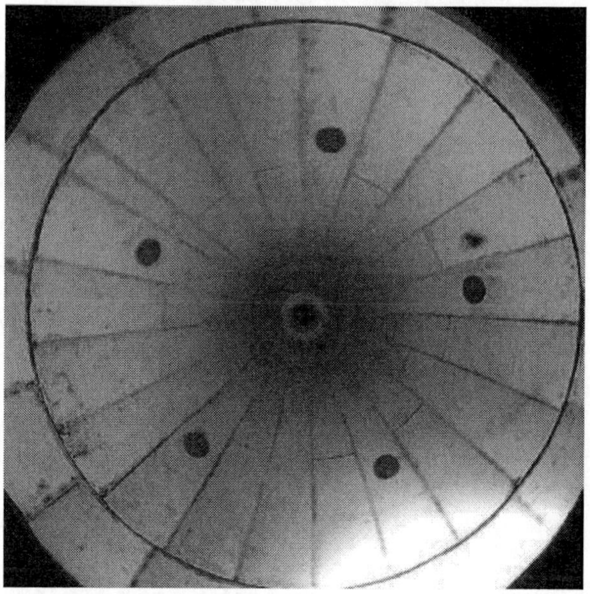

Figure 22: Internal view of a cyclone lined with Multotec's green dot tiles.

- 360° VR inspection camera.
  By inspecting the internal wear surface of a cyclone with this camera, detailed reports can be generated, and problem areas can be identified and addressed before corrective maintenance will be required. Below is an example of the type of information that can be generated with this tool:

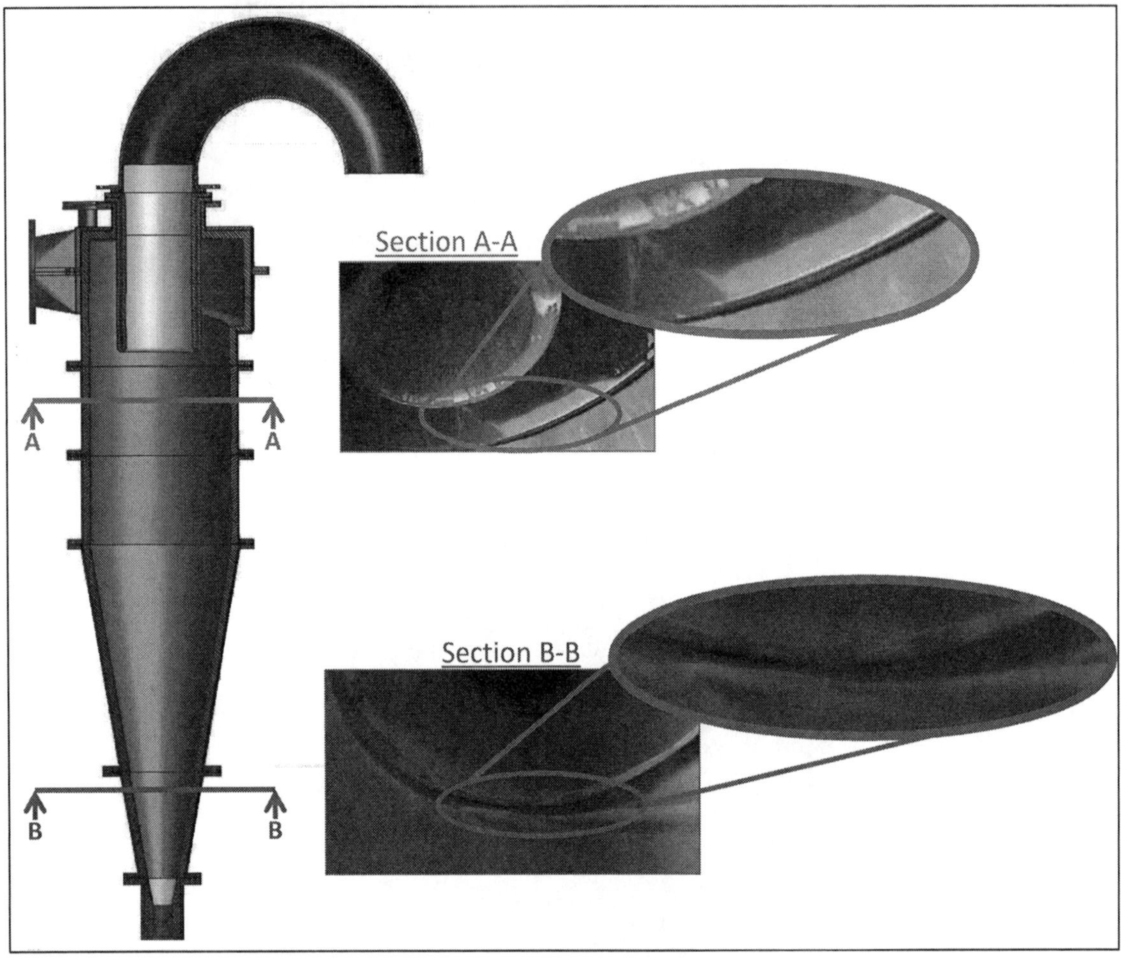

Figure 23: Internal inspection of a cyclone.

CONCLUSION

Cyclones are without a doubt at the heart of any coal processing plant. With lots of time and money invested in developing them to what we have today, they make a significant contribution to a plant's overall economic performance. Being a stationary piece of equipment with no moving parts, they are quite easy to maintain and keep at a satisfactory operating level. Even though more and more plants in South Africa need to beneficiate low grade coal reserves and the current coal prices are on a downward spiral, looking after the cyclones in the plant will ensure that they look after your bottom line.

REFERENCES

BRADLEY, D. 1965. The hydrocyclone. Oxford: Pergamon Press. 348 p.

BRETNEY, E. 1891. Water purifier. Patent: US 453,105. 5 p.

CLAASSEN, J.O. 2013. Yield improvement at a mid-sized coal mine in the Witbank coalfields. *The Journal of The Southern African Institute of Mining and Metallurgy,* 113(10):761-768.

DE KORTE, G.J. 2008. The influence of near-dense material on separation efficiency of dense-medium processes. *International Journal of Coal Preparation and utilization*, 28(2):69-93.

DE KORTE, G.J. 2012. Fine coal processing with dense medium cyclones. Paper read at the Fine Coal Symposium of the Coal preparation society of America in Pittsburgh on 23 to 25 October 2012. CSIR. 24 p.

DRIESSEN, M.G., KRŸGSMAN, C. & LEEMAN, J.N.J. 1951. Process for the separation of solids of different specific gravity and grain size. Patent: US 2,543,689. 4 p.

ENSLIN, F.H. 2017. Optimisation of a desliming cyclone in the Witbank coalfields – A case study. Paper presented at the Southern African Coal Processing Society's conference in Secunda on 22 to 24 August 2017. Multotec. 28 p.

HE, Y.B., & LASKOWSKI, J.S. 1995. Dense medium cyclone separation of fine particles. Part 1: The effect of medium split ratio on dense medium cyclone performance. *Coal Preparation*, 16:1-25.

HORSFALL, DW. 1987. The lows and highs of dense medium separation. Paper presented at the Australian Institute of Mining and Metallurgy's Dense Medium Operators Conference in Brisbane on 27 to 29 July 1987.

MAGWAI, M.K. & CLAASSEN, J.O. 2013. Near-gravity experience at Leeuwpan coal mine. Paper presented at the Southern African Coal Processing Society's conference in Secunda on 23 to 25 July 2013. Exxaro Resources. 18 p.

PEATFIELD, D. 2003. Coal and coal preparation in South Africa — A 2002 review. *The Journal of The Southern African Institute of Mining and Metallurgy,* 103(6):355-372.

REDDICK J.F., VON BLOTTNITZ H., KOTHUIS B., 2007, A cleaner production assessment of the ultra-fine coal waste generated in South Africa, *The journal of the Southern African Institute of Mining and Metallurgy,* 107:811-816.

SINGLETON, J.D. 2013. Development and evaluation of a dense media cyclone for the Southern African mineral and coal industries. Johannesburg: Wits. (Research report M.Sc) 53 p.

SMITH, V.C. 2000. Particle size estimation of hydrocyclone overflow. Cape Town: UCT. (Thesis M.Sc) 102 p.

SRIPRIYA, R., BANERJEE, P.K., RAO, P.V.T. DUTTA, A. & RAO, M.V.S. 2001. Critical evaluation of factors affecting the operation of dense medium cyclones treating medium coking coals. *Int. J. Miner. Process.*, 63:191–206.

TÜVRHEINLAND, 2019, EMIT Optimisation – Getting more out of existing equipment for less. https://www.risktec.tuv.com/risktec-knowledge-bank/asset-integrity-management/emit-optimisation-getting-more-out-of-existing-equipment-for-less/ Date of access: 4 Jun. 2019.

WIRTGEN SURFACE MINING TECHNOLOGY

Waylon Kukard
Sales Manager – Wirtgen South Africa

Environmentally friendly coal mining

The Shengli Mine, which extends over a total area of 340km², has become a reference project in China for the environmentally friendly mining of coal. The coal seam comprises layers of variable coal quality. A 2200 SM with a 3.8m wide cutting drum has been mining these layers selectively at depths of up to 25cm. It cuts up to 5000 tonnes in a single shift and with 97% of mined material having a particle size of smaller than 100mm. This enables a 10% more truck full factor per load and thus reduces truck transports by 5% - 10%. This results in fewer trucks, less fuel and maintenance for the same production volume. The 2200SM gives an impressive demonstration of how high-quality material can be mined without the huge environmental impacts caused by drilling and blasting.

Figure 1. Project Data:
Mining material – Coal
Unconfined compressive strength – 15 MPA
Surface miner – 2200 SM (1 machine)
Production rate per day – up to 5000 t

Figure 2. Selective mining ensures a high degree of purity of the material deposited using the windrowing method. Lower-quality coal can be detected and rejected more efficiently by the Wirtgen Surface Miner.

Figure 3. Due to the cutting technology all the material appears as a pre-crushed product, of which 97% is smaller than 100mm.

Selective Mining

One of the most important features of the Wirtgen Surface Miners is their ability to work selectively. Thin seams interspersed with intermediate rock layers can be cut precisely and economically, for example. Mining companies throughout the world are exploiting the advantages of Wirtgen Surface Miner technology with its extremely precise control of the cutting depth to extract high-quality mined material (ROM) in coal mines, phosphate mines and gypsum mines. Selective mining with Surface Miners ensures that thin seams of high-grade minerals and overburden can be mined separately.

Surface mining takes over from conventional coal mining

The New Acland Coal mine, located adjacent to the township of Acland, Queensland, Australia, is part of the Australian based New Hope Group. Since opening of the mine, coal and interburden have been extracted by means of bulldozers fitted with ripper teeth and wheel loaders. However, a sharp drop in coal prices required the management to look for alternative mining methods. A solution was soon found, the Wirtgen surface mining technology offers the ideal method for the challenging geological conditions in the mine, which comprise up to 27 mostly thin coal seams separated by layers of interburden. In comparison with conventional methods, the selective surface mining technology simplifies the mining process, thus reducing operating costs and supplying coal of higher quality to the coal processing plant.

6 Month trial phase completed successfully

Wirtgen surface mining has proven its worth as a tried-and-tested technology in numerous coal mining projects. It was no surprise, therefore, that the 6-month trial phase with defined target performance levels agreed upon between NAC and Wirtgen was a resounding success. The 4200SM achieved the projected cutting performance of up to 3000 t/h in the tough everyday mine environment. The point-attack cutting tools used were subjected to in-depth trials and optimized to fit the specific application as they were required to cut not only coal but also hard interburden material. Cooperation with the Wirtgen customer service, which was constantly involved in the trial, was also perfected at an early stage. This enabled many areas which typically offer room for improvement when introducing an entirely new technology to be identified and optimized quickly. Since all of the targeted performance were met during the trial phase, NAC decided to purchase the 4200SM.

The advantages of Wirtgen Surface mining

To maximise exploitation of the coal deposit, the high-performance 4200SM cuts and loads the material in a single machine pass, producing an even and easily trafficable surface on the mining pit floor. It was additionally determined during the trial phase that, being equal in performance, the surface miner replaces two bulldozers and one wheel loader. The resulting savings, mainly in fuel, lead to a significant reduction in operating costs. Additional marks in favour of the 4200SM include emission measurement results showing significantly reduced machine vibrations and noise levels. Paired with the ergonomically designed cabin, impacts on the operator's health are thus minimized particularly in the long run. A camera system designed exclusively for this operation enables the machine operator to conveniently differentiate between coal and interburden, thus significantly improving coal quality. The strict safety regulations and machine standards specified by the Australian coal industry were fully met by making the relevant machine modifications prior to shipment.

The cutting drum was specially designed for operation in soft rock and is ideally suited to mining not only coal but also the significantly harder interburden layers, up to 50MPa. The 4200SM mines coal in a selective process at exactly the specified cutting depth, producing a clean cutting surface and guaranteeing high degrees of purity. As a result, there are fewer truck travels required to transport the material to the coal processing plant for further processing and to return the separated waste to the mine. Compared with the mining method used to date, the 4200SM also produces a smaller percentage of fines (grain size <2mm) and a higher percentage of the end product fraction (grain size <38mm). In the final analysis, this also improves the efficiency of the coal processing plant.

Positive conclusion by the NAC management

"I am impressed with the economical, modern and environmentally friendly Wirtgen surface miner. NAC saves hard cash as soon as the miner is in operation," says Andrew McDonald, General Manager at NAC, praising the highly productive giant machine.

Figure 1. Project Data:
Mining material – Thermal coal (brittle)
Unconfined compressive strength – 20 MPA
Surface miner – 4200 SM (1 machine)
Production rate per hour – up to 3000 t

Figure 2. The challenging geological facts in New Acland suggest the use of the surface mining technology.

Figure 3. The high-performance conveyor loads high-quality thermal coal into a 200-tonne dump truck.

In summary

Wirtgen not only develop innovative machines of the highest quality. With our machine technology, we also constantly endeavour to keep environmental pollution as low as possible. A Wirtgen surface miner is a perfect example of this philosophy: it impresses with its environmentally friendly technology as it cuts, crushes and loads rock in a single operation. A tremendous advantage of selective mining is that the vibration less mining operation without drilling and blasting is accompanied by low levels of dust and noise. The low environmental impact also permits maximum exploitation of the deposit right up to the edge of residential areas. The selective mining of high-quality materials requires considerably less space than conventional mining methods. Our fuel-efficient, intelligently controlled engines comply with the strictest exhaust emission standards.

- WIRTGEN SURFACE MINING reduces four operating steps to one.
- STATE-OF-THE-ART ENGINE TECHNOLOGY reduces exhaust and noise emissions.
- ANTI-VIBRATION ENGINE SUPPORT reduces vibration and noise emissions.
- STATE-OF-THE-ART MACHINE CONTROL reduces exhaust and noise emissions.
- SOUNDPROOFING reduces noise emissions.

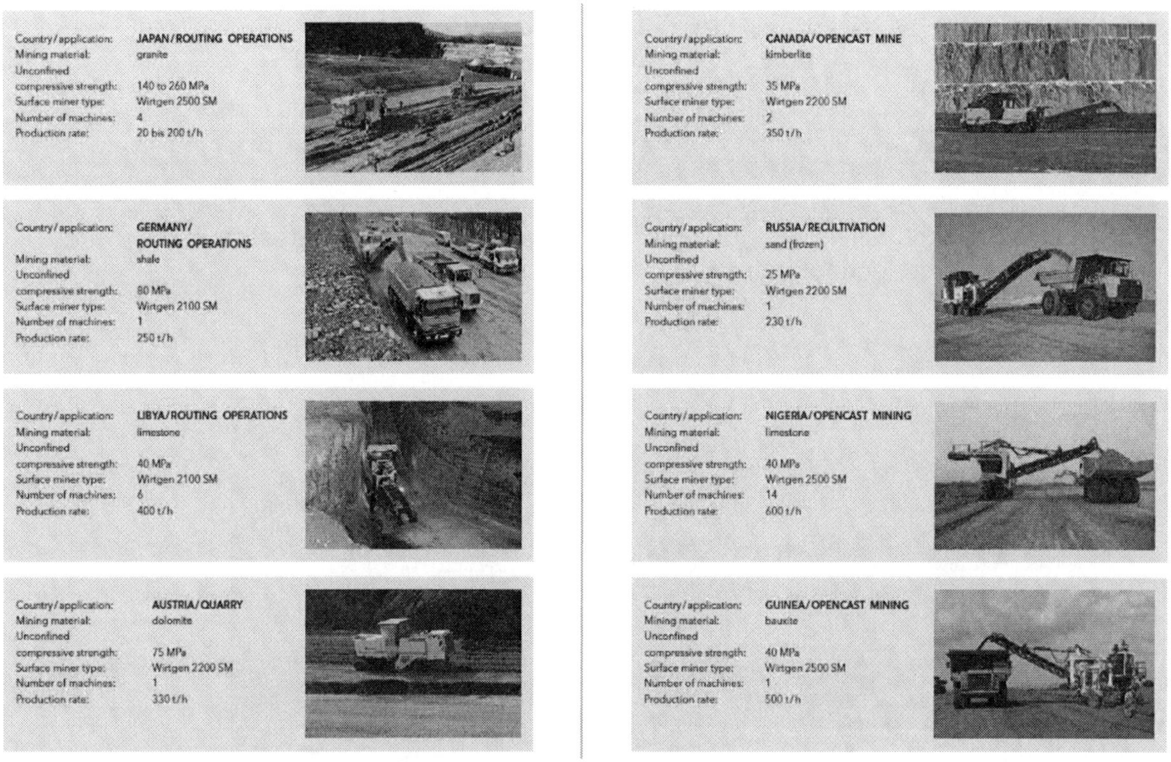

Figure 4. Other international reference projects.

AUTHOR INDEX